Microplastics and Soil Microbiome

Kanika Khanna • Sukhmeen Kaur Kohli
Renu Bhardwaj
Editors

Microplastics and Soil Microbiome

Recent trends and Future Prospects

Springer

Editors
Kanika Khanna
Department of Botanical and Environmental Sciences
Guru Nanak Dev University
Amritsar, Punjab, India

Sukhmeen Kaur Kohli
Institute of Science Education and Research
Tirupati, Andhra Pradesh, India

Renu Bhardwaj
Department of Botanical and Environmental Sciences
Guru Nanak Dev University
Amritsar, Punjab, India

ISBN 978-981-96-4977-8 ISBN 978-981-96-4978-5 (eBook)
https://doi.org/10.1007/978-981-96-4978-5

This Springer imprint is published by the registered company Springer Nature Singapore Pte Ltd.
The registered company address is: 152 Beach Road, #21-01/04 Gateway East, Singapore 189721, Singapore

Preface

Microplastics are the pollutants of emerging concern that have gained much more attention since the last few decades due to their adverse effect toward environment and living organisms. They have been present ubiquitously within the ecosystems including air, water, and soil. Moreover, they have also been observed to interact with biological components, soil microbiome, and other contaminants in the soil. The book has been designed with the aim to unravel the occurrence and distribution of microplastics in the environment, their associations with other pollutants, associations with microbiome, microbial degradation, and fate of microplastics within the environment. The biodegradation and bioaccumulation of microplastics through microbes and their enzymes have been focused. The mechanisms driving such interactions and their impact towards living populations and the environment are explored in the following book. Therefore, we have designed this book with the aim to explore interactions between microplastics and microorganisms (both fungi and bacteria) in soil environments along with following aspects: (1) effects of microplastics on microbial habitats via changes in soil physical, chemical, and biological properties; (2) effects of microplastics on soil microbial communities and functions; and (3) soil microbial-mediated plastic degradation with the likely mechanisms and potential remediation strategies. We aim to analyze the mechanisms driving these interactions and subsequent ecological effects, propose future directives for the study of microplastic in soils, and provide valuable information on the plastic bioremediation in contaminated soils.

Amritsar, Punjab, India — Kanika Khanna
Tirupati, Andhra Pradesh, India — Sukhmeen Kaur Kohli
Amritsar, Punjab, India — Renu Bhardwaj

Contents

Editors and Contributors

About the Editors

Kanika Khanna Dr. Kanika Khanna is working as a Quality Assurance Specialist at Donald Fine Foods, Alberta, Canada. Her research interests include microbiology, plant stress physiology, plant-microbe interactions, plant-nematode interactions, abiotic/biotic stresses, heavy metal toxicity, and plant growth and development. Dr. Khanna has extensive publications in reputed journals and has authored nearly 40 book chapters and 35 research articles and reviews. She has also been awarded many prestigious awards, Young Scholar Award (2019), Excellence Award (2020), Best Women Young Scientist Award (2021), and Appreciation Awards by MP of Canada (2019). She has also presented her research work in many international and national conferences and won Best Presenter Awards. Her Google Scholar Citations are 2520 with an h-index of 23 and i-10 index of 34. She has been an editorial member of several well-reputed journals of Elsevier, Springer, Wiley, Frontiers, and Taylor and Francis. Email: kanika.27590@gmail.com Department of Botanical and Environmental Sciences, Guru Nanak Dev University, Amritsar, Punjab, India

Sukhmeen Kaur Kohli Dr. Sukhmeen Kaur Kohli is working as an Assistant Professor at the School of Life Sciences, Central University, Hyderabad. Her interests include plant stress physiology, brassinosteroids, salicylic acid, abiotic/biotic stresses, heavy metal toxicity, and plant growth and development. Dr. Kohli has extensive publications in reputed journals and has authored nearly 40 book chapters and 45 research articles and reviews. She has also been awarded many prestigious awards, Research Excellence Award (2019), and Best Poster Awards (2019, 2018). She has also presented her research work in many international and national conferences and won Best Presenter Awards. Her Google Scholar Citations are 2127 with an h-index of 24 and i-10 index of 41. She has been an editorial member of many well-reputed journals of Elsevier, Springer, Wiley, Frontiers, and Taylor and Francis. Email: sukhmeenkohli@gmail.com School of Life Sciences, Central University, Hyderabad, India

Department of Botanical and Environmental Sciences, Guru Nanak Dev University, Amritsar, Punjab, India

Renu Bhardwaj Dr. Renu Bhardwaj is currently serving as Director Research, and a Distinguish Scientist, Professor in the Department of Botanical and Environmental Sciences, Guru Nanak Dev University, Amritsar, Punjab, India. She has more than 30 years of experience in teaching and research in the field of plant stress physiology, abiotic/biotic stress, biochemistry and molecular biology, plant metabolomics, ecophysiology, and phytoremediation. She has received various awards/honours from several scientific bodies such as Scientific and Environmental Research Institute, New Delhi, Linnean Society, London Society for Plant Research, Meerut, Foundation for Science and Environment (FSE), Kolkata, and International Foundation for Environment and Ecology. She has completed more than 16 major research projects funded by different funding agencies like UGC, CSIR, New Delhi, DST-SERC, DBT, GOI, New Delhi, GOI, IIN-India, BGCI & HSBC and MOEF, Govt. of India, New Delhi. She has successfully supervised 36 research scholars for doctorate degree. She has published more than 280 research papers in various journals of international and national repute and 87 book chapters in edited books by Wiley, Taylor and Francis Group, Elsevier, Springer, and Nova Science Publisher. She has organized more than 50 international conferences, seminars, and workshops and presented her research work in more than 100 international/national conferences. She is also a fellow of National Environmentalists Association (F.N.E.A). She has visited many countries such as Nepal, Czech Republic, Ukraine and Saint Petersburg, and Russia. Email: renubhardwaj82@gmail.com Director Research, Guru Nanak Dev University, Amritsar, Punjab, India

Professor, Department of Botanical and Environmental Sciences, Guru Nanak Dev University, Amritsar, Punjab, India

Contributors

Amritha Ajayakumar Department of Biochemistry and Molecular Biology, School of Biological Sciences, Central University of Kerala, Kasaragod, Kerala, India

Mohd Ali Department of Zoology, Guru Nanak Dev University, Amritsar, Punjab, India

Raj Bala Department of Botany and Environment Science, DAV University, Jalandhar, Punjab, India

Neelam Bhardwaj Department of Genetics and Plant Breeding, CSK HPKV, Palampur, HP, India

Renu Bhardwaj Department of Botanical and Environmental Sciences, Guru Nanak Dev University, Amritsar, Punjab, India

Madhu Chandel P.G. Department of Botany, Khalsa College, Amritsar, Punjab, India

Harish Changotra Department of Molecular Biology and Biochemistry, Guru Nanak Dev University, Amritsar, Punjab, India

Anu Bala Chowdhary Department of Botanical and Environmental Sciences, Guru Nanak Dev University, Amritsar, India

Anish Dua Aquatic Biology Laboratory, Department of Zoology Guru Nanak Dev University, Amritsar, Punjab, India

Harmanpreet Kaur Gill Aquatic Biology Laboratory, Department of Zoology Guru Nanak Dev University, Amritsar, Punjab, India

Neetu Jagota Department of Botany and Environment Science, DAV University, Jalandhar, Punjab, India

Shivam Jasrotia Department of Biosciences, Chandigarh University, Mohali, Punjab, India

Anureet Kaur Department of Zoology, School of Basic Sciences, Central University of Punjab, Bathinda, India

Gurvarinder Kaur Department of Botany, Punjabi University, Patiala, Punjab, India

Harpreet Kaur P.G. Department of Botany, Khalsa College, Amritsar, Punjab, India

Sukhmeen Kaur Kohli Department of Earth and Climate Sciences, Indian Institute of Science Education and Research, Tirupati, Andhra Pradesh, India

Deepak Kumar Department of Zoology,, Guru Nanak Dev University, Amritsar, Punjab, India

Vedna Kumari Department of Genetics and Plant Breeding, CSK HPKV, Palampur, HP, India

Isha Madaan Department of Botany, Punjabi University, Patiala, Punjab, India

Government College of Education, Jalandhar, Punjab, India

Ashok Kumar Madikonda Department of Biochemistry and Molecular Biology, School of Biological Sciences, Central University of Kerala, Kasaragod, Kerala, India

Rajesh Kumari Manhas Department of Microbiology, Guru Nanak Dev University, Amritsar, Punjab, India

H. Raja Naika Department of Environmental Science, School of Earth Science Systems (SESS), Central University of Kerala, Kasaragod, Kerala, India

V. Nareshkumar Department of Genetics and Plant Breeding, CSK HPKV, Palampur, HP, India

Puja Ohri Department of Zoology, Guru Nanak Dev University, Amritsar, Punjab, India

Nandigam Pardhipa Department of Earth and Climate Sciences, Indian Institute of Science Education and Research, Tirupati, Andhra Pradesh, India

Ripu Daman Parihar Department of Zoology, University of Jammu, Jammu, Jammu & Kashmir, India

Department of Earth and Climate Sciences, Indian Institute of Science Education and Research, Tirupati, Andhra Pradesh, India

Parkirti Department of Zoology, Guru Nanak Dev University, Amritsar, Punjab, India

Madhu Raina Department of Botany, Sri Venkateswara College, University of Delhi, New Delhi, Delhi, India

Ronika Department of Genetics and Plant Breeding, CSK HPKV, Palampur, HP, India

Priyanka Saini Department of Zoology, Deshbandhu College, University of Delhi, Delhi, India

Sandeep Department of Botany and Environment Science, DAV University, Jalandhar, Punjab, India

Faculty of Agricultural Sciences, DAV University, Jalandhar, Punjab, India

Ashish Sharma Department of Botany and Environment Science, DAV University, Jalandhar, Punjab, India

Ashutosh Sharma Faculty of Agricultural Sciences, DAV University, Jalandhar, Punjab, India

Indu Sharma Department of Life Sciences, University Institute of Sciences, Sant Baba Bhag Singh University, Jalandhar, Punjab, India

Roohi Sharma Department of Zoology, Guru Nanak Dev University, Amritsar, Punjab, India

Sonia Sharma P.G. Department of Botany, Khalsa College, Amritsar, Punjab, India

Sonika Sharma Faculty of Agricultural Sciences, DAV University, Jalandhar, Punjab, India

Sushant Sharma P.G. Department of Botany, Khalsa College, Amritsar, Punjab, India

Tarun Sharma Department of Agronomy, CSK HPKV, Palampur, HP, India

Anil Kumar Singh Department of Agriculture, Sant Baba Bhag Singh University, Khiala, Punjab, India

Geetika Sirhindi Department of Botany, Punjabi University, Patiala, Punjab, India

Suma C. Surendran Department of Biochemistry and Molecular Biology, School of Biological Sciences, Central University of Kerala, Kasaragod, Kerala, India

Ronika Thakur Department of Genetics and Plant Breeding, CSK HPKV, Palampur, HP, India

Raman Tikoria Department of Zoology, Guru Nanak Dev University, Amritsar, Punjab, India

Department of Zoology, Central University of Punjab, Ghudda Bathinda, Punjab, India

Vikram Department of Zoology, Guru Nanak Dev University, Amritsar, Punjab, India

Chapter 1
Microplastics in the Environment

Harpreet Kaur, Sonia Sharma, Madhu Chandel, Gurvarinder Kaur, Sushant Sharma, Isha Madaan, Geetika Sirhindi, and Renu Bhardwaj

Abstract Microplastics (MPs) have emerged as a consequence of industrialization, economic development, and improved living standards. These minuscule plastic fragments originate from the breakdown of larger plastic materials. As environmental pollutants, MPs pose significant risks to both flora and fauna. They are categorized into primary MPs—which are intentionally manufactured small plastic particles—and secondary MPs, which result from the degradation of larger plastic waste. Primary sources of MPs include plastic pellets, microbeads in personal care products, synthetic fibers from washing wastewater, sewage sludge, artificial turf, rubberized roads, and vehicle tire wear. On the other hand, secondary sources arise from the fragmentation of larger plastic waste such as discarded bags, bottles, fishing gear, agricultural plastic films, and municipal debris. Both types of MPs contribute to environmental contamination, adversely affecting ecosystems and public health. Sewage systems play a significant role in the spread of MPs. Studies indicate that 80–90% of microplastic particles in wastewater, such as fibers from synthetic clothing, accumulate in sewage sludge. These tiny plastic fragments have the potential to harbor pathogenic microorganisms, acting as vectors for disease transmission in the environment. Among the various pollutants, polyester fibers account for a substantial proportion of MPs found in oceans, with microfibers constituting over 85% of MP debris on shorelines globally. Although all primary sources have not been fully assessed, domestic laundry wastewater is believed to be one of the dominant contributors. Synthetic fibers and particles smaller than 5 mm can bypass filtra-

H. Kaur (✉) · S. Sharma · M. Chandel · S. Sharma
P.G. Department of Botany, Khalsa College, Amritsar, Punjab, India
e-mail: harpreetkaur@khalsacollege.edu.in

G. Kaur · G. Sirhindi
Department of Botany, Punjabi University, Patiala, Punjab, India

I. Madaan
Department of Botany, Punjabi University, Patiala, Punjab, India

Government College of Education, Jalandhar, Punjab, India

R. Bhardwaj
Department of Botanical and Environmental Sciences, GNDU, Amritsar, Punjab, India

K. Khanna et al. (eds.), *Microplastics and Soil Microbiome*,
https://doi.org/10.1007/978-981-96-4978-5_1

tion and solid separation systems in wastewater treatment plants, allowing them to enter aquatic ecosystems. The ingestion of MPs by marine organisms—including commercially important fish and crustaceans—is widespread, leading to contamination transfer within food chains. Additionally, sewage sludge containing MPs is frequently applied to agricultural land as fertilizer, leading to the accumulation of thousands of tons of MPs in soils each year. In waste management facilities, only a fraction of plastic waste is recycled or incinerated, while a significant portion is disposed of in landfills. Since plastic materials take centuries to degrade, they continuously release toxic compounds into soil and water systems, contributing to long-term ecological harm. Although the concern over marine MP pollution has gained significant attention, environmental experts suggest that MP contamination in terrestrial environments may be even more severe. Over the past few years, scientific investigations into MPs have increased, alongside growing public awareness and policy discussions aimed at addressing the issue. This chapter examines the current status of microplastic pollution, highlighting identification techniques, strategies to mitigate MP-related effects, and approaches to reduce their environmental impact. The discussion aims to bridge existing knowledge gaps in plastic detection and pollution management, contributing to research on sustainable solutions for microplastic contamination.

Keywords Microplastics · Sewage · Marine · Ecosystems · Pollution

1 Introduction

The twentieth century witnessed a significant surge in global plastic production, reaching approximately 370 million metric tons. Asia accounted for the largest share (49%), followed by North America (19%) and Europe (15%) (Plastics Facts Report, 2022). By 2019, global plastic production was estimated at 368 million metric tons, with nearly half originating from Asia (Tiseo 2021). Projections indicate that by 2050, approximately 12 billion metric tons of plastic waste could accumulate in landfills or infiltrate natural ecosystems (Geyer et al. 2017). However, in 2020, annual plastic production declined slightly to 367 million metric tons due to the impact of COVID-19 (Plastics Facts Report, 2022). Despite this temporary reduction, plastic production rebounded post-pandemic, largely driven by the increased demand for face masks and other single-use plastic products.

The widespread use of single-use plastics, valued for their chemical stability and efficiency, has exacerbated plastic pollution, significantly contaminating ecosystems and posing environmental and health hazards (Lusher 2015). Plastics do not fully degrade but instead fragment into progressively smaller particles, leading to the formation of microplastics (MPs) and nanoplastics (Hartmann et al. 2019). Discarded plastic waste persists in the environment for decades, breaking down into fibers, pellets, and tiny plastic fragments of varying sizes (Petersen and Hubbart 2021).

MPs are defined as plastic particles smaller than 5 mm, while nanoplastics exist at the nanoscale. These particles emerge from the fragmentation of larger plastics through processes such as biodegradation, abrasion, UV exposure, and hydrolysis (Hartmann et al. 2019; Mitrano and Wohlleben 2020). Studies by Lozano et al. (2021) have shown that MP weathering results in diverse structural forms, including fibers, pellets, spheres, foams, and films. Furthermore, MPs are composed of various synthetic polymers, such as polyvinyl chloride, polypropylene, and styrene (Guo et al. 2019).

MPs are transported through surface runoff and wind dispersal, allowing them to infiltrate various ecosystems, including river systems and marine environments (Liu et al. 2020a, b). Research has detected MPs in remote locations, such as the Antarctic marine system (Lacerda et al. 2019), the Arctic (Kanhai et al. 2020), and the Tibetan Plateau (Feng et al. 2021). These findings highlight the global distribution and persistence of MPs, underscoring their significant environmental impact.

Silva et al. (2020) affirmed that increased concentration of MPs pollution affects the growth and development of plants. Microplastics absorbed from the soil by plant roots and are translocated to the aboveground plant parts (Zhu et al. 2022) result in the formation of mechanical injuries, retardation of growth, imbalance in antioxidative activities, chromosomal aberrations due to genotoxicity, and increased levels of oxidative stress as studied by Lopez-Rojo et al. (2020). Some studies revealed that MPs accumulation is also reported in edible plant parts such as leaf, root, stem, and fruits and acts as a main source for MPs ingestion to human tract as studied by Oliveri Conti et al. (2020). Microplastics also contaminate crop end products that enter into the food chain and affect our ecosystem (Fig. 1.1), so this chapter summarized the ecological effects and occurrence of MPs in terrestrial and aquatic ecosystems.

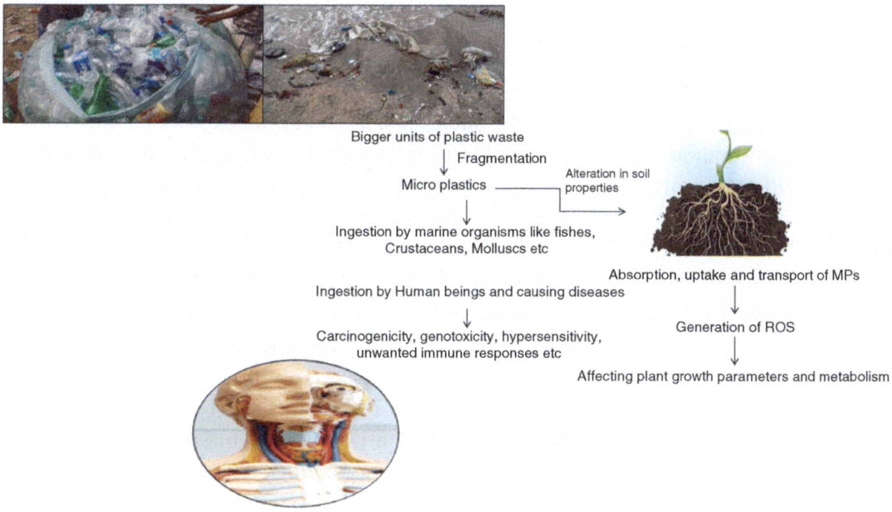

Fig. 1.1 Generalized mechanism of entry of microplastics in the human and plants environment

2 Classification and Sources of MPS

2.1 Primary Microplastics (MPs)

Primary MPs are defined as plastic particles ranging from 1 μm to 5 mm in size. These particles are intentionally added to various consumer and industrial products for specific functional purposes. In the personal care industry, MPs are used as exfoliants in skincare products, abrasives in soaps and cleansers, and carriers for pigments. Common examples include plastic pellets and granulated powders, which are frequently found in cosmetic and cleaning formulations (Lassen et al. 2015). Other sources of MP pollution include toothpaste, face washes, detergents, and shampoo packaging. Additionally, MPs are utilized in dental polishing agents, further contributing to their widespread environmental presence if not properly managed.

Beyond household and cosmetic products, primary MPs have extensive applications in industrial sectors. They are commonly used in oil and gas extraction, particularly in drilling fluids, as well as in paint removal from metal surfaces and engine cleaning processes (Derraik 2002). Industrially, MPs made from synthetic materials such as polystyrene, acrylic, and polyester are widely used in various manufacturing and production processes (Eriksen et al. 2013).

2.2 Secondary Microplastics (MPs)

Secondary MPs primarily originate from the degradation of larger plastic waste, including both household and industrial materials. A major contributor to secondary MP pollution is fibers released from synthetic textiles during washing, which significantly adds to environmental contamination (Eerkes-Medrano et al., Eerkes Medrano et al. 2015).

Various environmental processes, such as UV radiation exposure, mechanical stress, and photodegradation, lead to the fragmentation of discarded plastic materials, including shopping bags, fishing nets, and other plastic waste. These processes cause plastic debris to break down into smaller particles, classifying them as secondary MPs. Oceanic turbulence further accelerates this fragmentation in marine environments. Additionally, washing machines release synthetic fibers into wastewater, further exacerbating MP pollution (Eriksen et al. 2013).

The accumulation of secondary MPs in aquatic ecosystems is driven by continuous plastic waste disposal. Industries such as textile manufacturing, tire production, and paint industries contribute significantly to MP pollution (Lassen et al. 2015). Studies suggest that 15–31% of small plastic particles accumulate in oceanic environments. Among the primary contributors, synthetic fabric laundering accounts for approximately 35% of MP pollution, while tire wear from driving contributes 28%. Other sources include intentionally added MPs in personal care products, such as microbeads in facial scrubs (2%). Furthermore, the degradation of larger plastic items, including plastic bags, bottles, and fishing nets, is estimated to be responsible for 69–81% of MPs present in marine ecosystems.

3 The Potential Impacts of Microplastic

Plastic pollution was initially seen as an aesthetic sense pollution (Gregory 2009), but recent researches show marine animals could be negatively affected by the presence of MPs (Moore et al., Moore et al. 2001). Marine organisms could be affected mainly through the mandering of plastic, getting trapped by plastic and then plastic eating (Gregory 2009; Van Franeker et al. 2011). The microscopic size allows the bioavailability of plastic through ingestion tract (Betts 2008). A number of aquatic organisms have been observed to accumulate MPs in their bodies, including crabs.

4 Occurrence and Effects of MPs in Aquatic Habitats

The issue of plastic pollution in oceans has been extensively studied, with research highlighting its increasing severity (GESAMP 2015). Plastics are categorized based on size, with macroplastics referring to larger plastic debris, while microplastics (MPs) are defined as particles smaller than 5 mm (Thevenon et al. 2014). Since the widespread production of plastics, an estimated 8.3 billion tons have been manufactured, yet only 9% has been successfully recycled (Geyer et al. 2017). Additionally, between 5 and 13 million metric tons of plastic waste enter marine environments annually (Jambeck et al. 2015).

To effectively address this growing crisis, it is essential to understand the relationship between per capita plastic waste generation and the proportion of mismanaged plastics that ultimately contribute to ocean pollution (Jambeck et al. 2015). The adverse effects of MP contamination extend beyond environmental damage, posing significant socioeconomic challenges. Studies have documented widespread ingestion, entanglement, and physical injuries in marine organisms, as well as financial losses in tourism and maritime industries due to the degradation of coastal ecosystems (GESAMP 2015; Raynaud 2014).

The distribution of MPs in marine environments is largely influenced by their density. Low-density MPs, such as polypropylene, remain buoyant and are transported by ocean currents, leading to their accumulation in oceanic gyres and the formation of large plastic debris zones (Sebille et al. 2015). Conversely, denser MPs, such as acrylic, sink to the ocean floor, where they integrate into sediments and enter marine food chains (Woodall et al. 2014; Seltenrich 2015). This widespread contamination of marine ecosystems poses an escalating global challenge, raising serious concerns for ocean health and biodiversity conservation (Reed 2016).

5 Occurrence and Effects of MPS in Terrestrial Habitats

MPs enter terrestrial environments primarily through littering and the application of sewage sludge, which has been recognized as a significant reservoir for these particles (Barnes et al. 2009). While research on MP accumulation in soil ecosystems remains limited, existing studies have highlighted their presence in various terrestrial settings (Lambert et al. 2014). One proposed method for tracking past sludge applications involves using synthetic fibers as indicators of MP contamination in soil (Habib et al. 1998).

Zubris and Richards (2005) analyzed soil samples using polarized light microscopy and found that plastic fiber concentrations were significantly higher in areas with long-term sludge application compared to control sites. Their findings suggest that MP contamination in soil can persist for extended periods, raising concerns about its potential environmental impact.

6 MP Effects on Terrestrial Plants

Research on the effects of microplastics (MPs) on terrestrial plants remains limited, making it difficult to establish definitive conclusions. While some studies, such as that of Lozano and Rillig (2020), provide broader insights, most investigations have focused on individual plant species. Examples include studies on perennial ryegrass (Boots et al. 2019), common wheat (Judy et al. 2019), broad bean (Jiang et al. 2019), carrots, and spring onions (Lozano et al. 2021). These studies have primarily examined the effects of MPs in various soil types, including clay and loamy soils, under different exposure durations (ranging from 2 to 273 days) and MP concentrations (0.001–2% w/w).

The impact of MPs on plant growth appears to be species-specific. Boots et al. (2019) observed that exposure to acrylic polymers and polyester at 0.1% w/w negatively affected the germination of perennial ryegrass. In contrast, Judy et al. (2019) found no significant impact on common wheat germination when exposed to similar concentrations of polyester, polyethylene, and polyvinyl chloride, suggesting that plant responses to MPs vary across species. MPs can also influence different plant organs in distinct ways. For instance, perennial ryegrass exposed to polyethylene MPs exhibited increased root biomass, while shoot biomass remained unchanged (Boots et al. 2019).

The size of MP particles also plays a crucial role in their effects on plants. Jiang et al. (2019) reported that broad bean plants responded differently to polystyrene MP beads of varying sizes at identical concentrations (0.001%, 0.005%, and 0.01% w/w). Larger particles (5 μm) significantly reduced root length but did not cause genotoxic effects, whereas smaller particles (0.1 μm) induced genotoxicity without affecting root length. Further studies have explored the relationship between MPs and mycorrhizal fungi, which are essential for plant nutrient uptake. De Souza

Machado et al. (de Souza et al. 2018) investigated the morphological effects of MPs on plant-fungal symbiosis and found that different polymer types—polyamide, polyester, high-density polyethylene, polypropylene, polystyrene, and polyethylene terephthalate—had shape-specific effects on mycorrhizal colonization in Allium fistulosum. Among these, polyester fibers significantly enhanced mycorrhizal colonization around roots, likely due to alterations in soil structure that improved spring onion bulb yield.

These findings highlight how MPs can disrupt species interactions and plant growth, posing potential risks to the sustainability of agricultural ecosystems. Further research is needed to fully understand these effects and develop mitigation strategies for MP contamination in soil environment.

7 MPS and Human Health

MPs in seafood are responsible for major hazard to human strength. As seafood is a significant part of the human food, MPs contamination of the intestine poses a severe hazard of dispersal to other parts of the human alimentary canal. Kor and Mehdinia (2020) affirmed that ingestion and endocytosis are common route for the entry of MPs to human body. Toxicological impacts of MPs may adversely affect the fish metabolic activities, which are the main contemplation of humans who are nonvegetarian and eat fish as a component of their mealtime. More research are needed to check the realistic MPs and pollutant levels in the ecosystem (Neves et al. 2015).

8 Toxicological Effects in Human Health

In humans, the accumulation of MPs (MPs) in tissues has been associated with arterial swelling and blockages (Wang et al. 2016). Karbalaei et al. (2018) reported that in vitro studies have identified MP accumulation in the gills, stomach, and metabolic organs of crabs and fish, leading to microbial colonization and pollutant transfer (Manzoor et al. 2020). The severity of these impacts largely depends on an individual's exposure levels and biological sensitivity (Prata et al. 2020).

The introduction of MPs into the food chain has been linked to disruptions in antioxidative activity, cytotoxic effects, and the potential translocation of MPs to various organs (Galloway 2015). Due to their persistent nature, MPs remain in both the environment and biological systems for extended periods. Long-term exposure in organisms has been associated with continuous stress, inflammation, uncontrolled cell proliferation, immune dysfunction, and, in severe cases, mortality (Smith et al. 2018). Studies have also suggested that individuals with MP exposure show a significantly higher prevalence of gastric disorders compared to healthy individuals (Yan et al. 2021).

Specific MPs, such as polystyrene, have been found to inhibit the proliferation of Caco-2 cells over time (Wu et al. 2020). Additionally, polystyrene exposure has been linked to mitochondrial membrane potential disturbances (Wu et al. 2019). MPs also function as carriers for various microbial species (Kirstein et al. 2016), and their ability to absorb or release chemical compounds is influenced by environmental conditions (Crawford and Quinn 2017).

9 Oxidative Stress and Cytotoxicity

MPs contribute to oxidative stress by generating reactive oxygen species (ROS), which are produced as part of the body's inflammatory response (Valavanidis et al. 2013). Research by Forte et al. (2016) demonstrated that exposure to plastic nanoparticles can trigger inflammatory reactions in human gastric cells and A549 lung cells. Additionally, polyethylene particles have been shown to induce the release of pro-inflammatory cytokines, including IL-1, IL-6, and TNF-α, which are associated with the development of metabolic disorders such as type 2 diabetes, hypertension, and obesity-related conditions (Green et al. 1998; Popko et al. 2010).

Studies in animal models have suggested that macrophages can absorb MPs, potentially leading to immune system disturbances (Yacobi et al. 2008). Nanoparticles (NPs) are also known to influence cellular mechanisms by altering cell fate, inducing mutations, facilitating intercellular communication, and modifying structural properties at the nanoparticle interface (Behzadi et al. 2017). Research by Thubagere and Reinhard (2010) identified polystyrene plastics as cytotoxic to intestinal cells. However, strong recommendations for assessing cytotoxicity in in vitro studies remain inconclusive (Wu et al. 2019).

10 Carcinogenicity

Exposure to plastic materials has been associated with an increased risk of cancer. Prata (2018) suggested that prolonged inflammation and irritation resulting from MP ingestion could contribute to cancer development by inducing DNA damage. Similarly, Chang (2010) reported that oxidative stress and chronic cellular irritation caused by nanoplastics can trigger the release of pro-inflammatory agents. This inflammatory response may stimulate vascular changes, potentially facilitating tumor formation and progression.

11 Biodegradation Strategies for MP Pollution Control

11.1 Extraction of MPs from Aquatic and Terrestrial Environments

The removal of MPs from water, soil, and sediment is a crucial step in their biodegradation. This process requires the collection of plastic debris along with plastic-degrading microbes and enzymes under controlled conditions, such as optimal temperature, pH, humidity, and nutrient availability. Once extracted, these plastic particles can undergo biological depolymerization and fermentation to produce valuable chemical compounds (Delacuvellerie et al. 2019). Wastewater treatment plants (WWTPs) are identified as major pathways for the spread of secondary MPs into marine and terrestrial ecosystems.

11.2 Mitigating MP Contamination in the Environment

Conventional methods for MP disposal, including incineration, landfilling, and recycling, have been widely used. However, recent advancements emphasize biotransformation and bioremediation as promising approaches to tackling MP pollution. Researchers are exploring innovative biotechnological solutions to enhance the efficiency of these methods (Blair Espinoza 2019).

11.3 Biodegradation of Synthetic MPs

MPs are primarily composed of synthetic polymers that are resistant to natural degradation and insoluble in water. Given the diversity of plastic polymers present in the environment, biodegradation under assisted conditions has become an area of interest. Biotechnological interventions such as biostimulation, bioaugmentation, and enzymatic degradation are being explored to accelerate the breakdown of MPs (Miri et al. 2022).

11.4 Advancements in MP Biodegradation Technologies

Despite progress in plastic biodegradation research, there is still no consensus on the most efficient biotechnological method for MP breakdown. Since MPs consist of various polymer types, a combination of advanced biodegradation techniques may be required for effective removal from specific ecosystems. The degradation rate is also influenced by the surface area of MPs, with smaller particles degrading more rapidly than larger plastic items such as water bottles and cosmetic packaging (Miri et al. 2022).

12 Conclusion

The widespread production and utilization of MPs have significantly contributed to the accumulation of plastic debris in aquatic ecosystems. MPs originate from multiple sources, not solely from weathering-related fragmentation and surface degradation of plastics on shorelines but also from industrial waste, which remains a major contributor. Due to their persistent nature, even a small fraction of MPs in water bodies poses a substantial risk to aquatic organisms. These microscopic plastic particles are easily ingested by various aquatic species, including micro- and nano-plankton, facilitating the transfer of toxic substances across different trophic levels. The extent of this transfer depends on factors such as MP availability and retention time within organisms. Additionally, the internalization of plastic nanoparticles through endocytosis by micro- and nanofauna can lead to harmful toxic effects. Since aquatic organisms form the foundation of marine food webs, any disruption at this level may have widespread ecological consequences. Given the rising levels of MPs in global water bodies, it is essential to assess their long-term environmental impact and develop strategies to mitigate their effects. Future research should focus on understanding the ecological implications of MP pollution and identifying solutions to minimize its adverse effects on aquatic ecosystems.

Author Contributions All authors contributed equally to the manuscript.

References

Barnes DK, Galgani F, Thompson RC, Barlaz M (2009) Accumulation and fragmentation of plastic debris in global environments. Philos Trans R Soc Lond B 364:1985–1998

Behzadi S, Serpooshan V, Tao W, Hamaly MA, Alkawareek MY, Dreaden EC, Brown D, Alkilany AM, Farokhzad OC, Mahmoudi M (2017) Cellular uptake of nanoparticles: journey inside the cell. Chem Soc Rev 46(14):4218–4244. https://doi.org/10.1039/c6cs00636a. PMID: 28585944. PMCID: PMC5593313

Betts K (2008) Why small plastic particles may pose a big problem in the oceans. Environ Sci Technol. https://doi.org/10.1021/es802970v

Blair Espinoza RM (2019) MPs in wastewater treatment systems and receiving waters. University of Glasgow. Thesis https://doi.org/10.5525/gla.thesis.76781

Boots B, Russell CW, Green DS (2019) Effects of MPs in soil ecosystems: above and below ground. Environ Sci Technol 53(19):11496–11506. https://doi.org/10.1021/acs.est.9b03304

Chang C (2010) The immune effects of naturally occurring and synthetic nanoparticles. J Autoimmun 34(3):J234–J246. https://doi.org/10.1016/j.jaut.2009.11.009

Crawford CB, Quinn B (2017) The interactions of MPs and chemical pollutants. MP Pollutants 1:131–157. https://doi.org/10.1016/b978-0-12-809406-8.00006-2

de Souza MAA, Kloas W, Zarfl C, Hempel S, Rillig MC (2018) MPs as an emerging threat to terrestrial ecosystems. Glob Chang Biol 24(4):1405–1416. https://doi.org/10.1111/gcb.14020

Delacuvellerie A, Cyriaque V, Gobert S, Benali S, Wattiez R (2019) The plastisphere in marine ecosystem hosts potential specific microbial degraders including Alcanivorax borkumensis as a key player for the low-density polyethylene degradation. J Hazard Mater 380:120899. ISSN 0304-3894. https://doi.org/10.1016/j.jhazmat.2019.120899

Derraik JG (2002) The pollution of the marine environment by plastic debris: a review. Mar Pollut Bull 44(9):842–852

Eerkes Medrano D, Thompson RC, Aldridge DC (2015) MPs in freshwater systems: a review of the emerging threats, identification of knowledge gaps and prioritisation of research needs. Water Res 75:63–82

Eriksen M, Mason S, Wilson S, Box C, Zellers A, Edwards W, Farley H, Amato S (2013) MP pollution in the surface waters of the Laurentian Great Lakes. Mar Pollut Bull 77(1–2):177–182

Feng S, Lu H, Liu Y (2021) The occurrence of MPs in farmland and grassland soils in the Qinghai-Tibet plateau: different land use and mulching time in facility agriculture. Environ Pollut 279:116939. https://doi.org/10.1016/j.envpol.2021.116939

Forte M, Iachetta G, Tussellino M, Carotenuto R, Prisco M, De Falco M et al (2016) Polystyrene nanoparticles internalization in human gastric adenocarcinoma cells. Toxicol In Vitro 31:126–136. https://doi.org/10.1016/j.tiv.2015.11.006

Galloway TS (2015) Micro and nano-plastics and human health. In: Marine anthropogenic litter. Springer, Cham, pp 343–366. https://doi.org/10.1007/978-3-319-16510-3_13

GESAMP S (2015) Fate and effects of MPs in the marine environment: a global assessment. IMO/FAO/UNESCO-IOC/UNIDO/WMO/IAEA/UN/UNEP/ UNDP Joint Group of Experts on the Scientific Aspects of Marine Environmental Protection). Rep Stud, GESAMP, 96

Geyer R, Jambeck JR, Law KL (2017) Production, use, and fate of all plastics ever made. Sci Adv 3(7):e1700782

Green T, Fisher J, Stone M, Wroblewski BM, Ingham E (1998) Polyethylene particles of a 'critical size' are necessary for the induction of cytokines by macrophages in vitro. Biomaterials 19(24):2297–2302. https://doi.org/10.1016/s0142-9612(98)00140-9

Gregory MR (2009) Environmental implications of plastic debris in marine settings entanglement, ingestion smothering, hangers on, hitch hiking and alien invasions. Philos Trans R Soc Lond Ser B Biol Sci 364(1526):2013–2025. https://doi.org/10.1098/rstb.2008.0265

Guo X, Liu Y, Wang J (2019) Sorption of sulfamethazine onto different types of MPs: a combined experimental and molecular dynamics simulation study. Mar Pollut Bull 145:547–554. https://doi.org/10.1016/j.marpolbul.2019.06.063

Habib D, Locke DC, Cannone LJ (1998) Synthetic fibers as indicators of municipal sewage sludge, sludge products, and sewage treatment plant effluents. Water Air Soil Pollut 103:1–8

Hartmann NB, Huffer T, Thompson RC, Hassellov M, Verschoor A, Daugaard AE et al (2019) Are we speaking the same language? Recommendations for a definition and categorization framework for plastic debris. Environ Sci Technol 53:1039–1047

Jambeck JR, Geyer R, Wilcox C, Siegler TR, Perryman M, Andrady A, Narayan R, Law KL (2015) Plastic waste inputs from land into the ocean. Science 347:768–771. https://doi.org/10.1126/science.1260352

Jiang X, Chen H, Liao Y, Ye Z, Li M, Klobucar G (2019) Ecotoxicity and genotoxicity of polystyrene MPs on higher plant *Vicia faba*. Environ Pollut 250:831–838. https://doi.org/10.1016/j.envpol.2019.04.055

Judy JD, Williams M, Gregg A, Oliver D, Kumar A, Kookana R, Kirby JK (2019) MPs in municipal mixed-waste organic outputs induce minimal short to long-term toxicity in key terrestrial biota. Environ Pollut 252:522–531. https://doi.org/10.1016/j.envpol.2019.05.027

Kanhai LDK, Gardfeldt K, Krumpen T et al (2020) MPs in sea ice and seawater beneath ice floes from the Arctic Ocean. Sci Rep 10:5004. https://doi.org/10.1038/s41598-020-61948-6

Karbalaei S, Hanachi P, Walker TR, Cole M (2018) Occurrence, sources, human health impacts and mitigation of MP pollution. Environ Sci Pollut Res 25(36):36046–36063. https://doi.org/10.1007/s11356-018-3508-7

Kirstein IV, Kirmizi S, Wichels A, Garin Fernandez A, Erler R, Loder M et al (2016) Dangerous hitchhikers? Evidence for potentially Pathogenic Vibrio Spp. on MP particles. Mar Environ Res 120:1–8. https://doi.org/10.1016/j.marenvres.2016.07.004

Kor K, Mehdinia A (2020) Neustonic MP pollution in the Persian Gulf. Mar Pollut Bull 150:110665. https://doi.org/10.1016/j.marpolbul.2019.110665

Lacerda ALDF, Rodrigues LDS, van Sebille E, Rodrigues FL, Ribeiro L, Secchi ER et al (2019) Plastics in sea surface waters around the Antarctic Peninsula. Sci Rep 9(1):1–12. https://doi.org/10.1038/s41598-019-40311-4

Lambert S, Sinclair CJ, Boxall AB (2014) Occurrence, degradation and effect of polymer-based materials in the environment. Rev Environ Contamin Toxicol 227:1–53

Lassen C, Hansen SF, Magnusson K, Hartmann NB, Jensen PR, Nielsen TG, Brinch A (2015) MPs: occurrence, effects and sources of releases to the environment in Denmark. The Danish Environmental Protection Agency, pp 1–208

Liu K, Wang X, Song Z, Wei N, Li D (2020a) Terrestrial plants as a potential temporary sink of atmospheric MPs during transport. Sci Total Environ 742:140523. https://doi.org/10.1016/j.scitotenv.2020.140523

Liu K, Wang X, Song Z, Wei N, Ye H, Cong X et al (2020b) Global inventory of atmospheric fibrous MPs input into the ocean: an implication from the indoor origin. J Hazard Mater 400:123223. https://doi.org/10.1016/j.jhazmat.2020.123223

Lopez-Rojo N, Perez J, Alonso A, Correa-Araneda F, Boyero L (2020) MPs have lethal and sub-lethal effects on stream invertebrates and affect stream ecosystem functioning. Environ Pollut 259:113898. https://doi.org/10.1016/j.envpol.2019.113898. Epub 2019 Dec 30

Lozano YM, Rillig MC (2020) Effects of MP fibers and drought on plant communities. Environ Sci Technol 54(10):6166–6173. https://doi.org/10.1021/acs.est.0c01051

Lozano YM, Lehnert T, Linck LT, Lehmann A, Rillig MC (2021) MP shape, polymer type, and concentration affect soil properties and plant biomass. Front Plant Sci 12:169. https://doi.org/10.3389/fpls.2021.616645

Lusher A (2015) MPs in the marine environment: distribution, interactions and effects. In: Marine anthropogenic litter. Springer, Cham, pp 245–307. https://doi.org/10.1007/978-3-319-16510-3_10

Manzoor J, Sharma M, Sofi IR, Dar AA (2020) Plastic waste environmental and human health impacts. In: Handbook of research on environmental and human health impacts of plastic pollution. IGI Global, Hershey, pp 29–37. https://doi.org/10.4018/978-1-5225-9452-9.ch002

Miri S, Saini R, Davoodi SM, Pulicharla R, Brar SK, Magdouli S (2022) Biodegradation of MPs: better late than never. Chemosphere 286:131670., ISSN 0045-6535. https://doi.org/10.1016/j.chemosphere.2021.131670

Mitrano DM, Wohlleben W (2020) MP regulation should be more precise to incentivize both innovation and environmental safety. Nat Commun 11:5324. https://doi.org/10.1038/s41467-020-19069-1

Moore CJ, Moore SL, Leecaster MK, Weisberg SB (2001) A comparisonof plastic and plankton in the North PacificCentral Gyre. Mar Pollut Bull 42(12):1297–1300. https://doi.org/10.1016/S0025-326X(01)00114-X

Neves D, Sobral P, Ferreira JL, Pereira T (2015) Ingestion of MPs by commercial fish off the portuguese coast. Mar Pollut Bull 101(1):119–126. https://doi.org/10.1016/j.marpolbul.2015.11.008

Oliveri Conti G, Ferrante M, Banni M, Favara C, Nicolosi I, Cristaldi A, Fiore M, Zuccarello P (2020) Micro- and nano-plastics in edible fruit and vegetables. The first diet risks assessment for the general population. Environ Res 187:109677. https://doi.org/10.1016/j.envres.2020.109677. Epub 2020 May 20

Petersen F, Hubbart JA (2021) The occurrence and transport of MPs: the state of the science. Sci Total Environ 758:143936–143948

Plastics- the Facts (2022). https://plasticseurope.org/wp-content/uploads/2021/12/Plastics-the-Facts-2021-web-fnal.pdf
Popko K, Gorska E, Stelmaszczyk-Emmel A, Plywaczewski R, Stoklosa A, Gorecka D, Pyrzak B, Demkow U (2010) Proinflammatory cytokines Il-6 and TNF-α and the development of inflammation in obese subjects. Eur J Med Res 15(Suppl 2):120–122. https://doi.org/10.1186/2047-783x-15-s2-120. PMID: 21147638; PMCID: PMC4360270
Prata JC (2018) Airborne MPs: consequences to human health? Environ Pollut 234:115–126. https://doi.org/10.1016/j.envpol.2017.11.043
Prata JC, da Costa JP, Lopes I, Duarte AC, Rocha-Santos T (2020) Environmental exposure to MPs: an overview on possible human health effects. Environ Sci Total Environ 702:134455. https://doi.org/10.1016/j.scitotenv.2019.134455
Raynaud J (2014) United Nations Environment Programme, Valuing plastic: the business case for measuring, managing and disclosing plastic use in the consumer goods industry. https://wedocs.unep.org/20.500.11822/9238
Reed C (2016) Plastic age: how it's reshaping rocks, oceans and life. New Scientist, Feature 28 January 2015
Sebille EV, Wilcox C, Lebreton L, Maximenko N, Hardesty BD, van Franeker JA, Eriksen M, Siegel D, Galgani F, Law KL (2015) A global inventory of small floating plastic debris. Environ Res Lett 10:124006. https://doi.org/10.1088/1748-9326/10/12/124006
Seltenrich N (2015) New link in the food chain? Marine plastic pollution and seafood safety. Environ Health Perspect 123:A34–A41. https://doi.org/10.1289/ehp.123-a34
Silva ALP, Prata JC, Walker TR, Campos D, Duarte AC, Soares AMVM et al (2020) Rethinking and optimising plastic waste management under COVID-19 pandemic: policy solutions based on redesign and reduction of single-use plastics and personal protective equipment. Sci Tot Environ 742. https://doi.org/10.1016/j.scitotenv.2020.140565
Smith M, Love DC, Rochman CM, Neff RA (2018) MPs in seafood and the implications for human health. Curr Environ Health Rep 5(3):375–386. https://doi.org/10.1007/s40572-018-0206-z
Thevenon F, Caroll C, Sousa J (2014) Plastic debris in the oceans. (IUCN). The characterization of marine plastics and their environmental impacts, situation analysis report. IUCN, Gland, 52 pp
Thubagere A, Reinhard BM (2010) Nanoparticle-induced apoptosis propagates through hydrogen-peroxide-mediated bystander killing: insights from a human intestinal epithelium in vitro model. ACS Nano 4(7):3611–3622. https://doi.org/10.1021/nn100389a
Tiseo L (2021) Global plastic production 1950–2018. Statista
Valavanidis A, Vlachogianni T, Fiotakis K, Loridas S (2013) Pulmonary oxidative stress, inflammation and cancer: Respirable particulate matter, fibrous dusts and ozone as major causes of lung carcinogenesis through reactive oxygen species mechanisms. Int J Environ Res Public Health 10(9):3886–3907. https://doi.org/10.3390/ijerph10093886
Van Franeker JA, Blaize C, Danielsen J, Fairclough K, Gollan J, Guse N et al (2011) Monitoring plasticingestion by the northern fulmar Fulmarusglacialis in the North Sea. Environ Pollut 159(10):2609–2615. https://doi.org/10.1016/j.envpol.2011.06.008
Wang J, Tan Z, Peng J, Qiu Q, Li M (2016) The behaviors of MPs in the marine environment. Mar Environ Res 113:7–17. https://doi.org/10.1016/j.marenvres.2015.10.014
Woodall LC, Sanchez Vidal A, Canals M, Paterson GLJ, Coppock R, Sleight V, Calafat A, Rogers AD, Narayanaswamy BE, Thompson RC (2014) The deep sea is a major sink for MP debris. Open Sci 1:140317. https://doi.org/10.1098/rsos.140317
Wu B, Wu X, Liu S, Wang Z, Chen L (2019) Size-dependent effects of polystyrene MPs on cytotoxicity and efflux pump inhibition in human Caco-2 cells. Chemosphere 221:333–341. https://doi.org/10.1016/j.chemosphere.2019.01.056
Wu S, Wu M, Tian D, Qiu L, Li T (2020) Effects of polystyrene microbeads on cytotoxicity and transcriptomic profiles in human Caco-2 cells. Environ Toxicol 35(4):495–506
Yacobi NR, DeMaio L, Xie J, Hamm Alvarez SF, Borok Z, Kim KJ et al (2008) Polystyrene nanoparticle trafficking across alveolar epithelium. Nanomedicine 4(2):139–145. https://doi.org/10.1016/j.nano.2008.02.002

Yan Z, Liu Y, Zhang T, Zhang F, Ren H, Zhang Y (2021) Analysis of MPs in human feces reveals a correlation between fecal MPs and inflammatory bowel disease status. Environ Sci Technol. https://doi.org/10.1021/acs.est.1c03924
Zhu J, Zhang Y, Xu Y, Wang L, Wu Q, Zhang Z, Li L (2022) Effects of MPs on the accumulation and neurotoxicity of methyl mercury in zebra fish larvae. Mar Environ Res 176:105615
Zubris KA, Richards BK (2005) Synthetic fibers as an indicator of land application of sludge. Environ Pollut 138:201–211

Chapter 2
Microbial Colonization and Degradation of Microplastics

Indu Sharma, Sandeep, Neetu Jagota, Ashish Sharma, Raj Bala, Sonika Sharma, and Ashutosh Sharma

Abstract Plastics are one of the most harmful synthetic materials that are commonly used because of their strength, flexibility, and low cost. These widely used plastic polymers are nonreactive, resilient, and nonbiodegradable and, thus, accumulate in the environment, causing harm to the flora and fauna. About half of the plastic production is single-use plastic, which causes an imbalance in the ecosystem and poses hazardous risks to animals. Because of their single-use and casual disposal by consumers, plastic waste generation has increased exponentially in the atmosphere leading to its deposition, especially in oceans in the form of microplastics (MPs). The conventional methods of waste management such as recycling, incineration, and landfilling are costly and ineffective for the degradation of MPs. Therefore, better alternative methods that are eco-friendly, cost-effective, and efficient for the degradation of MPs debris are optimum solutions for the mitigation of MP pollution. Various biological approaches involving microorganisms can potentially colonize and enzymatically degrade MPs into environmentally friendly and less harmful products. Hence, the present chapter, in the light of available literature, highlights the potential implications of microorganisms in colonization and degradation of MPs.

Keywords Plastic polymer · Polypropylene · Plastisphere · Biodegradation · Plastic pollution · Biodeterioration · Bio-fragmentation · Mineralization

---

I. Sharma
Department of Life Sciences, University Institute of Sciences, Sant Baba Bhag Singh University, Jalandhar, Punjab, India

Sandeep
Department of Botany and Environment Science, DAV University, Jalandhar, Punjab, India

Faculty of Agricultural Sciences, DAV University, Jalandhar, Punjab, India

N. Jagota · A. Sharma · R. Bala
Department of Botany and Environment Science, DAV University, Jalandhar, Punjab, India

S. Sharma · A. Sharma (✉)
Faculty of Agricultural Sciences, DAV University, Jalandhar, Punjab, India
e-mail: ashutosh10110@davuniversity.org

K. Khanna et al. (eds.), *Microplastics and Soil Microbiome*,
https://doi.org/10.1007/978-981-96-4978-5_2

1 Introduction

A plethora of organic pollutants at different concentrations have been accumulated in the atmosphere due to reckless human interference. Some emerging pollutants introduced through anthropogenic activities, which are present in very low concentrations, but have become a serious threat to the environment due to their persistent nature, are referred to as micropollutants. These micropollutants come from a wide range of products including PPCPs (personal care products), PFAs (perfluoroalkyl and poly-fluoroalkyl substances), PCBs (polychlorinated biphenyls), PAHs (PAHs), steroid hormones, nanomaterials, pesticides, synthetic materials, and plastics and are found in groundwater, freshwater, soil, marine ecosystem, sediments, and dust (Kaur et al. 2024b). Plastic is one of the world's most often used synthetic materials because of its strength, lightness, low density, high ductility, and affordability (Chae and An 2018). According to Krueger et al. (2015), plastics can be categorized based on their physical and chemical characteristics, such as density, molecular weight, crystallinity, and functional groups, as well as the additives they include, such as plasticizers, antioxidants, and flame retardants. Polyethylene (PE), polypropylene (PP), polystyrene (PS), polyethylene terephthalate (PET), and polyvinyl chloride (PVC) are the primary categories of "nonbiodegradable" plastics. The use of plastics now is roughly 20 times greater than it was 50 years ago, and it is predicted that in the next 20 years, both production and use will double (Urbanek et al. 2018). Most plastic wastes released into the environment are chemically stable, corrosion-resistant, and challenging for microorganisms to break down (Yang et al. 2020). A serious environmental problem has arisen as a result of this (Wright et al. 2020). Moreover, Chen et al. (2021) reported that only 21% of plastic has been recycled or burned (UNEP 2018), with the remaining 79% decomposing into small pieces known as "microplastics" (less than 5 mm in diameter).

The accumulation of microplastic (or MP) in the environment presents several issues and concerns that have a negative impact on soil, water, creatures, and human health (Lamichhane et al. 2023). Additionally, MPs can impact fish reproduction and aquatic plant growth in seas, among other issues (Guo and Wang 2019). According to Vivekanand et al. (2021), MPs are ubiquitous pollutants that have an adverse influence on many ecosystems, including animals, the environment, food safety, ecological disruption, and water and air contamination (Fig. 2.1). Due to their durability and resistance to degradation, these microplastics accumulate in the environment over time. They can cause environmental degradation overall, have an effect on wildlife, and disturb natural ecosystems (Karbalaei et al. 2018). The whole structure and functioning of ecosystems may be impacted, as well as nutrient cycling, animal behavior, and eating habits (Ma et al. 2020). The biodiversity, ecological resilience, and general ecosystem health may all be negatively impacted by these changes. In conclusion, MP pollution has grown to be a problem of global proportions as a result of the nearly universal detection of MPs around the world (Bakir et al. 2014). Due to rapid production, easy availability, and "throw-away" consumer culture, the majority of waste plastic is not recycled. The reckless use and

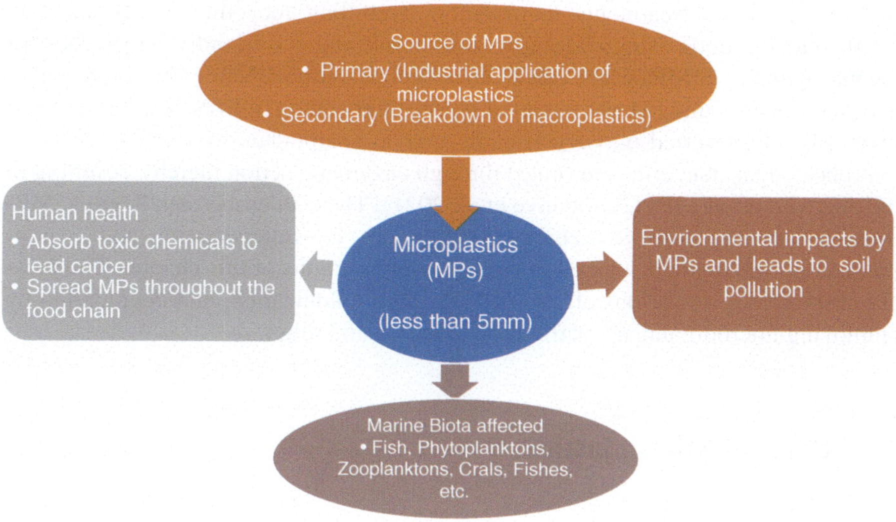

Fig. 2.1 Sources and effects of microplastic (MP) on the ecosystem

mismanaged waste disposal of plastic have resulted in its deposit in water bodies such as lakes, rivers, and ultimately oceans. The plastic that reaches the marine (ocean) environments has been reported to be quickly colonized by microorganisms, resulting in the formation of a unique community known as the plastisphere (Wright et al. 2020a). In this community, MP becomes biofouled, and microbial biofilms of plastics are developed owing to its microbial colonization. The plastisphere plays a critical role in the relationship between MPs and their surrounding environment, affecting ecotoxicity, aquatic ecosystem, and food webs, and the ultimate fate of MPs inaquatic environments.

For the removal of MPs from water bodies, various conventional wastewater treatment approaches like primary and secondary processes have been employed (Talvitie et al. 2017). However, large volumes of MPs were constantly discharged into aquatic bodies thereby confirming only partial removal of MPs from wastewaters. Besides these traditional approaches, advanced oxidation processes (AOPs) are effective techniques used to remove micropollutants from water (Kaur et al. 2024a). These processes include ozonation, electrochemical AOPs, photocatalysis, and Fenton-based AOPs. The AOPs offer advantages over conventional waste removal methods, making them a promising tool for micropollutant removal and valuable for environmental remediation. AOPs are reported to degrade recalcitrant pollutants from stabilized (less biodegradable) leachate and diethyl phthalate (phthalate esters or plasticizers) to some extent (Mohan et al. 2019). Another study reported the use of waste recycling methods, i.e., hydrothermal processing and solvent extraction for plastic waste segregation and recycling (Reimonn et al. 2019). Moreover, an array of cleanup mitigation strategies has been proposed for plastic removal, but it is still impossible to remove the deposited plastic waste through these approaches including chemical treatments, incineration, landfilling, recycling, etc. (Prata et al. 2019). Therefore, devising alternative sustainable and eco-friendly approaches for the management of MPs waste can be an optimum solution. Numerous microorganisms have been found to have potential for the degradation of MPs. The MPs can be ingested and converted into eco-friendly carbon compounds by some insects and microbes (algae, bacteria, and fungi) through enzymatic action thereby, resulting in the degradation of MPs (Amobonye et al. 2021). These microbes can degrade MPs through three-step process: (a) biodeterioration, (b) bio-fragmentation, and (c) mineralization, which are discussed in the following sections of this chapter. The present chapter reviews various effects of MPs, their colonization, and degradation by employing microorganisms, particularly bacteria and fungi.

2 Effects of Microplastics

Due to their ubiquity and possible damage to the entire ecosystem, microplastics, small plastic garbage particles with a diameter of less than 5 millimeters, have grown to be an increasingly problematic issue (Blackburn and Green 2022). In addition to microbeads in personal care products like toothpaste, microfibers from synthetic clothing, and fragments from the breakdown of larger plastic items used as cleaning abrasives, cosmetics, polymer-carriers for drug delivery, sandblasting agents, plastic-coated fertilizers, etc. are all examples of MPs that can be found in various forms as a primary source of microplastics (Yee et al. 2021). Upon the breakdown of the bulk plastic debris, secondary MPs are formed on their own. It refers to those that have fallen within the less than 5 mm limit due to mechanical weathering or photo 40 deterioration over time (Hartmann et al. 2019). Concerns have been expressed concerning the possible effects of microplastics on human health and the health of other living things due to their presence in the environment, including bodies of water (oceans), freshwater, sediments, soils, and live species (Blackburn and Green 2022). The impacts of microplastics on human health and the health of other living things (Fig. 2.1) are as follows:

2.1 Effects of Microplastics on Human Health

The ingestion of microplastics (MPS) is deleterious to human health. The majority of health problems result from MPs that have been enhanced by passing through the human body (Amobonye et al. 2021). MPs have both chemical and physical effects on people, which are the two main effects of the particle. The chemical characteristics of MP are a reflection of both the polymer's and the additive's chemical nature. Polyethylene (PE), polypropylene (PP), and polystyrene (PS) particles make up the majority of the contaminating MP. The additives include inert or reinforcing fillers, plasticizers, antioxidants, UV stabilizers, lubricants, dyes, and flame retardants, which give the plastic particular qualities (Campanale et al. 2020). According to Gasperi et al. (2018), the chemical effects of dyes and plasticizers may be poisonous, mutagenic, and carcinogenic to humans. Phthalates, a type of plasticizer, are frequently employed to give flexibility to plastics, because they cannot chemically bond to the polymer and are therefore discharged into the environment. They have been demonstrated to be present in human urine (Jornet-Martínez et al. 2015), household dust (Bamai, et al. 2014a), and breastfeeding (Högberg et al. 2008).

Size, shape, elasticity, shear strength, and surface charge of the particles are among the physical characteristics of MP. MP can be broken down into fibers, grains, granules, fragments, films, and foams based on shape. Because of their large surface-to-volume ratios, MP is an effective adsorbent of microbes and pollutants (Wright et al. 2020). Microplastics can enter human systems through the consumption of polluted food and water. An issue for the food web is the potential ingestion of MPs and subsequent contaminants. It has been discovered in a variety of foods, including seafood, salt, water bottles, and others. They can build up in the body and could result in inflammation, ingestion, prosthesis, organ damage, and endocrine system disturbance. Microplastics, which can absorb harmful compounds from the environment, have also been found in the air. Daily exposure to oxidizing pollutants causes oxidative stress, inflammation, and carcinogenesis in the lungs (Valavanidis et al. 2013). It is challenging to link the rise in respiratory diseases to inhaled MPs because of the negative health consequences that synthetic fibers have effects on human lungs (Gasperi et al. 2018). According to Ribeiro et al. (2019), MPs ingestion frequently happens as a result of aquatic life, seabirds, and wildlife ingesting humans. The average annual consumption of MPs was determined to be between 39,000 and 52,000 particles based on the average food intake (Cox et al. 2019).

2.2 *Effects of Microplastics on Other Living Beings*

Consumption and accumulation of microplastics throughout the ecosystem have an impact on both terrestrial and marine species. These contaminants can be consumed by marine species like fish and marine mammals directly (from media) or indirectly (via prey), i.e., passed down the food chain (Yong et al. 2020). Mollusks, ascidia, and zooplankton are examples of organisms that can acquire MP most easily. This can result in physical harm, gastrointestinal obstruction, decreased feeding effectiveness, and probable poisoning from the chemicals linked to the microplastics. Microplastics can have an impact on marine species' reproductive and developmental processes, resulting in decreased fertility, stunted growth, and changed behavior due to extensive exposure along the entire food chain, which affects different species. In organisms in greater tropical elevations, it can accumulate. The use of sewage sludge or the usage of plastic mulch in agriculture, both of which contaminate the soil, can also cause microplastics to build up in the soil. This might have an impact on terrestrial organisms as well as soil quality and plant nutrient uptake (Yong et al. 2020).

2.3 Effect of Microplastics on the Water Body

MPs have detrimental effects on the natural environment as well as major harmful effects on marine creatures in rivers, lakes, sediments, and seas. First, harmful plasticizers that have an adverse impact on the environment are released into the sea. Second, MPs and persistent organic matter can interact to create composite pollutants with stronger toxicological effects (Ahmad et al. 2020). Thirdly, interaction with heavy metals may alter the surface structure of MPs, causing them to become charged in seawater and increasing their toxicity. After consuming MPs, marine animals may collect harmful chemicals, which harm marine habitats' biodiversity (Nabi et al. 2022). They may have an impact on the survival, reproduction, and behavior of aquatic organisms, which may result in population loss. Food chain disruptions or the extinction of specific species can have a ripple impact on the ecosystem as a whole.

2.3.1 Invertebrates

According to Haegerbaeumer et al. (2019), the effects of MPs on mortality in invertebrates revealed three key modifications that contributed to the higher mortality in polychaetes. Oysters that were exposed to microsized polystyrene displayed decreased egg production, oocyte quality, and sperm motility (Sussarellu et al. 2016). According to Haegerbaeumer et al. (2019), MP exposure in invertebrates led to a decrease in feeding activity and fertility. It also slowed larval growth and development and increased oxygen consumption and reactive form production.

2.3.2 Fish

When MPs are exposed to fish, lipid peroxidation levels rise in the brain, gills, and spinal muscles. Acetylcholinesterase activity in the brain is also stimulated (Barboza et al. 2020). It has been demonstrated that MP absorption harms fish reproduction and interferes with their eating and motor behavior. Along with proof that prenatal exposure to MP impacts fish growth, maternal transfer of MP to offspring has also been confirmed (Pitt et al. 2018a). The extent of MP accumulation in tissues and the severity of the accompanying histological and biochemical alterations are made worse by increasing dosages of MP (Pannetier et al. 2020).

2.3.3 Zebrafish

It has been demonstrated that the MP nanoparticles affect fish motility, resulting in either hyperactivity or weakness, raising cortisol levels, lowering glucose levels, and speeding up the heartbeat, inflammatory responses, and liver fatty degeneration.

The nanoparticles interfere with synaptic transmission by suppressing acetylcholinesterase activity and inhibiting the expression of cytoskeleton genes. They also affect the antioxidant system by inhibiting glutathione reductase in the brain, muscles, and testes of adult fish and their offspring (Pitt et al. 2018b; Brun et al. 2019).

2.4 Effect of Microplastics on Land/Soil

Microplastics can contaminate soil from a number of sources, including the use of plastic mulch in agriculture, the application of sewage sludge, and the decomposition of plastic waste. The most heavily polluted places include landfills, cities, and beaches, with agricultural ecosystems also being badly impacted (Wang et al. 2019). Microplastics in the soil can affect species that live there, such as earthworms, springtails, and soil mites, which are essential to the breakdown and cycling of nutrients in the soil. These organisms may have physical obstructions, poor eating, and decreased reproductive success as a result of ingesting microplastics, thus disturbing soil ecosystems (Ng et al. 2018).

Despite the short-term advantages of plastic materials, their long-term implications on soil quality and crop yield are now receiving more attention (Boots et al. 2019). Microplastics interact with and absorb organic pollutants such as polychlorinated biphenyls (PCBs), dioxins, dichlorodiphenyltrichloroethane (DDT), and polycyclic aromatic hydrocarbons (PAHs) as well as trace elements (Bradney et al. 2019). Microplastics can operate as a vector for contaminants to be transmitted within and across different ecological compartments due to their surface interactions (Bank and Hansson 2019).

2.4.1 Mammals

MPs predominantly impact the digestive system of terrestrial mammals (Table 2.1). The colon epithelial barrier in mice is damaged by the buildup of 0.5–50 mm PS granules, which also reduces mucus production, inhibits the expression of tight-junction proteins, and changes the composition of the intestinal bacteria (Qiao et al. 2021). The presence of MP in the liver and kidneys of rodents resulted in inflammation, oxidative stress, a decline in the relative mass of the liver, and abnormalities in the metabolism of lipids and energy (Wang et al. 2021a, b).

3 Colonization of Microorganisms on the Plastic Body

In addition to physical support, nutritional availability, and habitats that let microorganisms withstand adverse environmental conditions, MPs can benefit colonizing microbial populations (Ahmad et al. 2020). As soon as the material enters the

Table 2.1 Effects of microplastics on human health and other living beings

S. no.	Name of MPs	Effects	References
1.	Plasticizers (phthalates)	Biological effects in utero that are associated with a short pregnancy duration	Latini et al. (2003)
2.	Phthalates	Asthma and allergies in children	Bamai et al. (2014b)
3.	Bisphenol-A (BPA)	Reproductive toxicants, being associated with adverse birth outcomes	Peretz et al. (2014)
4.	Plasticizers (BPA and phthalates)	Disrupt animal development and reproduction by affecting thyroid hormones, mitochondrial cholesterol transport, and increasing fatty acid oxidation	Mathieu-Denoncourt et al. (2015)
5.	Polyvinyl chloride (PVC)	Toxic effects at a cellular level, looking at cytotoxicity, oxidative stress, and cell viability	Schirinzi et al. (2017)
6.	di-2-ethylhexyl phthalate (DEHP) used in polyvinyl chloride plastics (PVC)	Affects liver, reproductive tract, kidneys, lungs, and heart, especially in developing animals. However, DEHP poses a relatively low risk of hepatic cancer in humans	Tickner et al. (2001)
7.	Phthalates	Cause oxidative stress in mammals, which can harm various systems like the reproductive, respiratory, endocrine, circulatory, and central nervous systems	Brassea-Pérez et al. (2022)

ecosystem, MPs connect to the rough or smooth surface with inorganic particles, organic matter, and microbes, colonizing MP surfaces and creating MP biofilms known as the "plastisphere" (Zettler et al. 2013). The "plastisphere" biofilm is made up of a variety of bacteria, fungi, viruses, archaea, algae, and protozoans that are found in the plastisphere's microbial communities (Wright et al. 2020a). According to Flemming (1998), the colonization of these bacteria can lead to the development of biofilms, which are complex ecosystems predominantly made up of microbes, cell secretions, organic and inorganic particles, etc. The structure and functionality of MPs can be harmed by biofilms in a number of ways, such as by disguising surface features, destroying additives, secreting enzymes that change or degrade MPs, and releasing metabolic byproducts (Miao et al. 2019).

There are typically four phases to the biofilm-mediated breakdown of MPs, which is more complicated than what is seen in pure bacterial and fungal cultures. The microorganisms attach to an MP's surface in the first step and change its surface characteristics (such as hydrophilicity and adhesion). The increased leaching of chemicals and monomers from the MPs during the second stage is caused by cellular and enzyme breakdown. In the third step, biologically derived enzymes or radials attack additives and MPs, causing both embrittlement and a loss of mechanical stability. The MPs are penetrated by accumulated water and microbial filaments in the fourth step, which causes MPs to decompose and be used further by microbes (Flemming 1998). MPs serve as a unique habitat that can take in nutrients to encourage the growth of microorganisms and have a significant impact on the microbial communities and functions they are linked with. Additionally, MPs are attacked and destroyed by biofilms through the action of microbes and enzymes, which they use as adhesion media, carbon sources, and energy sources.

According to Yang et al. (2020), the development of the microbial community in MP biofilms is a process of temporal succession that can be broken down into many phases. The pioneer microbes can create microbial assemblages and cover the greatest amount of plastic surface area during the early colonization stage. These generalist organisms multiply without competition, growing faster than the others to cover the surface (Wright et al. 2020). They also create a favorable environment for resource recycling and improved adhesion of other bacteria (Onda and Sharief 2020). As the biofilm develops, bacteria, viruses, and eukaryotic microorganisms begin to irreversibly bind to the plastic surface within days. Pili, protein adhesion, and the formation of complex extracellular polymeric substances (EPS) are responsible for achieving this (Dussud et al. 2018). EPS also helps cells adhere to surfaces and maintain cell-cell connections. According to Dang and Lovell (2015), the secondary metabolites may enhance quorum sensing and the antibacterial effect to resist rival microorganisms. Adsorption, desorption, and fragmentation of chemicals linked to microplastics are among the metabolic events these processes catalyze (Urbanek et al. 2018). Finally, the continuous recruitment, loss, and replacement of species transform the structure into a hotspot for diverse interacting communities (Onda and Sharief 2020), resulting in a mature plastisphere.

3.1 Bacterial Colonization on Plastic Bodies

Cyanobacteria and bacterial members of the Gammaproteobacteria and Alphaproteobacteria are the early dominant pioneer communities in the marine and estuarine plastisphere (Bryant et al. 2016; Oberbeckmann et al. 2015). Alphaproteobacteria (Pollet et al. 2018) and Bacteroidetes (Lee et al. 2008) are two examples of distinct microbial communities whose abundance increased quickly on the plastisphere as they matured. Bacteroidetes were always significantly more abundant at later stages (Hadad et al. 2005). Rhodobacteraceae, on the other hand, arrived in MPs first, were persistent, and eventually took over (Pollet et al. 2018). Rhodospirillaceae, which includes purple sulfur bacteria, may fix N_2 and so offer "fertilizer" to other members of the community. Additionally, in the later phases of the MPs colonization, Flavobacteria, Planctomycetaceae, and Phyllobacteriaceae were prevalent (Pinto et al. 2019). The primary bacteria abundant in the plastisphere are typically Proteobacteria, Bacteroidetes, Firmicutes, and Cyanobacteria. According to Woodall et al. (2018), this stands out from a variety of different substrates, including metal, rubber, glass, cloth, and seawater. Here are a few bacteria, including the following:

3.1.1 Cyanobacteria (Blue-Green Algae)

Typically, cyanobacteria and diatoms work together to contribute net primary production on MPs, which supplies nutrients to the plastisphere's heterotrophs. Numerous studies have demonstrated that Cyanobacteria constitute the plastisphere's predominant candidates, particularly in the early phases. For instance, samples of PE, PET, and PP were discovered to contain the species *Phormidium*, which was frequently present in the microbial mat on different substrates. This genus was found to be a significant component of a community in the North Atlantic that was colonized by plastic (Zettler et al. 2013), and it was also found to be capable of hydrocarbon degradation (Antić et al. 2006). Additionally, *Phormidium*, a filamentous cyanobacterium, was present on plastics but not in seawater samples (Zettler et al. 2013).

3.1.2 Proteobacteria (Gram-Negative Bacteria)

Compared to populations in seawater, Proteobacteria make up the plastisphere's most enriched phylum. The classes Alphaproteobacteria, Gammaproteobacteria, Betaproteobacteria, and Deltaproteobacteria of the phylum Proteobacteria are particularly concentrated on plastic materials. For instance, the majority of the plastisphere's inhabitants belonged to the family Rhodobacteraceae, a subclass of the Alphaproteobacteria. According to Dang et al. (2008), Roseobacter, a member of the Rhodobacteraceae family, is the main colonizer of diverse substrates in marine environments. Rhodobacteraceae actually plays a significant role in the formation of biofilms by reflecting the production of extracellular polymeric substance (EPS), which facilitates the settlement of other microorganisms (Amaral-Zettler et al. 2020). It is also abundant during all phases of colonization. Numerous *Vibrio* species are also considered hazardous and dangerous hitchhikers on MPs, which aids in their widespread dispersal throughout the marine environment. When compared to natural sediment, *Vibrio* spp. are typically the first to colonize the plastisphere, growing quickly and dominating the environment (Kesy et al. 2019).

3.1.3 Bacteroidetes

The second most prevalent bacterial phylum on plastic surfaces is Bacteroidetes. This is a crucial component and a major contributor to the development of biofilms (Pollet et al. 2018). It has been established that the class Flavobacteria colonizes a number of MPs, including PS, PE, PP, and PET. For instance, according to Amin et al. (2012), the family Flavobacteriaceae is a keystone taxon in biofilms that eat exudates from dinoflagellates and diatoms. According to Oberbeckmann et al. (2016, 2018), flavobacteriaceae were found more frequently on PET particles compared to glass beads and PE and PS compared to wooden pellets. Numerous reports of the genus *Lewinella* and the order Sphingobacteriales colonizing the plastisphere have been made (Ogonowski et al. 2018). Following outdoor incubation in North Seawater, Sphingobacteriales were shown to be prevalent in both short-term (60 days) and long-term (5 months) MP samples and capable of metabolizing polycyclic aromatic hydrocarbons (PAHs; Kirstein et al. 2019).

3.1.4 Firmicutes

On marine PS samples, Firmicutes were among the most numerous populations that were enriched (Syranidou et al. 2019). In particular, the genus *Bacillus* has demonstrated a strong and widespread ability to biodegrade a variety of MPs, including PE, PP, PS, PVC, and PET.

3.2 *Fungi Colonization on Plastic Bodies*

According to their morphologies, fungi can be classified as yeasts, filamentous fungi, or dimorphic fungi. They are eukaryotic organisms lacking chlorophyll. All fungi play a part in the mineralization of ambient micropollutants, but filamentous fungi like *Fusarium oxysporum*, *Aspergillus* species, *Purpureocillium lilacinum*, and *Fusarium falciforme* are particularly important. This may be related to filamentous fungi's exceptional capacity to adapt to shifting environmental conditions and to thrive in environments that are challenging for other microbes (Ferreira et al. 2020). Fungi can adhere to plastic surfaces and utilize them as a source of carbon and energy, because they are heterotrophic organisms. It releases a variety of digestive enzymes (such as protease, lipase, laccase, esterase, manganese peroxidase, and unspecific peroxygenases) and proteins (hydrophobins) that adhere to the polymer surface, penetrate, and break down macromolecules into a small organic compound that eventually releases CO_2, H_2O, and CH_4 as a by-product (Othman et al. 2021).

Nevertheless, the fungus has distinct strategies for removing pollutants, such as biomineralization, biosorption, and biotransformation (Harms et al. 2011). The endophytic fungus has a symbiotic relationship with the host in addition to secreting digestive enzymes, which causes additional degradation of the polymer. Fungi can potentially create biofilms in the marine ecosystem to break down microplastics. When different fungus communities that are geographically close to one another release chemicals (extracellular matrix) and establish stable consortia on the surface of the micropolymer, these biofilms are created (Sánchez et al. 2022). The structural and functional characteristics of microplastics are impacted by biofilms, which also break down additives and eventually produce metabolic by-products (Debroy et al. 2022). Enzymes play a role in the microbial degradation of plastic. The polymeric structure of plastic is specifically targeted by fungi's unique enzyme system, which breaks it down into its monomer and releases CO_2 and water under aerobic settings and CH_4 under anaerobic conditions (Zhang et al. 2021).

Numerous intracellular and extracellular enzymes found in fungi can catalyze a number of biochemical processes, including the breakdown of microplastics. The cytochrome P_{450} family (CYP), Phase I enzyme epoxidase, and Phase II enzyme transferases are involved in the intracellular enzyme system that is essential for the detoxification and adaptation of the fungus (Shin et al. 2018). Phase II enzymes are in charge of the conjugation activities, whereas Phase I enzymes are in charge of oxidation (Schwartz et al. 2018). The large family of heme-containing monooxygenases known as CYP is capable of catalyzing a wide range of reactions, including the metabolism of aliphatic, alicyclic, and aromatic compounds, which in turn results in their epoxidation, hydroxylation, dealkylation, sulfoxidation, desulfurization, deamination, dehalogenation, and N-oxide reduction. In contrast to plants, mammals, or bacteria, fungi have a multiplicity of these enzymes according to Shin et al. (2018). The depolymerization of lengthy polymers and heterogeneous reactions are both dependent on extracellular enzymes, which have also been extensively researched (Shin et al. 2018). Due to their broad reactivity range, these enzymes also promote surface functionalization, increasing mineralization (Amobonye et al. 2021). The hydrolytic system, which generates the hydrolases needed to break down polymers, and the noteworthy oxidation system, which is linked to the breakdown of the polymer structure, make up the extracellular enzymatic system. The class II peroxidases (manganese peroxidase, lignin peroxidase, and versatile peroxidase), laccase, dye-decolorizing peroxidase, and unspecific peroxygenases make up the majority of the oxidative system. This oxidative system is thought to be an effective instrument for cleaning up the environment, because it can oxidize a wide variety of substrates. By using hydrogen peroxide as an electron-accepting co-substrate in redox reactions, epoxidation, aromatic preoxygenation, sulfoxidation, or other methods mediated by unspecific peroxygenases, the oxidative system transfers electrons from the organic substrates to the molecular oxygen state (Karich et al. 2017).

Hydrophobins are surface hydrophobic proteins that are made by fungi and range in length from 70 to 350 amino acids (Wösten and Scholtmeijer 2015). These have a role in the development of filamentous fungi's hyphae, fruiting bodies, and spores. They also help the hyphae adhere to the hydrophobic plastic surface. On the interface of the hydrophobic and hydrophilic faces, the hydrophobins self-assemble to form monolayers, which are amphipathic films. Based on the assembly of the monolayer, different morphologies are obtained (Kulkarni et al. 2017). These proteins act as biosurfactants that encourage substrate mobility and bioavailability, making them an essential tool for bioremediation.

4 Biodegradation: Degradation of Microplastic by Microbes

The process by which microorganisms ingest and break down plastic pieces into readily available carbon sources by altering the functional group structure, molecular weight, tensile properties, and other characteristics of the plastics is known as the biodegradation of plastics. Microorganisms, MP characteristics, and environmental conditions all have an impact on the microbial breakdown of MPs (Zhai et al. 2023). According to Ammala et al. (2011), the breakdown of MPs is an integrated process that combines physicochemical and microbiological degradation variables in various settings. Although MPs are less vulnerable to microbial attack than other degradable materials, microbial degradation has significantly impacted the transformation of MPs (Rujnić-Sokele and Pilipović 2017). By supporting microbial colonization and growth and acting as a carbon source, MPs create a distinct ecological niche for microorganisms. Microorganisms are naturally able to adapt to practically any environment and have the capacity to break down a wide range of substances, including MPs (Brooks et al. 2011). In order to facilitate natural bioremediation and influence the cleaning of natural ecosystems without having negative effects, the use of microbes to degrade MPs will enhance biodegradation without endangering the environment (Qi et al. 2017). This makes it a promising and environmentally safe strategy (Shah et al. 2008).

Multiple metabolic events are involved in the microbial breakdown of MPs, and various MPs degrade in different ways. The chemical structure and molecular weight of MP, the type of microbes present, as well as other environmental parameters, all have a role in the microbe-mediated breakdown of MP. According to Yuan et al. (2020), MPs' many properties, including their mobility, crystallinity, molecular weight, kinds of functional groups and substituents, and plasticizers or additives that have been added, all have a significant impact on how quickly they degrade. Both must first be transformed into monomers during the degradation process before they can be mineralized. Compared to the pore size of the cell membrane, the MP particle size is considerably bigger. Therefore, MPs must first be depolymerized into smaller monomers in order to be absorbed and biodegraded within microbial cells, since they are too big to pass through cellular membranes. In a process that often involves hydrolysis, the most significant reaction for the destruction of MPs, microbial enzymes, and microorganisms both depolymerize MPs. In contrast to MP hydrolysis, oxidative degradation is a process that can change both hydrolyzable and nonhydrolyzable polymers. The enzyme attaches to the MPs and initiates hydrolytic cleavage during the hydrolysis process. Different microorganisms have depolymerase enzymes that can either work intracellularly or extracellularly to break down MPs (Yuan et al. 2020). An endogenous carbon store is hydrolyzed by collecting microorganisms during intracellular breakdown. Since extracellular enzymes from microorganisms break down MPs into short chains or smaller molecules (e.g., oligomers, dimers, and monomers), which are small enough to pass through semipermeable membranes, extracellular degradation involves the use of an exogenous carbon source and is not always carried out by the accumulating microorganisms. Depolymerization is the term used to describe this process. Through a process known as mineralization, these short-chain molecules are then transformed into end products (for example, CO_2, H_2O, or CH_4), which can then be used as carbon and energy sources (Gu 2003). The production of enzymes that catalyze the breakdown of MP chains into monomers starts the process of MP degradation caused by microbes. The microbial-mediated breakdown of MPs depends on microbial enzymes.

4.1 Role of Bacteria in the Decomposition of MPs

Of all organisms, bacteria major category of microorganisms and are the most prevalent (Bakir et al. 2014a). Bacteria typically inhabit soil, water, and the atmosphere, and many species are widely known for their capacity to break down contaminants. *Achromobacter*, *Acinetobacter*, *Arthrobacter*, *Aspergillus* (fungus), *Bacillus*, *Comamonas*, *Delftia*, *Micrococcus*, *Nesiotobacter*, *Paenibacillus*, *Pseudomonas*, *Rahnella*, *Staphylococcus*, and *Stenotrophomonas* are some of the microorganisms that can break down MPs (Table 2.2). *Ideonella* and *Saccharomonospora*, which were isolated from compost and a PET bottle recycling facility, respectively, are bacteria that can break down PET (Kawai et al. 2014). It has been shown that *Pseudomonas*, *Bacillus*, and *Achromobacter* degrade PVC (Giacomucci et al. 2019). In tests, *Aspergillus* was able to break down PU (Khan et al. 2017). Additionally, the sludge composting pool isolates *Thermus*, *Bacillus*, and *Geobacillus* destroyed PS, indicating that hyperthermophilic composting was a potential method for removing MPs from organic wastes (Chen et al. 2020).

High-impact polystyrene (HIPS)-degrading bacteria *Bacillus* and *Pseudomonas* were found in landfills (Mohan et al. 2016). Different soil samples yielded isolates of *Aspergillus*, *Bacillus*, *Pseudomonas*, *Staphylococcus*, and *Streptococcus*, all of which were capable of decomposing PET and PS (Asmita et al. 2015). Additionally, PS and PE could both be broken down by a *Rhodococcus* bacterium that was isolated from soil-burying PE mulch (Mor and Sivan 2008). Additionally, PS debris was home to the marine polychaete *Marphysa sanguinea*, which was able to destroy PS particles (Jang et al. 2018). *Pseudomonas*, which was isolated from plastics submerged in the Bay of Bengal for several days, was discovered to have a high capacity for degrading PP, LDPE, and HDPE (Sudhakar et al. 2007). The high hydrophobicity of *Pseudomonas*' cell surface, which encourages adherence to the polymer and the development of biofilms that aid in breakdown, appears to play a role in the organism's capacity for degradation. Extracellular polysaccharide release is crucial to promoting LDPE breakdown as well. The effects of degradation, however, differ depending on the type of plastic (Shah et al. 2008). Biofilm, which depends on the influence of nutritional circumstances for its production, has a significant impact on biodegradation.

The variation in glucose and ammonium sulfate concentrations revealed a change in *Pseudomonas* surface hydrophobicity (Tribedi and Sil 2014). The efficient production of a biofilm in a polyethylene succinate film was thus encouraged by the increased cell surface hydrophobicity, and degradation was positively connected with it (Tribedi and Sil 2014). In contrast, biofilm growth was reduced, and PE breakdown was low in the presence of marine sediments, which were rich in organic carbon (Nauendorf et al. 2016). As a result, environmental and dietary factors that

Table 2.2 Microbial degradation of microplastics (MPs) by bacteria

S. no.	Microbes	Microplastictype	Source	Effect	References
1.	*Bacillus*	Polypropylene (PP)	Municipal landfill Sediment	Weight reduced by 14.7% and diameter less by 22.6% in 60 days	Park and Kim (2019)
2.	*Paenibacillus*	Polypropylene (PP)	Municipal landfill Sediment	Weight reduced by 14.7% and diameter less by 22.6% in 60 days	Park and Kim (2019)
3.	Souda	High-density polyethylene (HDPE)	Souda bay	Loss of the weight 8% in 60 days	Tsiota et al. (2018)
4.	Agios	High-density polyethylene (HDPE)	Souda bay	Loss of the weight 18% in 60 days	Tsiota et al. (2018)
5.	*Bacillus simplex*	Low-density polyethylene (LDPE)	Earthworm gut	Reduction in weight in 21 days	Huerta Lwanga et al. (2018)
6.	*Enterobacter* sp. and *Pantoea* sp. (microbial consortia)	LDPE	Plastic garbage processing areas	Loss in weight: 81 and 38% in strip and pellet in 120 days	Skariyachan et al. (2016)
7.	*Brevibacillus borstelensis*		Coastal regions	Weight reduced by 11.4% in 30 days	Mohanrasu et al. (2018)
8.	*Microbulbifer hydrolyticus* IRE-31		Marine pulp mill Wastes rich in Lignin	Change comes in surface of MPs	Li et al. (2020)
9.	*Brevibacillus* spp.	PP, LDPE, and HDPE	Sewage treatment Plants and waste Management landfills	Loss the weight from 37.2 to 45.3% in 140 days	Skariyachan et al. (2018)
10.	*Bacillus*	PP	Mangrove	Reduce the weight by 4%	Auta et al. (2018)
11.	*Rhodococcus*	PP	Mangrove	Reduce the weight by 6.4%	Auta et al. (2018)
12.	*Bacillus gottheilii*	PE, PET, PP, PS	Mangrove	Reduce the weight respectively 6.2, 3.0, 3.6, 5.8%	Auta et al. (2017)
13.	*Aspergillus* sp.	PP	Sea water	Reduce the weight by 22% in 42 days	Sarkhel et al. (2020)
14.	*Vibrio* sp.	PP	Sea water	Reduce weight by 355 in 42 days	Sarkhel et al. (2020)

encourage the development of biofilms on plastic polymers are significant stimuli for *Pseudomonas* to degrade synthetic plastics (Wilkes and Aristilde 2017). One microorganism's harmful metabolite may serve as a substrate for the growth of another, neutralizing the toxic effects of the first (Yuan et al. 2020). The activity and tolerance for the treatment of contaminants, on the other hand, were increased by symbiotic relationships in consortia (Yuan et al. 2020).

4.2 Role of Fungi in the Decomposition of MPs

Fungi have the ability to cling to and use MP, in addition to bacteria (Mitik-Dineva et al. 2009). The hydrophobicity of MPs can be reduced by fungi by promoting the synthesis of various chemical bonds in MPs, such as carbonyl, carboxyl, and ester functional groups. Fungi can also encourage the oxidation and movement of chemicals (Chen et al. 2016). Studies on the fungal-mediated breakdown of MPs have been conducted recently, and the findings indicate that fungi can utilize MPs as a carbon source. Given their variety and efficient enzymatic system, fungi are advantageous in this regard for the breakdown of microplastics. The plastic polymers are broken down into monomers by extracellular and intracellular enzymes in fungi, which also create carbon dioxide and water in aerobic settings and methane in anaerobic situations. In addition, fungi release hydrophobins, surface proteins that facilitate substrate mobility, and bioavailability throughout the bioremediation process.

The microbial breakdown of microplastics involves a number of metabolic processes. The chemical makeup and molecular weight of the microplastics as well as the ambient factors (pH, temperature, salinity, etc.) affect how fungal-regulated microplastic breakdown occurs. Additionally, a number of microplastic characteristics, including crystallinity, mobility, functional groups, and type of plasticizer used, can aid in the breakdown of these materials. Currently, the precise method by which a fungus breaks down microplastic has not been well studied, and as a result, the associated pathways have not yet been sufficiently clarified. However, a potential degradation mechanism of microplastics can be suggested based on the information currently known on the fungal breakdown of similar polymers. The polymer is often broken down into monomers during degradation, and these monomers are then mineralized. The pore size of the cell membrane is less than the size of microplastics. Depolymerizing the microplastic into monomers is therefore required in order for the cells to readily absorb them (Yuan et al. 2020). To convert hydrolyzable and non-hydrolyzable polymers into monomers, the process commonly comprises hydrolysis and, in contrast, oxidative destruction. The process of degradation can take place either intracellularly, using an endogenous carbon source for hydrolysis, or externally, using an exogenous carbon source. After decomposition, the short-chain compounds are finally mineralized and utilized as sources of energy and carbon (Chandra and Singh 2020). Degradation of microplastics caused by fungi can take place in the following phases:

(a) *Biodeterioration*: The first process, known as biodeterioration, entails the surface breakdown of the polymer caused by the combined physical and chemical action of fungi and abiotic elements (Anjana et al. 2020). Microplastic loses some of its resistance as a result of the fungus colonizing its surface. Due to the hydrophobic nature of microplastics, fungus develops hydrophobins, which are surface-active proteins, to interact with the surface (Nauendorf et al. 2016). Proteins called hydrophobins range in length from 70 to 350 amino acids. These have a role in the development of hyphae, fruiting bodies, and the spores produced by filamentous fungi that help the hyphae adhere to the plastic surface. The potential for deterioration is increased by biofilm, since it is particularly adept at enhancing surface interaction. The mechanical characteristics of the polymer are reduced as a result of the fungal hyphae adhering to its surface and releasing extracellular enzymes (Sánchez 2020).

(b) *Bio-fragmentation*: The enormous microplastics are now being depolymerized into smaller oligomers, monomers, or dimers (Miri et al. 2022). The hydrolysis of glycosidic, ester, and peptide bonds by fungal enzymes, either intracellularly or extracellularly, is required for the depolymerization of microplastics (Temporiti et al. 2022). The bigger microplastics are first reduced by fungus-secreted extracellular enzymes into simpler molecules that may pass through the fungus' semipermeable membrane before being further reduced by fungus-secreted intracellular enzymes into short chains. After fragmentation, the lower molecular weight molecules are taken up by the fungus, and the various membrane transport mechanisms can provide both active and passive transport for the same (Ru et al. 2020).

(c) *Mineralization*: Finally, by mineralization, these obtained short chains are converted into carbon dioxide, water, or methane, either aerobically or anaerobically. The fungus then uses these compounds as a source of carbon or energy (Montazer et al. 2019). Fungal enzymes play a big part in how quickly microplastics break down. Recent research has reported on the function of fungal enzymes in the breakdown of microplastics. Zhang et al. (2020) revealed the potential of the enzyme laccase-like multicopper oxidase in polyethylene remediation and isolated a polyethylene-degrading fungus, *A. favus*, from the intestines of the wax moth *Galleria mellonella*. The role of the enzyme serine hydrolase, which was isolated from the fungus *Pestalotiopsis microspora*, in the biodegradation of the microplastic polyurethane was identified by Russell et al. (2011) in a different investigation. Devi et al. (2015) confirmed that extracellular enzymes play a role in the biodegradation of HDPE microplastic by *A. tubingensis* VRKPTI and *A. favus* VRKPT2. Overall, the results show that fungal enzymes play a major role in the breakdown of microplastics (Table 2.3). To fully grasp the future of degradative pathways, substantial research must be conducted on the fungal enzymes that are responsible for the degradation of microplastics, because they are not totally understood.

Table 2.3 Fungi associated with microbial degradation of MPs

S. no.	Microbes	Microplastic type	Source	Effect	References
1.	*Zalerion maritimum*	PE pellets	Marine	Decreases the mass and size in 28 days	Paço et al. (2017)
2.	*Aspergillus flavus*	PE	Seawater	Weight loss by 16.2%	Alshehrei (2017)
3.	*Aspergillus fumigatus*			Weight loss by 20.5% in 30 days	
4.	*Penicillium oxalicum*	LDPE and HDPE sheets	Waste disposal land soil sample	Loss in weight after 60 days	Ojha et al. (2017)
5.	*Penicillium griseofulvum*	PE and PU	Shoreline of lake	Show the potential of plastic for few days	Brunner et al. (2018)
6.	*Cladosporium cladosporioides*	PU			
7.	*Leptosphaeria* sp.	PE and PU			
8.	*Xepiculopsis graminea*				
9.	*Trichoderma viride*	LDPE	Waste disposal land soil sample	Weight loss 5.13% and tensile strength reduce by 58% in 45 days	Munir et al. (2018)
10.	*Aspergillus nomius*			Weight loss 6.63% and tensile strength reduce by 40% in 45 days	
11.	*Aspergillus oryzae*	LDPE	Waste disposal land soil sample	Weight reduce by 36.4 ± 5.53%in 112 weeks	Muhonja et al. (2018)
12.	*Aspergillus* sp.	PU	Waste disposal land soil sample	Change in melting point and 15–20% Wt. loss in 28 days	Osman et al. (2018)
13.	*Penicillium* sp.	PU	Waste from terrestrial land	Degradation of PU in 60 days	Magnin et al. (2019)
14.	*Aspergillus terreus*	PE	Mangrove dumpsite	Weight loss 37.94 ± 3.06% and tensile strength reduce	Sangale et al. (2019)
15.	*Aspergillus sydowii*			Weight loss 41.82 ± % and tensile strength reduce in 60 days	
16.	*Aspergillus flavus*	HDPE	Guts of wax moth	Decreased Mn, carbonyls group and	Zhang et al. (2020)

5 Conclusions and Future Prospects

Plastics are one of the most harmful and nonbiodegradable synthetic materials that are commonly used because of their strength, flexibility, and low cost. The widely used plastic polymers include PE (polyethylene), LDPE (low-density polyethylene), HDPE (high-density polyethylene), PS (polystyrene), and PP (polypropylene) accumulate in the environment, causing harm to the flora and fauna. Nearly 50% of the plastic production is single-use plastic, which causes severe environmental damage. One of the states (Karnataka) in India has 201 plastic manufacturing units, with 120 in Bengaluru alone, producing 77,247 tons each year (Skariyachan et al. 2018). This incessant plastic production frequency is causing an imbalance in the ecosystem and posing hazardous risks to humans. The excessive mass production and perpetual use of plastic have resulted in plastic pollution that has badly affected the quality of life, especially in water bodies. Because of its long-lasting and nonbiodegradable nature, the management of plastic waste has become a global concern. The disposal of plastic waste through conventional methods including landfills, incineration, and recycling is ineffective and also poses several ecological risks. Therefore, some eco-friendly microbial and biotechnological interventions would be better alternatives to current waste disposal approaches. An array of microorganisms can break down waste materials. Further studies revealed that microorganisms when used alone or in combination have been found effective in breaking down plastics including microplastics. Therefore, several strains of bacteria and fungi as well as the use of microbial consortia can be a highly cost-effective and environmentally friendly approach to managing plastic waste. Further studies on the formulation of effective and novel microbial strains for the disposal of plastic waste at a large scale can be a viable solution to manage plastic pollution. In aquatic environments, microbes can colonize the plastic and form a community called as plastisphere. The plastisphere directly interacts with plastic, and the surrounding aquatic environment affects microbial communities, aquatic food webs, and the fate of microplastic and its ecotoxic potential. More studies are required to investigate the interaction of the microbes-plastic-ocean interface and the degradation of plastic particles in the plastisphere. However, further research is much warranted to fully understand the biofilms in plastisphere and identify numerous viable plastic degradation solutions and mechanisms. Furthermore, sustainable, integrated, and eco-friendly approaches and international cooperation including national governance are crucial for the reduction of plastic release in wastewater, consumer education, and improvements in plastic's life cycle management.

Acknowledgments Authors are thankful to the administration of Sant Baba Bhag Singh University and DAV University, Jalandhar, for providing all the necessary infrastructural facilities.

References

Ahmad M, Li JL, Wang PD, Hozzein WN, Li WJ (2020) Environmental perspectives of microplastic pollution in the aquatic environment: a review. Mar Life Sci Technol 2:414–430

Bamai YA, Araki A, Kawai T, Tsuboi T, Saito I, Yoshioka E, Kanazawa A, Tajima S, Shi C, Tamakoshi A, Kishi R (2014a) Associations of phthalate concentrations in floor dust and multi-surface dust with the interior materials in Japanese dwellings. Sci Total Environ 468:147–157

Bamai YA, Shibata E, Saito I, Araki A, Kanazawa A, Morimoto K, Nakayama K, Tanaka M, Takigawa T, Yoshimura T, Kishi R (2014b) Exposure to house dust phthalates in relation to asthma and allergies in both children and adults. Sci Total Environ 485:153–163

Alshehrei F (2017) Biodegradation of low density polyethylene by fungi isolated from Red Sea water. Int J Curr Microbiol App Sci 6(8):1703–1709

Amaral-Zettler LA, Zettler ER, Mincer TJ (2020) Ecology of the plastisphere. Nat Rev Microbiol 18(3):139–151

Amin SA, Parker MS, Armbrust EV (2012) Interactions between diatoms and bacteria. Microbiol Mol Biol Rev 76(3):667–684

Ammala A, Bateman S, Dean K, Petinakis E, Sangwan P, Wong S, Yuan Q, Yu L, Patrick C, Leong KH (2011) An overview of degradable and biodegradable polyolefins. Prog Polym Sci 36(8):1015–1049

Amobonye A, Bhagwat P, Singh S, Pillai S (2021) Plastic biodegradation: frontline microbes and their enzymes. Sci Total Environ 759:143536

Anjana K, Hinduja M, Sujitha K, Dharani G (2020) Review on plastic wastes in marine environment–biodegradation and biotechnological solutions. Mar Pollut Bull 150:110733

Antić MP, Jovancicevic B, Vrvić MM, Schwarzbauer J (2006) Petroleum pollutant degradation by surface water microorganisms (8 pp). Environ Sci Pollut Res 13:320–327

Asmita K, Shubhamsingh T, Tejashree S (2015) Isolation of plastic degrading micro-organisms from soil samples collected at various locations in Mumbai, India. Int Res J Environ Sci 4(3):77–85

Auta HS, Emenike CU, Fauziah SH (2017) Screening of *Bacillus* strains isolated from mangrove ecosystems in Peninsular Malaysia for microplastic degradation. Environ Pollut 231:1552–1559

Auta HS, Emenike CU, Jayanthi B, Fauziah SH (2018) Growth kinetics and biodeterioration of polypropylene microplastics by *Bacillus* sp. and *Rhodococcus* sp. isolated from mangrove sediment. Mar Pollut Bull 127:15–21

Bakir A, Rowland SJ, Thompson RC (2014) Transport of persistent organic pollutants by microplastics in estuarine conditions. Estuar Coast Shelf Sci 140:14–21

Bakir A, Rowland SJ, Thompson RC (2014a) Enhanced desorption of persistent organic pollutants from microplastics under simulated physiological conditions. Environ Pollut 185:16–23

Bank MS, Hansson SV (2019) The plastic cycle: a novel and holistic paradigm for the Anthropocene. Environ Sci Technol 53(13):7177–7179

Barboza LGA, Lopes C, Oliveira P, Bessa F, Otero V, Henriques B, Raimundo J, Caetano M, Vale C, Guilhermino L (2020) Microplastics in wild fish from North East Atlantic Ocean and its potential for causing neurotoxic effects, lipid oxidative damage, and human health risks associated with ingestion exposure. Sci Total Environ 717:134625

Blackburn K, Green D (2022) The potential effects of microplastics on human health: what is known and what is unknown. Ambio 51(3):518–530

Boots B, Russell CW, Green DS (2019) Effects of microplastics in soil ecosystems: above and below ground. Environ Sci Technol 53(19):11496–11506

Bradney L, Wijesekara H, Palansooriya KN, Obadamudalige N, Bolan NS, Ok YS, Rinklebe J, Kim KH, Kirkham MB (2019) Particulate plastics as a vector for toxic trace-element uptake by aquatic and terrestrial organisms and human health risk. Environ Int 131:104937

Brassea-Pérez E, Hernández-Camacho CJ, Labrada-Martagón V, Vázquez-Medina JP, Gaxiola-Robles R, Zenteno-Savín T (2022) Oxidative stress induced by phthalates in mammals: state of the art and potential biomarkers. Environ Res 206:112636

Brooks AN, Turkarslan S, Beer KD, Yin Lo F, Baliga NS (2011) Adaptation of cells to new environments. Wiley Interdiscip Rev Syst Biol Med 3(5):544–561

Brun NR, Van Hage P, Hunting ER, Haramis APG, Vink SC, Vijver MG, Schaaf MJ, Tudorache C (2019) Polystyrene nanoplastics disrupt glucose metabolism and cortisol levels with a possible link to behavioural changes in larval zebrafish. Commun Biol 2(1):382

Brunner I, Fischer M, Rüthi J, Stierli B, Frey B (2018) Ability of fungi isolated from plastic debris floating in the shoreline of a lake to degrade plastics. PLoS One 13(8):e0202047

Bryant JA, Clemente TM, Viviani DA, Fong AA, Thomas KA, Kemp P, Karl DM, White AE, DeLong EF (2016) Diversity and activity of communities inhabiting plastic debris in the North Pacific Gyre. MSystems 1(3):e00024–e00016

Campanale C, Massarelli C, Savino I, Locaputo V, Uricchio VF (2020) A detailed review study on potential effects of microplastics and additives of concern on human health. Int J Environ Res Public Health 17(4):1212

Dang H, Lovell CR (2015) Microbial surface colonization and biofilm development in marine environments. Microbiol Mol Biol Rev 80(1):91–138

Chae Y, An YJ (2018) Current research trends on plastic pollution and ecological impacts on the soil ecosystem: a review. Environ Pollut 240:387–395

Chandra P, Singh DP (2020) Microplastic degradation by bacteria in aquatic ecosystem. In: Microorganisms for sustainable environment and health. Elsevier, pp 431–467

Chen Y, Stemple B, Kumar M, Wei N (2016) Cell surface display fungal laccase as a renewable biocatalyst for degradation of persistent micropollutants bisphenol A and sulfamethoxazole. Environ Sci Technol 50(16):8799–8808

Chen Z, Zhao W, Xing R, Xie S, Yang X, Cui P, Lü J, Liao H, Yu Z, Wang S, Zhou S (2020) Enhanced in situ biodegradation of microplastics in sewage sludge using hyperthermophilic composting technology. J Hazard Mater 384:121271

Chen HL, Nath TK, Chong S, Foo V, Gibbins C, Lechner AM (2021) The plastic waste problem in Malaysia: management, recycling and disposal of local and global plastic waste. SN Appl Sci 3:1–15

Cox KD, Covernton GA, Davies HL, Dower JF, Juanes F, Dudas SE (2019) Human consumption of microplastics. Environ Sci Technol 53(12):7068–7074

Dang H, Li T, Chen M, Huang G (2008) Cross-ocean distribution of Rhodobacterales bacteria as primary surface colonizers in temperate coastal marine waters. Appl Environ Microbiol 74(1):52–60

Debroy A, George N, Mukherjee G (2022) Role of biofilms in the degradation of microplastics in aquatic environments. J Chem Technol Biotechnol 97(12):3271–3282

Dussud C, Meistertzheim AL, Conan P, Pujo-Pay M, George M, Fabre P, Coudane J, Higgs P, Elineau A, Pedrotti ML, Ghiglione JF (2018) Evidence of niche partitioning among bacteria living on plastics, organic particles and surrounding seawaters. Environ Pollut 236:807–816

Ferreira JA, Varjani S, Taherzadeh MJ (2020) A critical review on the ubiquitous role of filamentous fungi in pollution mitigation. Curr Pollut Rep 6:295–309

Flemming HC (1998) Relevance of biofilms for the biodeterioration of surfaces of polymeric materials. Polym Degrad Stab 59(1–3):309–315

Gasperi J, Wright SL, Dris R, Collard F, Mandin C, Guerrouache M, Langlois V, Kelly FJ, Tassin B (2018) Microplastics in air: are we breathing it in? Curr Opin Environ Sci Health 1:1–5

Giacomucci L, Raddadi N, Soccio M, Lotti N, Fava F (2019) Polyvinyl chloride biodegradation by *Pseudomonas citronellolis* and *Bacillus flexus*. New Biotechnol 52:35–41

Gu JD (2003) Microbiological deterioration and degradation of synthetic polymeric materials: recent research advances. Int Biodeterior Biodegradation 52(2):69–91

Guo X, Wang J (2019) The chemical behaviors of microplastics in marine environment: a review. Mar Pollut Bull 142:1–14

Hadad D, Geresh S, Sivan A (2005) Biodegradation of polyethylene by the thermophilic bacterium *Brevibacillus borstelensis*. J Appl Microbiol 98(5):1093–1100

Haegerbaeumer A, Mueller MT, Fueser H, Traunspurger W (2019) Impacts of micro-and nano-sized plastic particles on benthic invertebrates: a literature review and gap analysis. Front Environ Sci 7:425457

Harms H, Schlosser D, Wick LY (2011) Untapped potential: exploiting fungi in bioremediation of hazardous chemicals. Nat Rev Microbiol 9(3):177–192

Hartmann NB, Hüffer T, Thompson RC, Hassellöv M, Verschoor A, Daugaard AE, Rist S, Karlsson T, Brennholt N, Cole M, Wagner M (2019) Are we speaking the same language?

Recommendations for a definition and categorization framework for plastic debris. Environ Sci Technol 53(3):1039–1047

Högberg J, Hanberg A, Berglund M, Skerfving S, Remberger M, Calafat AM, Filipsson AF, Jansson B, Johansson N, Appelgren M, Håkansson H (2008) Phthalate diesters and their metabolites in human breast milk, blood or serum, and urine as biomarkers of exposure in vulnerable populations. Environ Health Perspect 116(3):334–339

Lwanga EH, Thapa B, Yang X, Gertsen H, Salánki T, Geissen V, Garbeva P (2018) Decay of low-density polyethylene by bacteria extracted from earthworm's guts: a potential for soil restoration. Sci Total Environ 624:753–757

Jang M, Shim WJ, Han GM, Song YK, Hong SH (2018) Formation of microplastics by polychaetes (*Marphysasanguinea*) inhabiting expanded polystyrene marine debris. Mar Pollut Bull 131:365–369

Jornet-Martínez N, Antón-Soriano C, Campíns-Falcó P (2015) Estimation of the presence of unmetabolized dialkyl phthalates in untreated human urine by an on-line miniaturized reliable method. Sci Total Environ 532:239–244

Karbalaei S, Hanachi P, Walker TR, Cole M (2018) Occurrence, sources, human health impacts and mitigation of microplastic pollution. Environ Sci Pollut Res 25:36046–36063

Karich A, Ullrich R, Scheibner K, Hofrichter M (2017) Fungal unspecific peroxygenases oxidize the majority of organic EPA priority pollutants. Front Microbiol 8:288117

Kaur H, Kaur R, Jagota N, Singh S, Sharma A, Sharma A (2024a) Process of removing organic micropollutants using advanced oxidation techniques. In: Organic micropollutants in aquatic and terrestrial environments. Cham, Springer, pp 295–313

Kaur R, Kaur H, Singh S, Jagota N, Sharma A, Sharma A (2024b) Organic micropollutants in the urban soils: technological advances and management concerns. In: Organic micropollutants in aquatic and terrestrial environments, pp 47–63

Kawai F, Oda M, Tamashiro T, Waku T, Tanaka N, Yamamoto M, Mizushima H, Miyakawa T, Tanokura M (2014) A novel Ca^{2+}-activated, thermostabilized polyesterase capable of hydrolyzing polyethylene terephthalate from *Saccharomonosporaviridis* AHK190. Appl Microbiol Biotechnol 98:10053–10064

Kesy K, Oberbeckmann S, Kreikemeyer B, Labrenz M (2019) Spatial environmental heterogeneity determines young biofilm assemblages on microplastics in Baltic Sea mesocosms. Front Microbiol 10:444967

Khan S, Nadir S, Shah ZU, Shah AA, Karunarathna SC, Xu J, Khan A, Munir S, Hasan F (2017) Biodegradation of polyester polyurethane by *Aspergillus tubingensis*. Environ Pollut 225:469–480

Kirstein IV, Wichels A, Gullans E, Krohne G, Gerdts G (2019) The plastisphere–uncovering tightly attached plastic "specific" microorganisms. PLoS One 14(4):e0215859

Krueger MC, Harms H, Schlosser D (2015) Prospects for microbiological solutions to environmental pollution with plastics. Appl Microbiol Biotechnol 99:8857–8874

Kulkarni S, Nene S, Joshi K (2017) Production of hydrophobins from fungi. Process Biochem 61:1–11

Lamichhane G, Acharya A, Marahatha R, Modi B, Paudel R, Adhikari A, Parajuli N (2023) Microplastics in environment: global concern, challenges, and controlling measures. Int J Environ Sci Technol 20(4):4673–4694

Latini G, De Felice C, Presta G, Del Vecchio A, Paris I, Ruggieri F, Mazzeo P (2003) In utero exposure to di-(2-ethylhexyl) phthalate and duration of human pregnancy. Environ Health Perspect 111(14):1783–1785

Lee JW, Nam JH, Kim YH, Lee KH, Lee DH (2008) Bacterial communities in the initial stage of marine biofilm formation on artificial surfaces. J Microbiol 46:174–182

Li J, Kim HR, Lee HM, Yu HC, Jeon E, Lee S, Kim DH (2020) Rapid biodegradation of polyphenylene sulfide plastic beads by Pseudomonas sp. Sci Total Environ 720:137616

Ma H, Pu S, Liu S, Bai Y, Mandal S, Xing B (2020) Microplastics in aquatic environments: toxicity to trigger ecological consequences. Environ Pollut 261:114089

Magnin A, Hoornaert L, Pollet E, Laurichesse S, Phalip V, Avérous L (2019) Isolation and characterization of different promising fungi for biological waste management of polyurethanes. Microb Biotechnol 12(3):544–555

Mathieu-Denoncourt J, Wallace SJ, de Solla SR, Langlois VS (2015) Plasticizer endocrine disruption: highlighting developmental and reproductive effects in mammals and non-mammalian aquatic species. Gen Comp Endocrinol 219:74–88

Miao L, Wang P, Hou J, Yao Y, Liu Z, Liu S, Li T (2019) Distinct community structure and microbial functions of biofilms colonizing microplastics. Sci Total Environ 650:2395–2402

Miri S, Saini R, Davoodi SM, Pulicharla R, Brar SK, Magdouli S (2022) Biodegradation of microplastics: better late than never. Chemosphere 286:131670

Mitik-Dineva N, Wang J, Truong VK, Stoddart P, Malherbe F, Crawford RJ, Ivanova EP (2009) Escherichia coli, Pseudomonas aeruginosa, and Staphylococcus aureus attachment patterns on glass surfaces with nanoscale roughness. Curr Microbiol 58:268–273

Mohan AJ, Sekhar VC, Bhaskar T, Nampoothiri KM (2016) Microbial assisted high impact polystyrene (HIPS) degradation. Bioresour Technol 213:204–207

Mohan S, Mamane H, Avisar D, Gozlan I, Kaplan A, Dayalan G (2019) Treatment of diethyl phthalate leached from plastic products in municipal solid waste using an ozone-based advanced oxidation process. Materials 12(24):4119

Mohanrasu K, Premnath N, Prakash GS, Sudhakar M, Boobalan T, Arun A (2018) Exploring multi potential uses of marine bacteria; an integrated approach for PHB production, PAHs and polyethylene biodegradation. J Photochem Photobiol B Biol 185:55–65

Montazer Z, Habibi Najafi MB, Levin DB (2019) Microbial degradation of low-density polyethylene and synthesis of polyhydroxyalkanoate polymers. Can J Microbiol 65(3):224–234

Mor R, Sivan A (2008) Biofilm formation and partial biodegradation of polystyrene by the actinomycete Rhodococcus ruber: biodegradation of polystyrene. Biodegradation 19:851–858

Muhonja CN, Makonde H, Magoma G, Imbuga M (2018) Biodegradability of polyethylene by bacteria and fungi from Dandora dumpsite Nairobi-Kenya. PLoS One 13(7):e0198446

Munir E, Harefa RSM, Priyani N, Suryanto D (2018, March) Plastic degrading fungi *Trichoderma viride* and *Aspergillus nomius* isolated from local landfill soil in Medan. In: IOP conference series: earth and environmental science, vol 126, No. 1. IOP Publishing, p 012145

Nabi G, Ahmad S, Ullah S, Zada S, Sarfraz M, Guo X, Wanghe K (2022) The adverse health effects of increasing microplastic pollution on aquatic mammals. J King Saud Univ Sci 34(4):102006

Nauendorf A, Krause S, Bigalke NK, Gorb EV, Gorb SN, Haeckel M, Wahl M, Treude T (2016) Microbial colonization and degradation of polyethylene and biodegradable plastic bags in temperate fine-grained organic-rich marine sediments. Mar Pollut Bull 103(1–2):168–178

Ng EL, Lwanga EH, Eldridge SM, Johnston P, Hu HW, Geissen V, Chen D (2018) An overview of microplastic and nanoplastic pollution in agroecosystems. Sci Total Environ 627:1377–1388

Oberbeckmann S, Löder MG, Labrenz M (2015) Marine microplastic-associated biofilms–a review. Environ Chem 12(5):551–562

Oberbeckmann S, Kreikemeyer B, Labrenz M (2018) Environmental factors support the formation of specific bacterial assemblages on microplastics. Front Microbiol 8:308230

Oberbeckmann S, Osborn AM, Duhaime MB (2016) Microbes on a bottle: substrate, season and geography influence community composition of microbes colonizing marine plastic debris. PLoS One 11(8):e0159289

Ogonowski M, Motiei A, Ininbergs K, Hell E, Gerdes Z, Udekwu KI, Bacsik Z, Gorokhova E (2018) Evidence for selective bacterial community structuring on microplastics. Environ Microbiol 20(8):2796–2808

Ojha N, Pradhan N, Singh S, Barla A, Shrivastava A, Khatua P, Rai V, Bose S (2017) Evaluation of HDPE and LDPE degradation by fungus, implemented by statistical optimization. Sci Rep 7(1):39515

Onda DFL, Sharief KM (2020) Identification of microorganisms related to microplastics. In: Handbook of Microplastics in the Environment, pp 1–34

Osman M, Satti SM, Luqman A, Hasan F, Shah Z, Shah AA (2018) Degradation of polyester polyurethane by *Aspergillus* sp. strain S45 isolated from soil. J Polym Environ 26:301–310
Othman AR, Hasan HA, Muhamad MH, Ismail NI, Abdullah SRS (2021) Microbial degradation of microplastics by enzymatic processes: a review. Environ Chem Lett 19:3057–3073
Paço A, Duarte K, da Costa JP, Santos PS, Pereira R, Pereira ME, Freitas AC, Duarte AC, Rocha-Santos TA (2017) Biodegradation of polyethylene microplastics by the marine fungus *Zalerion maritimum*. Sci Total Environ 586:10–15
Pannetier P, Morin B, Le Bihanic F, Dubreil L, Clérandeau C, Chouvellon F, Van Arkel K, Danion M, Cachot J (2020) Environmental samples of microplastics induce significant toxic effects in fish larvae. Environ Int 134:105047
Park SY, Kim CG (2019) Biodegradation of micro-polyethylene particles by bacterial colonization of a mixed microbial consortium isolated from a landfill site. Chemosphere 222:527–533
Peretz J, Vrooman L, Ricke WA, Hunt PA, Ehrlich S, Hauser R, Padmanabhan V, Taylor HS, Swan SH, VandeVoort CA, Flaws JA (2014) Bisphenol A and reproductive health: update of experimental and human evidence, 2007–2013. Environ Health Perspect 122(8):775–786
Pinto M, Langer TM, Hüffer T, Hofmann T, Herndl GJ (2019) The composition of bacterial communities associated with plastic biofilms differs between different polymers and stages of biofilm succession. PLoS One 14(6):e0217165
Pitt JA, Kozal JS, Jayasundara N, Massarsky A, Trevisan R, Geitner N, Wiesner M, Levin ED, Di Giulio RT (2018b) Uptake, tissue distribution, and toxicity of polystyrene nanoparticles in developing zebrafish (*Danio rerio*). Aquat Toxicol 194:185–194
Pitt JA, Trevisan R, Massarsky A, Kozal JS, Levin ED, Di Giulio RT (2018a) Maternal transfer of nanoplastics to offspring in zebrafish (*Danio rerio*): a case study with nanopolystyrene. Sci Total Environ 643:324–334
Pollet T, Berdjeb L, Garnier C, Durrieu G, Le Poupon C, Misson B, Briand JF (2018) Prokaryotic community successions and interactions in marine biofilms: the key role of Flavobacteriia. FEMS Microbiol Ecol 94(6):fiy083
Prata JC, Silva ALP, Da Costa JP, Mouneyrac C, Walker TR, Duarte AC, Rocha-Santos T (2019) Solutions and integrated strategies for the control and mitigation of plastic and microplastic pollution. Int J Environ Res Public Health 16(13):2411
Qi X, Ren Y, Wang X (2017) New advances in the biodegradation of poly (lactic) acid. Int Biodeterior Biodegradation 117:215–223
Qiao J, Chen R, Wang M, Bai R, Cui X, Liu Y, Wu C, Chen C (2021) Perturbation of gut microbiota plays an important role in micro/nanoplastics-induced gut barrier dysfunction. Nanoscale 13(19):8806–8816
Reimonn G, Lu T, Gandhi N, Chen WT (2019) Review of microplastic pollution in the environment and emerging recycling solutions. J Renewable Mater 7(12):1251–1268
Ribeiro F, O'Brien JW, Galloway T, Thomas KV (2019) Accumulation and fate of nano-and microplastics and associated contaminants in organisms. TrAC Trends Anal Chem 111:139–147
Ru J, Huo Y, Yang Y (2020) Microbial degradation and valorization of plastic wastes. Front Microbiol 11:507487
Rujnić-Sokele M, Pilipović A (2017) Challenges and opportunities of biodegradable plastics: a mini review. Waste Manag Res 35(2):132–140
Russell JR, Huang J, Anand P, Kucera K, Sandoval AG, Dantzler KW, Hickman D, Jee J, Kimovec FM, Koppstein D, Strobel SA (2011) Biodegradation of polyester polyurethane by endophytic fungi. Appl Environ Microbiol 77(17):6076–6084
Sánchez C (2020) Fungal potential for the degradation of petroleum-based polymers: an overview of macro-and microplastics biodegradation. Biotechnol Adv 40:107501
Sánchez MC, Velapatiño A, Llama-Palacios A, Valdés A, Cifuentes A, Ciudad MJ, Collado L (2022) Metataxonomic and metabolomic evidence of biofilm homeostasis disruption related to caries: An in vitro study. Mol Oral Microbiol 37(2):81–96
Sangale MK, Shahnawaz M, Ade AB (2019) Potential of fungi isolated from the dumping sites mangrove rhizosphere soil to degrade polythene. Sci Rep 9(1):5390

Devi RS, Kannan VR, Nivas D, Kannan K, Chandru S, Antony AR (2015) Biodegradation of HDPE by *Aspergillus* spp. from marine ecosystem of Gulf of Mannar, India. Mar Pollut Bull 96(1–2):32–40

Sarkhel R, Sengupta S, Das P, Bhowal A (2020) Comparative biodegradation study of polymer from plastic bottle waste using novel isolated bacteria and fungi from marine source. J Polym Res 27:1–8

Schirinzi GF, Pérez-Pomeda I, Sanchís J, Rossini C, Farré M, Barceló D (2017) Cytotoxic effects of commonly used nanomaterials and microplastics on cerebral and epithelial human cells. Environ Res 159:579–587

Schwartz M, Perrot T, Aubert E, Dumarçay S, Favier F, Gérardin P, Morel-Rouhier M, Mulliert G, Saiag F, Didierjean C, Gelhaye E (2018) Molecular recognition of wood polyphenols by phase II detoxification enzymes of the white rot *Trametes versicolor*. Sci Rep 8(1):8472

Shah AA, Hasan F, Hameed A, Ahmed S (2008) Biological degradation of plastics: a comprehensive review. Biotechnol Adv 26(3):246–265

Shin J, Kim JE, Lee YW, Son H (2018) Fungal cytochrome P450s and the P450 complement (CYPome) of Fusarium graminearum. Toxins 10(3):112

Skariyachan S, Manjunatha V, Sultana S, Jois C, Bai V, Vasist KS (2016) Novel bacterial consortia isolated from plastic garbage processing areas demonstrated enhanced degradation for low density polyethylene. Environ Sci Pollut Res 23:18307–18319

Skariyachan S, Patil AA, Shankar A, Manjunath M, Bachappanavar N, Kiran S (2018) Enhanced polymer degradation of polyethylene and polypropylene by novel thermophilic consortia of *Brevibacillus* sps. and *Aneurinibacillus* sp. screened from waste management landfills and sewage treatment plants. Polym Degrad Stab 149:52–68

Sudhakar M, Trishul A, Doble M, Kumar KS, Jahan SS, Inbakandan D, Viduthalai RR, Umadevi VR, Murthy PS, Venkatesan R (2007) Biofouling and biodegradation of polyolefins in ocean waters. Polym Degrad Stab 92(9):1743–1752

Sussarellu R, Suquet M, Thomas Y, Lambert C, Fabioux C, Pernet MEJ, Huvet A (2016) Oyster reproduction is affected by exposure to polystyrene microplastics. Proc Natl Acad Sci 113(9):2430–2435

Syranidou E, Karkanorachaki K, Amorotti F, Avgeropoulos A, Kolvenbach B, Zhou NY, Fava F, Corvini PFX, Kalogerakis N (2019) Biodegradation of mixture of plastic films by tailored marine consortia. J Hazard Mater 375:33–42

Talvitie J, Mikola A, Koistinen A, Setälä O (2017) Solutions to microplastic pollution–removal of microplastics from wastewater effluent with advanced wastewater treatment technologies. Water Res 123:401–407

Temporiti MEE, Nicola L, Nielsen E, Tosi S (2022) Fungal enzymes involved in plastics biodegradation. Microorganisms 10(6):1180

Tickner JA, Schettler T, Guidotti T, McCally M, Rossi M (2001) Health risks posed by use of Di-2-ethylhexyl phthalate (DEHP) in PVC medical devices: a critical review. Am J Ind Med 39(1):100–111

Tribedi P, Sil AK (2014) Cell surface hydrophobicity: a key component in the degradation of polyethylene succinate by *Pseudomonas* sp. AKS2. J Appl Microbiol 116(2):295–303

Tsiota P, Karkanorachaki K, Syranidou E, Franchini M, Kalogerakis N (2018) Microbial degradation of HDPE secondary microplastics: preliminary results. In: Proceedings of the international conference on microplastic pollution in the Mediterranean Sea. Springer, pp 181–188

UNEP (2018) Single-use plastics: a roadmap for sustainability. Retrieved from: https://www.unenvironment.org/resources/report/single-use-plastics-roadmap-sustainability

Urbanek AK, Rymowicz W, Mirończuk AM (2018) Degradation of plastics and plastic-degrading bacteria in cold marine habitats. Appl Microbiol Biotechnol 102:7669–7678

Valavanidis A, Vlachogianni T, Fiotakis K, Loridas S (2013) Pulmonary oxidative stress, inflammation and cancer: respirable particulate matter, fibrous dusts and ozone as major causes of lung carcinogenesis through reactive oxygen species mechanisms. Int J Environ Res Public Health 10(9):3886–3907

Vivekanand AC, Mohapatra S, Tyagi VK (2021) Microplastics in the aquatic environment: challenges and perspectives. Chemosphere 282:131151

Wang J, Li Y, Lu L, Zheng M, Zhang X, Tian H, Wang W, Ru S (2019) Polystyrene microplastics cause tissue damages, sex-specific reproductive disruption and transgenerational effects in marine medaka (*Oryziasmelastigma*). Environ Pollut 254:113024

Wang YL, Lee YH, Hsu YH, Chiu IJ, Huang CCY, Huang CC, Chia ZC, Lee CP, Lin YF, Chiu HW (2021a) The kidney-related effects of polystyrene microplastics on human kidney proximal tubular epithelial cells HK-2 and male C57BL/6 mice. Environ Health Perspect 129(5):057003

Wang X, Bolan N, Tsang DC, Sarkar B, Bradney L, Li Y (2021b) A review of microplastics aggregation in aquatic environment: influence factors, analytical methods, and environmental implications. J Hazard Mater 402:123496

Wilkes RA, Aristilde L (2017) Degradation and metabolism of synthetic plastics and associated products by *Pseudomonas* sp.: capabilities and challenges. J Appl Microbiol 123(3):582–593

Woodall LC, Jungblut AD, Hopkins K, Hall A, Robinson LF, Gwinnett C, Paterson GL (2018) Deep-sea anthropogenic macrodebris harbours rich and diverse communities of bacteria and archaea. PLoS One 13(11):e0206220

Wösten HA, Scholtmeijer K (2015) Applications of hydrophobins: current state and perspectives. Appl Microbiol Biotechnol 99:1587–1597

Wright SL, Ulke J, Font A, Chan KLA, Kelly FJ (2020) Atmospheric microplastic deposition in an urban environment and an evaluation of transport. Environ Int 136:105411

Wright RJ, Erni-Cassola G, Zadjelovic V, Latva M, Christie-Oleza JA (2020a) Marine plastic debris: a new surface for microbial colonization. Environ Sci Technol 54(19):11657–11672

Yang Y, Liu W, Zhang Z, Grossart HP, Gadd GM (2020) Microplastics provide new microbial niches in aquatic environments. Appl Microbiol Biotechnol 104:6501–6511

Yee MSL, Hii LW, Looi CK, Lim WM, Wong SF, Kok YY, Tan BK, Wong CY, Leong CO (2021) Impact of microplastics and nanoplastics on human health. Nano 11(2):496

Yong CQY, Valiyaveettil S, Tang BL (2020) Toxicity of microplastics and nanoplastics in mammalian systems. Int J Environ Res Public Health 17(05):1509

Yuan J, Ma J, Sun Y, Zhou T, Zhao Y, Yu F (2020) Microbial degradation and other environmental aspects of microplastics/plastics. Sci Total Environ 715:136968

Zettler ER, Mincer TJ, Amaral-Zettler LA (2013) Life in the "plastisphere": microbial communities on plastic marine debris. Environ Sci Technol 47(13):7137–7146

Zhai X, Zhang XH, Yu M (2023) Microbial colonization and degradation of marine microplastics in the plastisphere: a review. Front Microbiol 14:1127308

Zhang J, Gao D, Li Q, Zhao Y, Li L, Lin H, Bi Q, Zhao Y (2020) Biodegradation of polyethylene microplastic particles by the fungus Aspergillus flavus from the guts of wax moth *Galleria mellonella*. Sci Total Environ 704:135931

Zhang K, Hamidian AH, Tubić A, Zhang Y, Fang JK, Wu C, Lam PK (2021) Understanding plastic degradation and microplastic formation in the environment: a review. Environ Pollut 274:116554

Chapter 3
Microplastic in Aquatic Habitat: Impact and Sustainable Remediation Approaches

Mohd Ali, Ripu Daman Parihar, Shivam Jasrotia, Raman Tikoria, Deepak Kumar, Roohi Sharma, Parkirti, Vikram, and Puja Ohri

Abstract When plastic waste is dumped in the water, it eventually breaks down into small pieces. These plastic pieces, which fall into the category of microplastics (MPs), are typically 0.05–5 mm in size, and because of their small size, aquatic species regularly utilize them. As a result, extensive MPs infiltration poses a threat to already-existing life forms and is a global problem for the aquatic environment. MPs readily bind to other harmful compounds or metals, acting as a vector for the introduction of such toxic substances into living forms. The growing contaminants polyethylene, polypropylene, polystyrene, and other polymers are harmful to all kinds of organisms. Urban wastewater flushing is the primary entry point for MPs into aquatic environments. The current book chapter focuses on the source, origin, formation, impact, and remediation strategies of microplastics in aquatic habitats.

Keywords Microplastics · Aquatic habitat · Pollutants · Urban wastewater · Remediation strategies

M. Ali · D. Kumar · R. Sharma · Parkirti · Vikram · P. Ohri (✉)
Department of Zoology, Guru Nanak Dev University, Amritsar, Punjab, India

R. D. Parihar
Department of Zoology, University of Jammu, Jammu, Jammu & Kashmir, India

S. Jasrotia
Department of Biosciences, Chandigarh University, Mohali, Punjab, India

R. Tikoria
Department of Zoology, Guru Nanak Dev University, Amritsar, Punjab, India

Department of Zoology, Centra University Punjab, Ghudda Bathinda, Punjab, India

K. Khanna et al. (eds.), *Microplastics and Soil Microbiome*,
https://doi.org/10.1007/978-981-96-4978-5_3

1 Introduction

Plastics are known for being very flexible, light, mechanically stable, and having superior electrical and thermal insulating characteristics, making them tough and suitable for use in harsh environments (Hahladakis and Iacovidou 2018). The first synthetic plastic, Bakelite, was created in 1907, marking the beginning of the global plastic manufacturing industry (Klun et al. 2022). Yet, the rapid increase in global plastic manufacture did not occur until the 1950s. Due to its promise, affordable price, and appropriateness for a broad spectrum of business and industrial applications, plastics output increased by over 200 times over the ensuing 65 years, reaching more than 0.380 billion tons in 2015, or roughly 67% of the world's population (Klöckner et al. 2021; Kristanti et al. 2022). Practically every aspect of society uses plastic, including transportation, technological equipment, clothing, and components of packaging that transport goods like food and beverages (Rodrigues et al. 2019; Chia et al. 2021). After 2020, it is predicted that more over 400 million tons of plastic garbage would be produced annually due to the mass production and extensive use of plastics (Ali et al. 2021). Approximately, 76% of all plastic manufacturing is used up as waste, of which 9% is recycled, 12% is burned, and 79% is buried or discharged into the environment (Pan et al. 2022). Microplastics (MPs) with diameters smaller than 5 mm were generated by photodegradation, thermal oxidation and degradation, and possibly biodegradation of plastics of various sources, forms, and types (Duan et al. 2021; Sarma et al. 2022; Wojnowska-Baryła et al. 2022). These MP particles (˂5 mm) are easily absorbed by the biotic system, posing a major hazard to the earth's inhabitants. Ingestion of microplastic particles has negative effects on juvenile, adult, and planktonic aquatic organisms, including decreased food consumption, decreased fecundity, decreased hatchability, digestive system obstruction, behavioral changes, morphological and psychological changes, and even death (De Sá et al. 2018; Chen et al. 2021; Gola et al. 2021; Vo and Pham 2021). Nowadays, research is being focused on the sources, quantity, degradation, and interactions of MPs with their surface species in aquatic environments. The majority of the plastic trash that finds its way into the aquatic habitat comes from terrestrial sources (Chaukura et al. 2021). Domestic wastewater is a source of MPs particles and plastic fibers generated throughout human life, but because of their small size, it is hard to efficiently catch and remove these materials creating problems for standard wastewater treatment plant processes (Browne et al. 2011; Tang et al. 2020; Pan et al. 2022). With the wastewater from the wastewater treatment plants, MPs enter rivers or aggregate in river sediments before entering the marine environment, where they will remain for decades (Alimi et al. 2018; Drummond et al. 2022). Ocean currents can carry MPs a great distance; therefore, MPs are extensively dispersed throughout the ocean, sediments, and even the deep sea (Cunningham et al. 2020). MPs pollution has negative repercussions and have slowly moved from aquatic ecosystems to the rest of the environment (Ivleva et al. 2017; Guzzetti et al. 2018). Thus, it is crucial to create methods for MP removal, recycling, and degrading in order to reduce any potential ecological problems.

Therefore, the current chapter, which takes into account the negative effects of plastics and MPs, concentrates on the main sources of MPs, their harmful effects, and potential solutions to mitigate the pollution sustainably in aquatic habitat.

2 Microplastic Production

Since the 1960s, the production of polymeric plastic has significantly increased throughout the world and has taken over the consumer market (Olatunji 2022). Global production increased from 1.7 million tons in 1950 to 280 million metric tons in 2012 and 299 million metric tons in 2013, respectively (Duis and Coors 2016). By 2050, there will likely be 33 billion tons more plastic on the earth, according to estimates (Agamuthu 2018). Due to their high rates of manufacturing and longevity, it may be anticipated that some of these plastics will end up in the environment. Presently, recycling or landfill disposal accounts for half of all plastic production. The remainder eventually travels to an aquatic environment (Wurm et al. 2020). By solar UV radiation, mechanical abrasion, hydrolysis, biological deterioration, and disintegration, these aquatic plastics can gradually break down into micro- and nanoplastics (Sun et al. 2022). Micro- and nanoplastics are predicted to become more prevalent in the aquatic environment as a result of technological advancements in the plastics industry, especially in marine environments (Silva et al. 2018; Al-Thawadi 2020). Very few synthetic organic polymer particles known as microplastics have been discovered in sediments, wastewater effluents, and aquatic habitats (fresh and salt waters) all over the world (Al-Thawadi 2020). They are divided into primary and secondary microplastics, which are developing with time and are thought to be the most abundant kind of solid waste on Earth. (Ju et al. 2023; Priya et al. 2022; Zhang et al. 2022). For various industrial uses, primary microplastics are those particles that are consciously formed at the microscale. Primary microplastics are also used in cosmetics, detergents, and surface-cleaning products with abrasive beads, lubricants, pigments used as carriers, additives to manage the viscosity of hot glue, glossy magazines, coating materials to prevent fruit from bruising, water softeners, and pharmaceutical compounds as vectors (Awolesi et al. 2023). Additionally, personal care products contain microplastic components known as microbeads (Mdlalose and Chimuka 2022). In Germany, five categories of personal care products—shower gels and liquid soaps, cleansers for body care, skincare and sun protection products, dental hygiene products, and other body care items—use a combined 500 tons of polyethylene microparticles annually (Al-Thawadi 2020; Zhou et al. 2023). It is estimated that 3125 tons of polyethylene microparticles are used annually in cosmetic items sold in Europe (Essel et al. 2015; Aurisano et al. 2021). In Canada, microbeads are also added to cleaning products, such as scrubbers for rust or paint removal. They are also added to personal care products (Miraj et al. 2021; Al-Thawadi 2020. Microplastics are released into the environment through runoff and the degradation of macro- and mesoblastic by biotic and abiotic weathering (Mehra et al. 2020; Sridharan et al. 2021; Al-Thawadi

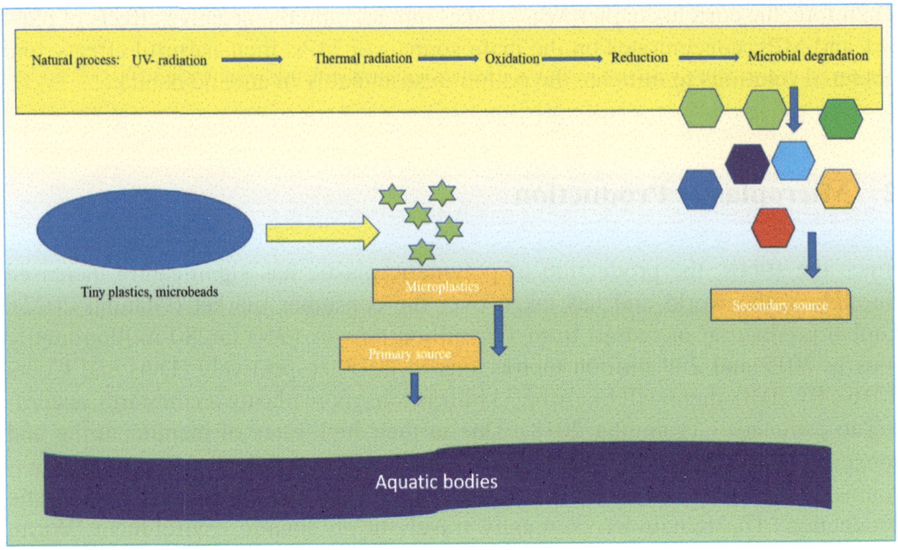

Fig. 3.1 Schematic illustration of microplastic production

2020). An autocatalytic pathway that starts with solar UV-B radiation and progresses through thermo-oxidative degradation can lead to the deterioration of polymers. Before photo-oxidative degradation begins in plastics exposed to the ocean, thermo-oxidative degradation may start the deterioration process (Andrady et al. 2022; Al-Thawadi 2020). Hydrolysis with water is another method of degrading plastics, but it is minor in comparison to photodegradation (Chu et al. 2022). However, solar UV light is the most effective way for plastic to degrade (Malankowska et al. 2021; Al-Thawadi 2020) (Fig. 3.1).

3 Origin and Sources of Microplastics in the Aquatic Ecosystem

3.1 Origin of Microplastic

Plastic may also be known by certain other nomenclatures like nanoplastics, microplastics, mesoplastics, and macroplastics. Today, as per adverse effects on human health, microplastics (MPs) are of significant concern which is of microscale (size ranging from 5 mm to 1 μm) with a polymeric structure (Horton and Barnes 2020; Danopoulos et al. 2020). Being more concerned about MPs, we need to know their origin and sources, so that we can well work upon remediation of the same. Studies reveal that contaminated water and food are one among the obvious source of MPs chasing our health (Danopoulos et al. 2020). MPs are being contributed to the environment by cosmetic and personal care products at a large scale (Anderson et al. 2016). MPs are found in the components of certain cleansers and exfoliants where they are just a substitute against our conventionally used ingredients like oats, almonds, and corns along with the deliberate use of MPs in shower gels, soaps, toothpaste, etc. (Meng et al. 2020a, b). Through every single application, an exfoliant is estimated to release up to 94,000 MPs (Anderson et al. 2016; Ding et al. 2021).

As per US Department of Commerce's National Oceanic and Atmospheric Administration (NOAA), plastic is found to be a significant aquatic junk particle (Westphalen and Abdelrasoul 2017). Since they are of small size and fragmented existence, it is difficult to trace back the origin of MPs. But it is necessary to understand their origin to just explore about the effective methods that will finally show a reduction in their entry into marine environments. There are certain common ways of their origin into the environment like drainage systems (Wagner et al. 2018), wastewater treatment plants (Murphy et al. 2016), dissolution from agricultural polyethylene foils, litter from ships and recreational events, washing and cleaning of cloths (Mintenig et al. 2017), fertilizer runoff, and car tire abrasions (Dubaish and Liebezeit 2013). Today, plastic waste can be found disturbing the marine life at a very large scale, and the origination of the same is terrestrial sources. Mentioning a few, wind or rain mobilize street litter into waterways, dumping of waste in an illegal or improper way where vehicles or containers may not be covered properly thereby, poorly managed dumping sites, amenities of plastic processing and manufacturing, sewer overflows and sewage treatment, fishing activities, and seashore disposal of solid waste (UNEP 2009). Approximately 80% of the plastics found in marine environments are originated from run-flow water via land waters to wastewater treatment plants (EIA 2015). Plastic wastes are not only documented from marine environments but from freshwater environments also (Eckert et al. 2018; Murphy and Quinn 2018; Peng et al. 2018; Su et al. 2018; Derraik 2002; Li et al. 2019).

Basically, microplastics are originated from plastics of different types, shapes, and sources after their decomposition into particles of size less than 5 mm in diameter via various processes like thermal oxidation, photodegradation, possible biodegradation, and thermal degradation (Andrady 2011; Briassoulis et al. 2019). MPs are small-sized particles having remarkable chemical stability and large specific surface area (Matjasic et al. 2021). As MPs originated in the food chain, there occurs attachment of certain toxic substances and pollutants to their surface which will ultimately be ingested by animals. Such MPs are ultimately threatening human and animal health (Baztan et al. 2014; Hermabessiere et al. 2017; Raddadi and Fava 2019). Microplastics are found to be of a size in between 5 and 100 mm (Zhang et al. 2020). Moreover, recent studies expressed them to be within a range of 5 mm–1 μm (Hartman et al. 2019). There are multiple contaminants that constitute collectively to give rise to the formation of microplastics (Rochman et al. 2019). Mainly there are two sources that have been reported so far responsible for the origin of MPs, viz., primary and secondary sources (Su et al. 2019), where primary sources involve production of MPs less than 5 mm in size (Yurtsever 2019) and secondary sources are produced in a size more than 5.0 mm (Shahnawaz et al.

2019). But processes like physical, chemical, or biological weathering led to the breakdown of such microplastics into smaller sizes even less than 5.0 mm (Naik et al. 2020).

According to a recent investigation about pollution of MPs, they are found to cause pollution in the agricultural soil (Ding et al. 2020). In wet agricultural land-use systems, MPs-coated urea fertilizers and irrigation have been found to be their main origin, since fertilizers and irrigation help the farmers in attaining best growth of crops in areas facing a problem of water scarcity (Lin et al. 2019; Shamshina et al. 2020). Enhanced crop growth can be witnessed by using humic acid polymer-coated fertilizers (Gil-Ortiz et al. 2020). Likewise, sewage sludge and plastic mulch are found to be the main origins of microplastics in a sweet potato farm in rainfed agricultural land-use systems (Huang et al. 2020). Not even the sewage sludge and plastic mulch are responsible for production of MPs on agricultural land, but improper disposal of littering, municipal garbage, and tire abrasion are other major origins of microplastics in the soil. On the other hand, atmospheric disposal of microplastics is the only main origin of microplastics in terrestrial environments like natural forests (Thinh et al. 2020; Wright et al. 2020). Upon being deposited on the soil's surface, there may occur vertical transmission of MPs into sublayers of the soil and ultimately found in deeper layers of the soil (Li et al. 2021a, b). Eventually, microplastics once come in contact with soil may enter the groundwater system by penetration into the pores of the soil or via leaching (Re 2019). Soil is not just a potential microplastics sink rather it is also a viable route to groundwater systems or aquifers (O'Connor et al. 2019). Since MPs in soil are found to be the origin of MPs in water, to know about their transport in soils is necessary to exactly assess their possible transfer to groundwater (Bigalke and Filella 2019). Studies also determined the origin of microplastics in groundwater after their recent identification in drinking water. Origin of MPs in groundwater may be due to contamination and pollution by inorganic compounds such as nitrates, etc. Human activities are found to be responsible for this like fishery, agriculture, certain aboveground activities, and wastewater treatment work (Raza and Lee 2019; Allé et al. 2021).

3.2 Source of Microplastic

In 2019, European Union's Scientific Advice Mechanism published that microplastic is everywhere in the environment (Vivekanand et al. 2021). Microplastics are slow-degrading contaminants that have characteristics like high stability, high potency of getting fragmented, capability to absorb the other contaminants, and long abode time (Padervand et al. 2020). Based on origin, sources of microplastics are of two types, i.e., primary and secondary sources (Fig. 3.2).

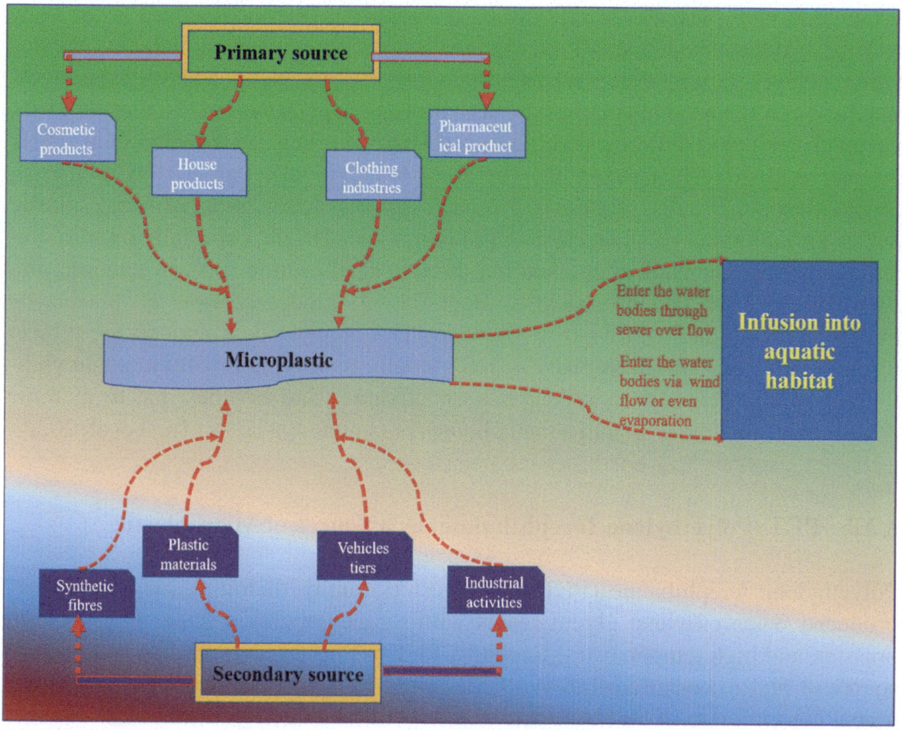

Fig. 3.2 Source and infusion of microplastic in to aquatic habitat

3.2.1 Primary Sources

Primary sources are the ones where microparticles are initially produced in the requisite dimensions, whereas in secondary sources, microplastic is produced by the breakdown or degradation of larger debris (Tursi et al. 2022). Primary sources of microplastic include medicines, textiles, body and facial scrubs, and manufactured pellets that are used for plastic and feedstock production (Rochman et al. 2015; Hernandez et al. 2017; Lechner et al. 2014).

3.2.2 Secondary Sources

Secondary sources of microplastics are usually derived by breaking down the plastic litter items of larger size, such as plastic nets, containers, films, and line fibers, via aging and weathering processes like hydrolysis, UV radiation, wave, and wind abrasion (Alimi et al. 2018). Primary microplastic is released from industrial and domestic wastewater and enters estuaries and rivers (Hidalgo-Ruz et al. 2012). In the aquatic environment, the largest microplastic source is secondary plastic (Eriksen et al. 2013). Moreover, its abundance increase due to the continuous accumulation of plastic debris gathering from various origins. Because of the large specific surface area and small microplastic size, it enters many aquatic organisms, which induces toxicity that percolates deep into the food chain. Microplastic ingestion has been reported in amphipods, lugworms, copepods, barnacles, mussels, fish, decapod crustaceans, turtles, and seabirds (Desforges et al. 2015; Nelms et al. 2018: Alomar et al. 2017; VanCauwenberghe et al. 2015; Abbasi et al. 2018; Caron et al. 2018). Secondary microplastic is produced by the degradation of larger plastic debris through processes like photodegradation via sunlight, especially by ultraviolet B radiation, mechanical degradation through sand friction and wave action, oxidative erosion or thermo-oxidative degradation, biodegradation of plastic by microorganisms, and hydrolysis by the seawater. Large plastic fragments are subject to these processes simultaneously or individually to varying environmental conditions. Usually, degradation accelerates on offshores and beaches due to extreme conditions compared to inland aquatic bodies or continental areas (Tursi et al. 2022).

3.2.3 PET (Polyethylene Terephthalate) as a Source of Microplastic

Polyethylene terephthalate (PET), a polyesters family thermoplastic polymer, is produced when ethylene glycol reacts with terephthalic acid under low vacuum pressure and high temperatures. Recently, PET has been recognized as an essential source of plastic because of all synthetic fibers, as it accounts for 70% of and has global consumption of 50 Mton/year (Castelvetro et al. 2020). Studies have shown that diet is also a minor source of PET in pets, because its traces have been detected in dog, cat feces, and food samples (Zhang et al. 2019).

3.2.4 Textile Industry as a Source of Microplastic

Due to population growth and fast fashion, consumption of textiles is significantly increasing. Fiber fragments (1 μm–5 mm) are released from home textiles and garments during wearing, washing, and drying. Every process (bleaching, brushing, and sanding) that weakens the fiber strength and interactions enhances the release of fiber fragments into water bodies which enters the food chain due to the consumption of sea table salt, sea creatures, and water (Periyasamy and Tehrani-Bagha 2022). From 2000 to 2020, globally, the fiber production of both natural and synthetic has increased tremendously from 57 million metric tons (MMT) to 111 MMT and has overtaken cotton from the market, accounting for about 65% of output worldwide (Pepper and Truscott 2021). Besides synthetic fibers (34.8%), the fiber fragments emission also takes place from tires (28.3%), landfilling, road markings, and fishing nets, which eventually enter the marine environment (Boucher and Friot 2017).

3.2.5 Vehicles as a Source of Microplastic

By 2030, plastic emissions will reach 53 million metric tons/per year because of the increase in global plastic usage (Borrelle et al. 2020). Studies showed that 42% of the total microplastics transported into the ocean are due to road wear particles, marking constituents, and tyres (Siegfried et al. 2017; Vogelsang et al. 2019; Dehghani et al. 2017). Road dust produced through the fragmentation of roadside debris, including plastic containers and packaging material, contributes to microplastic pollution in the environment (Monira et al. 2021; O'Brien et al. 2021; Zhang et al. 2017; Giustozzi et al. 2021).

3.2.6 Personnel Care Products as Source of Microplastic

Plastic-embedded microbeads make up to 10% weight of products and are used as viscosity enhancers, exfoliants, and sorbents in personal care products (Boucher and Friot 2017; Lassen et al. 2012). The single utility of these products releases many thousands of microbeads (Napper et al. 2015). The use of plastic in formulation, packing, and use of cosmetics has raised the alarm, because manufacturers intentionally incorporate it in their products (Leslie 2014). These plastic particles finally get washed away into the aquatic bodies and hence demand policy interventions (Boucher and Friot 2017).

3.2.7 Miscellaneous Sources of Microplastic

Other sources of microplastics are municipal sewage, wastewater treatment plants (WWTPs), building paints, detergents, and fertilizers (Carr et al. 2016). To sum up, microplastic has penetrated every daily activity of humans and every corner of the environment. Besides awareness among the scientific community, there is a need for mass awareness to save this planet from plastic pollution.

4 Factor Influencing the Bioavailability of Microplastic

The amount and concentration of microplastic (MP) found in marine environment around the world vary greatly. The percentage of total amount of chemicals in the environment that are available for an organism to absorb is known as the biological availability. The bioavailability of MP can be impacted by a variety of biotic and abiotic variables.

4.1 Color

Color of MPs may increase their bioavailability and their likeness to prey particles and may increase the likelihood that they will be consumed. Euphausiids and copepods, which are primarily black, red, and blue in color, are significant MP grazers in the coastal waters of the north eastern Pacific. Due to their similarity to prospective food items, particularly to visually oriented raptorial animals, the color of MPs may increase their bioavailability (Wright et al. 2013). The impact of color on zooplankton's consumption of MP has barely been studied. The majority of the MP discovered was blue (66%). Yet, numerous studies have employed light-colored MPs that a variety of plankton species are able to quickly consume (Cole et al. 2013, 2015; Cole and Galloway 2015); however, no interspecies difference in particle color was discovered.

4.2 Size

MPs are bioavailable, because they are small in size. MPs might be mistaken for prey by natural predators due to their small size during routine feeding activities. MPs are bioavailable, because they are of little size ranging from 0.5 to 816 m and are consumed by some plankton species (Cole and Galloway 2015). Due to their comparable size, MPs might be mistaken for a species' natural prey or passively swallowed during regular feeding behavior. In a study, smaller MPs (15 m) were swallowed by the copepod *Calanus finmarchicus* more frequently than bigger MPs (30 m), showing that smaller MPs had a higher bioavailability for this species (Vroom et al. 2017). Desforges et al. (2015) discovered that the copepod *Neocalanus cristatus* (length approximately: 8.5 mm) preferred to consume particles with a size of 556 m, while the euphausiid *Euphausia pacifica* (length approximately: 22 mm) preferentially ingested particles with a size of 816 m. This reflects the size disparity between the species and illustrates how these plastic particles become accessible to smaller-sized organisms as they weather and degrade. MP will eventually transform into nanoplastics (1 m), which are smaller than the human hair. Sun et al. (2017) stated that shrimp and copepod species in the South China Sea had consumed MPs that were 140 and 130 m in size, respectively, but once more, the size of the zooplankton was not specified. Although the prevalence and effects of nanosized particles on organisms are largely unknown, recent research has shown adverse effects on zooplankton feeding, behavior, and physiology (Bergami et al. 2016).

4.3 Chemical Composition and Polymer Density

MPs are expected to change in density and buoyancy regularly, though, as a result of transformational processes like biofouling and animal ingestion/egestion making them bioavailable to organisms at various layers in the water column. The density of plastic trash in the water column would be used to gauge its bioavailability. Zooplankton and planktivores suspension-feeding animals are likely to come into contact with lower-density MPs such as polyethylene (PE), at the sea surface (Wright et al. 2013). In planktivorous' surface waters, filter feeders and even suspension eaters are likely to come into contact with sustainable, lower-density plastics like polyethylene on their ocean faces (PE). For instance, PE food bags measuring 20 by 28 cm produced a well-formed biofilm within a week, and because of their neutral elasticity, they began to drain after 3 weeks submerged in the ocean (Lobelle and Cunliffe 2011). PVC, a high-density material, easily sinks and becomes accessible to benthic suspension and deposit feeders in contrast (Wright et al. 2013). Hence, a key quality of the MPs is their chemical composition. While polyethylene, polypropylene, and polyethylene terephthalate are all frequently found in the field, polystyrene (PS) is frequently employed in laboratory investigations (Zhou et al. 2022).

4.4 Shape

Microplastics can be directly disposed in the environment as cylindrical beads used in sewage treatment facilities, laundry detergent, and cosmetic goods (Napper and Thompson 2016). In contrast, laboratory-based research have primarily used MPs spherical beads (Cole et al. 2013; Lee et al. 2012; Cole and Galloway 2015). The majority of species readily absorbed the microbeads, demonstrating that a wide variety of taxa are bioavailable to this shape. A recent study by Vroom et al. (2017) examined the ingestion of MPs pieces (30 m) in addition to microbeads. They discovered that both juvenile and adult *Calanus finmarchicus* readily absorbed the particles. Moreover, sheepshead minnow (*Cyprinodon variegatus*) larvae readily consumed irregular polyethylene forms (6–350 m). The irregularly shaped MPs had a negative impact on the swimming behavior of the larvae compared to spherical MPs reducing both the speed and total distance covered Choi et al. (2018).

4.5 Abundance

In marine habitats, certain types of MPs are more prevalent in particular localities. In contrast to polyethylene and polypropylene, expanded polystyrene was found more prevalent in Eastern and Southeastern Asia, according to one study (Shahul

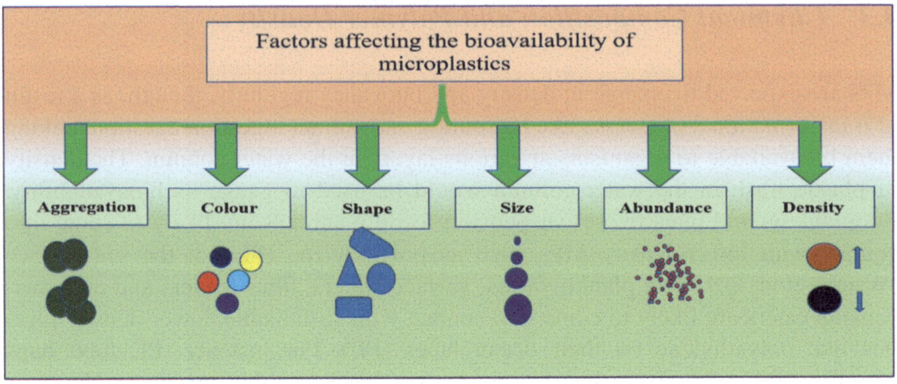

Fig. 3.3 Factor influencing the bioavailability of microplastic

Hamid et al. 2018). Moreover, it was found that seasonal change affected the amount of MP. The likelihood that MPs will be consumed by organisms increases with the number of MPs in a given environment (Kang et al. 2015).

4.6 Aggregation

Microplastics hydrophobic characteristics can cause them to collect and become part of marine aggregates like marine snow. Depending on the type of plastic, this results in an increase in overall particle size and may have an impact on density. They thus turn into bioavailable form to species of various sizes and those located in various strata of the water column. There is evidence that these aggregations can develop inside the digestive tract (Cole et al. 2013; Vroom et al. 2017). Many copepod species have been seen to collect microbeads in the anterior midgut before ingesting them in tightly packed feces pellets (Cole et al. 2013). Another species of copepod, *Calanus finmarchicus*, has aggregates of polystyrene shards (30 m) in the stomach (front and/or hind intestines) that, according to ocular inspection, occupied 30–90% of the entire gut (Vroom et al. 2017) (Fig. 3.3).

5 Implications of Microplastic on Aquatic Habitat

Currently, significant study focus has been paid to the possible environmental dangers and effects of microplastic particles (Duis and Coors 2016; Burns and Boxall 2018). With other problems like climate change, ozone depletion, and ocean acidification, there have been requests to designate microplastics, dangerous and a threat to the global borderline (Platjouw and Pozdnakova 2023). Microplastics in aquatic systems provide direct and indirect impacts on the environment and global health (Katare et al. 2021). Microplastics specifically have direct effects, whereas the discharge of possibly dangerous micropollutants and the transfer of viruses have indirect effects (Chaukura et al. 2021). Acquisition by aquatic animals like fish, mussels, shrimps, zooplanktons, and extended release of potentially harmful pollutants in case of biomagnification such as polybutylene terephthalate (PBTs) are among the effects that have been identified (Auta et al. 2017). The possible environmental dangers of microplastics in aquatic systems and future research goals have been investigated in a variety of evaluations (Shen et al. 2022). Studies focused on implications on animals, and aquatic ecosystems dominate research on the environmental health implications of microplastics, while those on human health are very scarce (Ahmed et al. 2023).

Many researches covering different species reproductive cycle, stage of growth, and genus have examined the ecological effects of MPs (Maga et al. 2022; Hou et al. 2021; Burns and Boxall 2018). A lot of research has focused on the effects of microplastics on physiology, the gastrointestinal tract, reproductive, growth, and longevity in aquatic species as well as on microplastics' role as carriers of contaminants, both inorganic and organic (Hodkovicova et al. 2022; Tian et al. 2022; Elgarahy et al. 2021). Fish, crustaceans and zooplanktivores are among the aquatic creatures that mistakenly consume microplastics as food (Bos 2019; Ziino et al. 2021). For instance, Whitefish (*Coregonus lavaretus*), Rockfish (*Sebastes schlegelii*), as well as zooplanktivores consume pre-exposed feed and consume microplastics via their gills (e.g., mussel *Mytilus edulis*) (Huuskonen et al. 2020; Yin et al. 2018). In addition, three different species of fish from the Spanish beaches were also found to consume microplastics, with red mullets having the highest concentration (33.3%) and dogfish having the lowest (20.8%) in their guts where these MPs ingested have terrible effect (Bellas et al. 2016). Following 60 and 90 days of exposure to PVC microplastics, histologic changes were observed in the distal intestine of European sea bass (*Dicentrarchus labrax*) (Nhamo et al. 2021). The gill function and ion exchange of the shore crab (*Carcinus maenas*) exposed to PSMPs in aqueous solution after an hour of exposure showed significant implications. The dose-dependent effects on oxygen uptake were seen, and they returned to normal after 16 h. In contrast, a large drop in hemolymph Na + ions and increase in Ca2+ ions was seen after a 24-h exposure (Watts et al. 2014, 2016).

The alteration was also observed in the growth phases and overall reproductive health of the organisms. As in case of sea urchin (*Lytechinus variegatus*) embryo development, the beach-stranded plastic pellets acted as carriers for pollutants and the toxicity of leached chemicals (Abuwatfa et al. 2021; Nobre et al. 2015). Another study used fluorescent microscopy to demonstrate that goldfish toxicity to microplastics varies with size and dosage that altered the immunity, gene expression, and histopathology (Abarghouei et al. 2021). The research reveals increased levels of small-sized MPs in the colon and hepatic system and increased number of bigger particles in the goldfish gills. Yet, previous studies that analyzed various tissues indicated that the organs with the overloads of stored small microplastics (PS) in decreasing order were the gills, intestine, and liver (Lu et al. 2016).

It has been established that the buildup of microplastics in the lungs might result in inflammation and oxidative stress (Akanyagne et al. 2021). For instance, microbiome impairment, oxidative stress, and immunoregulation were brought on by an accumulation of PE microplastics in crayfish (*Procambarus clarkii*). On being exposed for 21 days to PE, the crayfish underwent through oxidative stress (H_2O_2 and SOD level elevated), hepatopancreas encapsulated (lowered hemocytes), there was decline in phenoloxidase activity, and excessive mucus release was observed in the gut (Zhang et al. 2022). The rates of desorption of phenanthrene, di-2-ethylhexyl phthalate perfluorooctanoic acid, and dichlorodiphenyltrichloroethane on PE and PVC were 30 times faster under simulated gut settings than in oceans. There was also a rise under conditions that imitates warm-blooded life forms.

Another study measured the uptake from natural exposure routes to examine the impact of PE microplastics in sediment on the absorption of polychlorinated biphenyls (PCBs) by a marine lugworm (*Arenicola marina*) (Besseling et al. 2017). Moreover, pathogenic organisms, fecal contamination, and algal blooms can spread and survive in aquatic habitats where microplastics are present (Naik et al. 2020; Philipp et al. 2021), for example, bottles made up of PET being examined as suitable marine receptacles for planktonic and fecal bacteria (Grogan et al. 2021). LDPE and disposable plastics are colonized by *Enterococcus faecalis* and *Vibrio harveyi* under saltwater circumstances (Hchaichi et al. 2020). According to few recent studies, interactions between biofilms and plastic may modify the physicochemical properties of the particles, which may then influence the fate and effects of microplastics (He et al. 2022; Debroy et al. 2022). The ambient applicability of lab experiments can be increased by imitating real-world settings and microbiological controls, which are important drivers of biogeochemical cycles. Additional research is, though, needed to examine these interactions (Eich et al. 2015; Luo et al. 2022) (Table 3.1).

6 Remediation Strategies of Microplastics in the Aquatic Habitat

Microplastics (MP) include plastic having dimension less than 5 mm. Due to their global production and poor degradation, they are a great threat to the environment especially the aquatic environment (Laskar and Kumar 2019). Microplastics released into water are ingested directly or indirectly by aquatic animals like fish, crustaceans, molluscs, rotifers, worms, etc. and ultimately in the organisms feeding on these aquatic animals. The aquatic prey-predator relationship is totally affected by MPs as it induces toxicity in food web (Rakib et al. 2023). The major source of microplastic contamination in aquatic habitat includes domestic waste like microplastic fibers from washing machines, untreated water from sewage treatment plant, industries like textile industries, cosmetic industries, tire industries, mills, paint industries, etc. (Katare et al. 2021). Polyethylene (PE), polystyrene (PS), artificial organic compound polyethylene terephthalate (PET), polyamides (PA), and propylene (PP) are main constituents of plastic that contaminate various water bodies like rivers, streams, lakes, ponds, sea, oceans where aquatic living organisms ingest it,

Table 3.1 Implications of microplastics on aquatic habitat

Type of microplastic	Aquatic species	Size and duration	Impact	References
Polybrominated diphenyl ethers (PDBE)	*Allorchestescompressa* (Amphipod)	<5 mm, 36 hours	Posed bigger concern to marine amphipods that mistakenly ate microplastics (near about ≤45 pieces) contaminated with PBDE substances	Chua et al. (2014)
Triclosan, PBDE-47 and polyvinyl chloride	*Arenicola marina* (lugworms)	µm sized, 10 days	Triclosan reduces immunological response and longevity as well as the capacity to ingest and metabolize sediments. Triclosan absorption via PVC resulted in death (>55%), while PVC alone increased worms' vulnerability to oxidative stress by >30%	Browne et al. (2011)
UV-degraded PSMPs polystyrene	*Caenorhabditis elegans* (nematode)	0.1–100 µg/l, 10 days	DNA damage-dependent tissue necrosis and acute reproductive disruptions (associated gene expression for ced-3, ced-4, and ced-9 were drastically affected)	Chen et al. (2021)
Polystyrene	*Eisenia fetida*		Caused gut bacterial immunity alterations and damage of gut barriers. (*Aeromonas* and *Escherichia/Shigella*)	Li et al. (2022)
Polystyrene	*Caenorhabditis elegans* (nematode)	100 µg/ml, 3 days	Elevated production of reactive oxygen species (ROS), buildup of lipofuscin and gene expression associated with oxidative stress and caused heavy intestinal injury	Yu et al. (2020)
MPs	*C. elegans*	1 and 5 µm, 3 days	Reduced the activity of the skn-1 gene at the minimum exposure concentration, compromising nematode viability and intestinal performance.	Shang et al. (2020)
UV aged polystyrene	*Caenorhabditis elegans* (Nematode)	10–100 µg, 10 days	Affected dopamine, glutamate, and serotonin neurotoxicity via altered neuron differentiation, neurotransmitter contents, and gene related to neurotransmission	Chen et al. (2021)
Phthalates	Fish and mussels	nm, >48 days	Caused fish reproductive system malfunction and neuroendocrine disturbance	Bouwmeester et al. (2015)

(continued)

Table 3.1 (continued)

Type of microplastic	Aquatic species	Size and duration	Impact	References
Polystyrene	Mussels	8–10 μm, 14 days	Crab (*Carcinus maenas*) fed on mussels and retained the MPs in their foregut which is hazardous for human consumption	Watts et al. (2014)
Polystyrene	*Carcinus maenas* (shore crab)	8 μm, 24 hours	Effected branchial function and gas exchange, dose-dependent effect on oxygen consumption. During 24 h postexposure, a significant change was seen in ion exchange in hemolymph	Watts et al. (2016)
Polyethylene pellets	*Lytechinus variegates* (sea urchin)	Granules, 2 days	The toxicity of stuck plastic pellets was lesser than that of virgin plastic pellets, although both types of MPs promoted the abnormal embryonic development (reproductive toxicology) of sea urchins by 58.1% and 66.5%, respectively	Nobre et al. (2015)
Polyvinyl chloride	*Dicentrarchus labrax* (sea bass)	<5 mm, 60–90 days	Caused chronic histological changes in the distal intestine of European sea bass	Peda et al. (2016)
Polyvinyl chloride PUR (polyurethane)	*Daphnia magna*	59 μm, 21 days	The ecotoxocological effect of MPs altered the reproductive health of the zooplankton	Zimmermann et al. (2020)
Nylon	*Calanus finmarchicus* (copepod)	10–30 μm, 21 days	MPs caused juvenile copepods to prematurely molt and drastically reduced algal intake rates.	Cole et al. (2019)
MPs	*Dicentrarchus labrax* (sea bass)	0.69 mg/l, 4 days	Produce energy-related alterations, neurotoxicity, oxidative damage, and interacted with mercury bioaccumulation, rendering it unfit for human ingestion	Barboza et al. (2018)
BaP (Benzo(a)pyrene)	*Epnephelus coioides* (fish)	3×10^{-5} mol/l, 24 h	Increased the oxidative stress in primary liver cells through antioxidant enzyme assays and reduced the protein amount	Derakhshesh et al. (2019)

Polyvinyl chloride	*Karenia mikimotoi*	100 mg/l, 96 hours	Significant reduction in pigments, chlorophyll content, physical blockage of photosynthetic efficiency and ultimately the growth of dinoflagellate was impaired	Zhao et al. (2019)
Polystyrene	*Chlorella pyrenoidosa*	1 µm, 10 days	Markedly decreased the amount of photosynthesis pigment, extracellular polymeric materials, soluble proteins, caused oxidative stress and damaged cellular integrity	Cao et al. (2022)
Polyethylene (UV-altered)	*Scenedesmus vacuolatus*	<300 µm, 24 h	Algal biomass, cell proliferation, and photosynthetic activity decreased by PE	Rummel et al. (2022)
pMPs (pristine polystyrene)	*Phaeodactylum tricornutum*	<5 mm, 30 days	Prevented the growth of microalgae at large levels showing hormesis like effects	Chen et al. (2021)
Polystyrene	Zebra fish	0.5 and 50 µm diameter, 14 days	Resulted in endocrine abnormalities, metabolic disorders, oxidative stress, altered immunological responses, changes to the antioxidant enzyme system, and changes to their genes	Jin et al. (2018)
MPs	Mussels	300–1000 µm, 5 days	The ingested MPs posed a threat for human consumption	Gerigny et al. (2022)
MPs	*Caretta caretta* (sea turtle)	Stranded	Digestive tract contained fishing line and also had necrotic limbs. In addition, sea turtles suffered significant anemia, cellular necrosis in many tissues, and kidney damage, as per hematological and biochemical studies	Pagano et al. (2019)
MPs	*Pomatoschistusmicrops* (fish)	1–5 µm, 48 h	Reduction in the activity of acetylcholine esterase and underwent peroxidation	Ferreira et al. (2016)

(continued)

Table 3.1 (continued)

Type of microplastic	Aquatic species	Size and duration	Impact	References
MPs (esp. PP)	*Neophocaena asiaeorientalis sunameri* (porpoises)	<10 cm pellets	Trophic transfer of MPs posed threat to sea food consumption by mammals causing neurotoxicity and digestive tract implications	Xiong et al. (2018)
Polystyrene	*Calanus helgolandicus* (copepod)	20 μm, 24 h	Reduced fertility and reproduction efficiency without any discernible alterations in egg production	Cole et al. (2015)

and gradually the microplastic that carries many toxins enters into the food chain. Once ingested by aquatic microorganisms, these microplastics start accumulating in their digestive systems and disturb normal functioning of the body that sometimes leads to death (Wang et al. 2019). Microplastic in fishes reduces their feeding capacity, damages gastrointestinal tract structurally and functionally, reduces growth and nutrient levels in fishes, and also affects their immune system. Moreover, hydrophobicity and large surface area of microplastic enables it to concentrate toxic chemicals over it which enters into aquatic fauna and damage their body to greater extent (Wang et al. 2020a, b). A study on tadpoles of *Rana macrocnemis* and *Pelophylax* characterized the accumulated MPs from tadpoles in which polyacrylic, nylon, and PET were dominant ones (Karaoğlu and Gül 2020). MPs also affect lower tropical fauna of aquatic system. It reduces development of microalgal tissues and reduces its chlorophyll content. It also has some biochemical and physiological impact on phytoplankton (Rakib et al. 2023). Microplastic present in waste water also inhabits many kinds of pathogenic bacteria. Moreover, it can also act as antibiotic-resistance gene carrier (Sooriyakumar et al. 2022). There are still more toxic effects of microplastic yet to be discovered that's why microplastic has become a global problem to concern. Therefore, it is necessary to manage this issue by proper management of water resources including various strategies for microplastic remediation. These methods include the following:

6.1 *Membrane Bioreactor (MBR)*

MBR is one of the modern technologies and one of the most effective tertiary treatment methods of wastewater. It consists of a combination of activated sludge or bioreactor and filtration membrane known as membrane bioreactor. Its efficiency rate is more than 90%, and its pore size ranges from 0.1 to 0.5 micrometers for wastewater treatment (Ngo et al. 2019; Mishra et al. 2022). Its main characteristics include high-quality effluent, solid retention time, small footprint, and hydraulic retention time with complete separation. Though it consumes large energy while operating, MBR is still in great demand to treat wastewater of industries like textile dyeing, food processing, electroplating, pharmaceutical, coal chemical fine chemical industries, etc., because the membranes used for application of MBR have intrinsic filtration resistance of 10^{10}–10^{12} m^{-1}, also the membrane would have 3.6–360 $L/(m^2h)/kPa$ ranged high specific flux for more and better flow of water (Xiao et al. 2019).

6.2 *Rapid Sand Filters (RSF)*

It consists of layers of sand, and when water passes through it, the sand entraps the microplastic and produces clear water. It can remove particulate microplastic and plastic microfibers with 95.5 and 53.83%, respectively. The sand used in it can be a silica sand, quartz sand, or other sand each having their own characteristic efficiency to remove microplastic. It is effective in eliminating microplastic size ranging greater than 200 micrometers (Tirkey et al. 2022). Before passing sample through RSF, the given waste water sample is first treated in coagulation, flocculation, or pre-sedimentation unit (Sembiring et al. 2021). This RSF consisting granular limestone is cost-efficient, and it is an abundant natural material which can filter even nanoplastic like polystyrene which is thousand times smaller than microplastic (Li et al. 2023). Sand filtration can filter microplastic in tertiary treatment which was observed in a study on three wastewater treatment plants having microplastic in it with PE as the most abundant one (Funck et al. 2021). But microorganisms with size less than 0.01 mm can escape through sand filters while treatment of wastewater. Moreover, it also requires large quantity of water for washing, and it is of inconsistent quality which can be a major drawback of using sand filter sometimes (Marsono et al. 2022). However, using sand filter along with other methods like addition of a thin permeable layer of biochar can improve the efficiency of sand filters in removing microplastic from the wastewater (Hsieh et al. 2022).

6.3 Dissolved Air Floatation (DAF)

In this method, the air is dissolved in water at high atmospheric pressure. This water then enters into floatation column and forms bubbles which associate with microplastic ending with floatation and finally removal of microplastic from the system by increasing their buoyancy. DAF has different rate of removal efficiency for different kind of microplastic constituents like PTE, PE, PVC, nylon, etc.; the reason can be the different densities of different plastics affecting floatation (Esfandiari and Mowla 2021). The microbubbles produced by DAF have large surface area to volume ratio that enhances their efficiency to capture fine particles of PP, PA, PE, PET, PS, and PVF (Rajapakse et al. 2022; Zhu et al. 2022). In DAF, Fe-based and Al-based coagulants are also used under specified parameters like coagulant dosage, pH, rate of flow of water into flotation cell, and pressure whose PE reducing capacity was observed to be 70.56 and 96.10%, respectively (Esfandiari and Mowla 2021).

6.4 Coagulation

Coagulation is the common strategy to remove microplastic from aquatic bodies. Coagulation involves the mechanism to overcome repulsive energy barrier among interparticles by increasing its ionic strength. The presence of inorganic coagulants neutralizes the surface charge on colloidal system resulting in particle aggregation

(Lee et al. 2012). Different kinds of coagulants are available for coagulation purpose like polyaluminium chloride, polytitanium chloride, etc., but some coagulants are far better than others; for example, coagulation of microparticles using polymerized polyaluminum-titanium chloride composite coagulant (PATC) was a better method to remove PS during actual lake water treatment (Liu et al. 2023). In a study, polyaluminium chloride and ferric chloride were tested at particular pH as coagulants against PS-MPs in tertiary treatment of WWTP sample water, and results show that it was 99.4% efficient in removing microplastic. This represents that choice of right chemical as coagulant in right dose at right pH enhances efficiency rate to higher levels (Rajala et al. 2020). Another study showed that as compared to traditional coagulant such as magnesium hydroxide coagulant (MHC), the modified coagulant, magnetic magnesium hydroxide coagulant (MMHC) was found to be more efficient in removing MPs with increment up to 92% (Zhang et al. 2021a, b, c). While in another study, a dual-coagulant anionic polyacrylamide was used to enhance the efficiency of coagulant magnesium hydroxide to remove MPs (Li et al. 2022).

6.5 Biochar

Biochar is used as efficient contaminant adsorptions. It is highly porous and is obtained from biomass waste exposed at high temperature in the presence of limited oxygen. It is cost-efficient and environment-friendly microplastic remediation method (Abuwatfa et al. 2021). This carbonaceous porous product is obtained from various feedstocks that include cattle waste, wood waste, sewage sludge, and agricultural residue (Gopinath et al. 2021). A study shows that biochar obtained from hardwood and corn straw provides more than 95% efficiency in removing microplastic of size 10 micrometers by immobilizing it as microplastic get stuck and trapped in the grains or small chip of biochar (Wang et al. 2020a, b). Biochar is also available in modified forms like magnetic biochar having better efficiency as shown in a study where Zn-modified magnetic biochar and Mg-modified magnetic biochar result in more than 94% MPs removing efficiency because of competitive adsorption effect, while the adsorbed microplastics are degraded by Zn-/Mg-originated catalytic active sites (Wang et al. 2021a, b).

In similar way, other methods include the use of periphytic biofilm that decomposes many polymeric MPs such as PE, PP, and PET in the presence of nutrients like glucose, peptone, etc. (Dey et al. 2021). Magnetic nano-Fe_3O_4 magnetizes MPs by surface absorption followed by their removal with the help of suction of magnet (Shi et al. 2022).

Some other methods effective in remediation of microplastics from aquatic habitat are given in the Table 3.2.

Table 3.2 Remediation methods of microplastic with their pros and cons

Methods	Types of microplastic	Role	Pros	Cons	References
Floatation	PP, PVC, LDPE, PS, PLA, PBS, PBAT	Extraction of microplastic occurred and depended on density flotation	Easily available, environment-friendly, recyclable	Some floatation media were toxic for environment	Li et al. (2021a, b), Gohla et al. (2021)
Coagulation	Polyethylene	The dose of coagulant played an important role in flocculation and charge neutralization to remove plastic	Smaller the PE size, better was the removal	Post coagulation membrane fouling was induced	Ma et al. (2019)
Coagulation-flocculation-sedimentation (CFS)	Carboxylated polystyrene	Hydrophobicity and surface charge influence the microplastic	Removed a variety of microplastics	Pristine microplastics did not bind to coagulant efficiently	Xue et al. (2021)
Electrocoagulation	PE, PP, PMMA, and CA	Micro-coagulants were formed under electric field and settled down as sludge of microplastic	Efficiently removed fiber microplastic	Not much effective in removing grain microplastic	Shen et al. (2022)
Sand filtration	PS MPs	Sand provided surface for attachment of MPs on sand grain	The larger the MPs size, the better was its removal	Leaching of some toxic chemicals	Na et al. (2021)
Magnetic submarines	PS spheres	Their motion characterization helped in removal of microplastic from water	Removed MPs by noncontact method	In acidic solutions their velocity reduced due to acidic corrosion	Sun et al. (2020)
Ultrafiltration	PP, PET, PE	Water is passed though membrane for purification	Low energy consumption, high separation efficiency	Plastic formed cake layer and influenced fouling	Tadsuwan and Babel (2022), Malankowska et al. (2021)
Dynamic membrane technology	Plastic	Cake layer or dynamic membrane acted as supporting membrane and created barrier for microplastic	Cost-effective, compact setup, low supply of energy is required	Thick and dense layering lead to membrane fouling	Li et al. (2018)

Reverse osmosis	Polystyrene	Through semipermeable membrane, concentrated water was passed using high pressure to remove MPs	It dissolved solid particles larger than 0.1–1.0 nm in size and removes all colloidal matter	Membrane fouling was major drawback along with membrane density leading to low membrane productivity	Ziajahromi et al. (2017)
Membrane bioreactor	Polyethylene fragment, polyester, polypropylene, polyamide	Biodegradation of organic matter promoted by biocatalysts decreased solution complexity	High effluent quality. Easy scale-up and small footprint	Fouling was the major challenge	Maliwan et al. (2021)
Pulse clarification	Fibers and film fragments of PET, PS, PE, PP	A sludge jacket made by vacuum column acted as screening filter to remove microplastic	Turbidity reduction along with microplastic reduction occurred	Efficiency is moderate to remove microplastic	Sarkar et al. (2021)
Dissolved air floatation	Polyethylene, nylon 66, and polyethylene terephthalate	Floating flocs consisting microplastic was formed which was easy to remove	Targets suspended solids, highly safe, nontoxic	Proper investigation on its performance under different microplastic sizes, composition, etc. yet to be done	Wang et al. (2020)
Anaerobic membrane bioreactor	PS, PVC, PE, PET	Anaerobic digestion degraded organic matter and membrane technology separates microplastic	Its efficiency depended upon quality and nature of biomass waste, also it produced low sludge residual	Formation of cake layer or blockage of pores caused membrane fouling	Guo et al. (2020), Zhang et al. (2021)

(continued)

Table 3.2 (continued)

Methods	Types of microplastic	Role	Pros	Cons	References
Surface-modified microbubbles—CCGAs	Unsurface-coated polystyrene (PS) and carboxyl-modified poly- (methyl methacrylate) (PMMA)	Due to electrostatic attraction, the bubble carried the charged microplastic	Removed dissolved organic matter along with microscale microplastic	Bubble production required large energy	Zhang et al. (2021a, b, c)
Disk filters	PE, PET, and PP	Disk-shaped fibers formed filtration medium that acted as barrier for microplastic	Efferently removed microplastic with rate 98.5%	With time membrane pores got clogged	Hidayaturrahman and Lee (2019)
Membrane filtration	PVC, PE, and others	Water was passed under pressure through the filtration membrane	Showed 100% efficacy in removing microplastic	Fouling phenomenon was the main limitation	Akarsu et al. (2021)
Membrane bioreactor	Polypropylene	Effluent of wastewater sedimentation tank after filtering through mesh screen were in flown into MBRs	At lower concentration of MPs, MBR can run long without any fault	Membrane fouling, microbial community, extracellular polymeric substance accumulated with time	Wang et al. (2022)

7 Conclusion

Microplastics are now among the most frequent types of marine waste. The harmful effects of MPs on marine life are noticed by almost all species. Microplastic contamination is minimized by using modern biotechnology and an appropriate regulatory framework. Numerous studies have been conducted on the modification and removal of MPs from aquatic environments, including the creation of materials with adsorption employing a variety of natural products from plants and animals to accomplish effective MPs adsorption in water. Drinking water treatment plant (DWTPs) and wastewater treatment plant (WWTPs) play a significant role in the delivery of MPs to people, and they both contribute significantly to reducing MPs pollution. CFS technology, which can remove the majority of MPs in raw water, is the primary method of MP removal used by DWTPs. The majority of MPs can be removed by the typical treatment methods used by WWTPs, but due to the substantial volume of effluent produced by WWTPs, a significant proportion of MPs continue to enter the aquatic environment through these facilities. Comparing various secondary treatment technologies utilized in wastewater treatment procedures, the MBR technology generally offers the highest MP removal effectiveness. Therefore, in order to increase the effectiveness of MP removal from DWTPs and WWTPs, more cutting-edge treatment technologies must be developed.

References

Abarghouei S, Hedayati A, Raeisi M, Hadavand BS, Rezaei H, Abed-Elmdoust A (2021) Size-dependent effects of microplastic on uptake, immune system, related gene expression and histopathology of goldfish (Carassius auratus). Chemosphere 276:129977

Abbasi S, Soltani N, Keshavarzi B, Moore F, Turner A, Hassanaghaei M (2018) Microplastics in different tissues of fish and prawn from the Musa Estuary, Persian Gulf. Chemosphere 205:80–87

Abuwatfa WH, Al-Muqbel D, Al-Othman A, Halalsheh N, Tawalbeh M (2021) Insights into the removal of microplastics from water using biochar in the era of COVID-19: a mini review. Case Stud Chem Environ Eng 4:100151

Agamuthu P (2018) Marine debris, plastics, microplastics and nano-plastics: what next. Waste Manag Res 36(10):869–871

Ahmed AS, Billah MM, Ali MM, Bhuiyan MKA, Guo L, Mohinuzzaman M, Hossain MB, Rahman MS, Islam MS, Yan M Cai W (2023) Microplastics in aquatic environments: A comprehensive review of toxicity, removal, and remediation strategies. Sci Total Environ 876:162414

Akarsu C, Kumbur H, Kideys AE (2021) Removal of microplastics from wastewater through electrocoagulation-electroflotation and membrane filtration processes. Water Sci Technol 84(7):1648–1662

Akanyange SN, Lyu X, Zhao X, Li X, Zhang Y, Crittenden JC, Zhao H (2021) Does microplastic really represent a threat? A review of the atmospheric contamination sources and potential impacts. Sci Total Environ 777:146020

Ali SS, Elsamahy T, Al-Tohamy R, Zhu D, Mahmoud YAG, Koutra E, Metwally MA, Kornaros M, Sun J (2021) Plastic wastes biodegradation: mechanisms, challenges and future prospects. Sci Total Environ 780:146590

Alimi OS, FarnerBudarz J, Hernandez LM, Tufenkji N (2018) Microplastics and nanoplastics in aquatic environments: aggregation, deposition, and enhanced contaminant transport. Environ Sci Technol 52(4):1704–1724

Allé PH, Garcia-Muñoz P, Adouby K, Keller N, Robert D (2021) Efficient photocatalytic mineralization of polymethylmethacrylate and polystyrene nanoplastics by TiO2/β-SiC alveolar foams. Environ Chem Lett 19:1803–1808. https://doi.org/10.1007/s10311-020-01099-2

Alomar C, Sureda A, Capó X, Guijarro B, Tejada S, Deudero S (2017) Microplastic ingestion by Mullussurmuletus Linnaeus, 1758 fish and its potential for causing oxidative stress. Environ Res 159:135–142

Al-Thawadi S (2020) Microplastics and nanoplastics in aquatic environments: challenges and threats to aquatic organisms. Arab J Sci Eng 45(6):4419–4440

Anderson AG, Grose J, Pahl S, Thompson RC, Wyles KJ (2016) Microplastics in personal care products: exploring perceptions of environmentalists, beauticians and students. Mar Pollut Bull 113(1–2):454–460. https://doi.org/10.1016/j.marpolbul.2016.10.048

Andrady AL (2011) Microplastics in the marine environment. Mar Pollut Bull 62(8):1596e1605. https://doi.org/10.1016/J.marpolbul.2011.05.030

Andrady AL, Barnes PW, Bornman JF, Gouin T, Madronich S, White CC, White RG et al (2022) Oxidation and fragmentation of plastics in a changing environment; from UV-radiation to biological degradation. Sci Total Environ 158022:158022

Aurisano N, Weber R, Fantke P (2021) Enabling a circular economy for chemicals in plastics. Curr Opin Green Sustain Chem 31:100513

Auta HS, Emenike CU, Fauziah SH (2017) Distribution and importance of microplastics in the marine environment: a review of the sources, fate, effects, and potential solutions. Environ International 102:165–176

Awolesi O, Oni P, Arwenyo B (2023) Microplastics and nano-plastics: from initiation to termination. J Geosci Environ Protect 11(1):249–280

Barboza LGA, Vieira LR, Branco V, Figueiredo N, Carvalho F, Carvalho C, Guilhermino L (2018) Microplastics cause neurotoxicity, oxidative damage and energy-related changes and interact with the bioaccumulation of mercury in the European seabass, Dicentrarchus labrax (Linnaeus, 1758). Aqua Toxicol 195:49–57

Baztan J, Carrasco A, Chouinard O et al (2014) Protected areas in the Atlantic facing the hazards of micro-plastic pollution: first diagnosis of three islands in the Canary Current. Mar Pollut Bull 80(1e2):302e311. https://doi.org/10.1016/J.MARPOLBUL.2013.12.052

Bellas J, Martínez-Armental J, Martínez-Cámara A, Besada V, Martínez-Gómez C (2016) Ingestion of microplastics by demersal fish from the Spanish Atlantic and Mediterranean coasts. Mar Pollut Bull 109(1):55–60

Bergami E, Bocci E, Vannuccini ML, Monopoli M, Salvati A, Dawson KA, Corsi I (2016) Nano-sized polystyrene affects feeding, behavior and physiology of brine shrimp Artemia franciscana larvae. Ecotoxicol Environ Saf 123:18–25

Besseling E, Foekema EM, van den Heuvel-Greve MJ, Koelmans AA (2017) The effect of microplastic on the uptake of chemicals by the lugworm Arenicola marina (L.) under environmentally relevant exposure conditions. Environ Sci Technol 51(15):8795–8804

Bigalke M, Filella M (2019) Foreword to the research front on 'microplastics in soils.'. Environ Chem 16:1–2. https://doi.org/10.1071/ENv16n1_FO

Borrelle SB, Ringma J, Law KL, Monnahan CC, Lebreton L, McGivern A, Murphy E, Jambeck J, Leonard GH, Hilleary MA, Rochman CM (2020) Predicted growth in plastic waste exceeds efforts to mitigate plastic pollution. Science 369(6510):1515–1518

Boucher J, Friot D (2017) Primary microplastics in the oceans: a global evaluation of sources, vol 10. Iucn, Gland

Bouwmeester H, Hollman PC, Peters RJ (2015) Potential health impact of environmentally released micro-and nanoplastics in the human food production chain: experiences from nanotoxicology. Environ Sci Technol 49(15):8932–8947

Briassoulis D, Pikasi A, Briassoulis C, Mistriotis A (2019) Disintegration behaviour of bio-based plastics in coastal zone marine environments: a field experiment under natural conditions. Sci Total Environ 688:208. https://doi.org/10.1016/j.scitotenv.2019.06.129

Browne MA, Crump P, Niven SJ, Teuten E, Tonkin A, Galloway T, Thompson R (2011) Accumulation of microplastic on shorelines woldwide: sources and sinks. Environ Sci Technol 45(21):9175–9179

Burns EE, Boxall AB (2018) Microplastics in the aquatic environment: Evidence for or against adverse impacts and major knowledge gaps. Environ Toxicol Chem 37(11):2776–2796

Bos RP (2019) The association between stomach fullness and vertical migration behavior in deep-pelagic crustaceans and fishes in the gulf of Mexico, with notes on microplastic ingestion

Cao Q, Sun W, Yang T, Zhu Z, Jiang Y, Hu W, Yang H (2022) The toxic effects of polystyrene microplastics on freshwater algae Chlorella pyrenoidosa depends on the different size of polystyrene microplastics. Chemosphere 308:136135

Caron AG, Thomas CR, Berry KL, Motti CA, Ariel E, Brodie JE (2018) Ingestion of microplastic debris by green sea turtles (Chelonia mydas) in the Great Barrier Reef: validation of a sequential extraction protocol. Mar Pollut Bull 127:743–751

Carr SA, Liu J, Tesoro AG (2016) Transport and fate of microplastic particles in wastewater treatment plants. Water Res 91:174–182

Castelvetro V, Corti A, Bianchi S, Ceccarini A, Manariti A, Vinciguerra V (2020) Quantification of poly (ethylene terephthalate) micro-and nanoparticle contaminants in marine sediments and other environmental matrices. J Hazard Mater 385:121517

Chaukura N, Kefeni KK, Chikurunhe I, Nyambiya I, Gwenzi W, Moyo W, Nkambule TT, Mamba BB, Abulude FO (2021) Microplastics in the aquatic environment – the occurrence, sources, ecological impacts, fate, and remediation challenges. Pollutants 1(2):95–118

Chen X, Li X, Li Y (2021) Toxicity inhibition strategy of microplastics to aquatic organisms through molecular docking, molecular dynamics simulation and molecular modification. Ecotoxicol Environ Saf 226:112870

Chia RW, Lee JY, Kim H, Jang J (2021) Microplastic pollution in soil and groundwater: a review. Environ Chem Lett 19(6):4211–4224

Choi JS, Jung YJ, Hong NH, Hong SH, Park JW (2018) Toxicological effects of irregularly shaped and spherical microplastics in a marine teleost, the sheepshead minnow (Cyprinodon variegatus). Mar Pollut Bull 129(1):231–240

Chu S, Zhang B, Zhao X, Soo HS, Wang F, Xiao R, Zhang H (2022) Photocatalytic conversion of plastic waste: from photodegradation to photosynthesis. Adv Energy Mater 12(22):2200435

Chua EM, Shimeta J, Nugegoda D, Morrison PD, Clarke BO (2014) Assimilation of polybrominated diphenyl ethers from microplastics by the marine amphipod, Allorchestes compressa. Environ Sci Technol 48(14):8127–8134

Cole M, Galloway TS (2015) Ingestion of nanoplastics and microplastics by Pacific oyster larvae. Environ Sci Technol 49(24):14625–14632

Cole M, Lindeque P, Fileman E, Halsband C, Goodhead R, Moger J, Galloway TS (2013) Microplastic ingestion by zooplankton. Environ Sci Technol 47(12):6646–6655

Cole M, Lindeque P, Fileman E, Halsband C, Galloway TS (2015) The impact of polystyrene microplastics on feeding, function and fecundity in the marine copepod Calanus helgolandicus. Environ Sci Technol 49(2):1130–1137

Cole M, Coppock R, Lindeque PK, Altin D, Reed S, Pond DW, Booth AM (2019) Effects of nylon microplastic on feeding, lipid accumulation, and moulting in a coldwater copepod. Environ Sci Technol 53(12):7075–7082

Cunningham EM, Ehlers SM, Dick JT, Sigwart JD, Linse K, Dick JJ, Kiriakoulakis K (2020) High abundances of microplastic pollution in deep-sea sediments: evidence from Antarctica and the Southern Ocean. Environ Sci Technol 54(21):13661–13671

Danopoulos E, Twiddy M, Rotchell JM (2020) Microplastic contamination of drinking water: a systematic review. PLoS One 15:e0236838. https://doi.org/10.1371/journal.pone.0236838

De Sá LC, Oliveira M, Ribeiro F, Rocha TL, Futter MN (2018) Studies of the effects of microplastics on aquatic organisms: what do we know and where should we focus our efforts in the future? Sci Total Environ 645:1029–1039

Debroy A, George N, Mukherjee G (2022) Role of biofilms in the degradation of microplastics in aquatic environments. J Chem Technol Biotech 97(12):3271–3282

Dehghani S, Moore F, Akhbarizadeh R (2017) Microplastic pollution in deposited urban dust, Tehran metropolis, Iran. Environ Sci Pollut Res 24:20360–20371

Derraik GBJ (2002) The pollution of the marine environment by plastic debris: A review. Mar Pollut Bull 44:842–852

Derakhshesh N, Salamat N, Movahedinia A, Hashemitabar M, Bayati V (2019) Exposure of liver cell culture from the orange-spotted grouper, Epinephelus coioides, to benzo [a] pyrene and light results in oxidative damage as measured by antioxidant enzymes. Chemosphere, 226, 534-544

Desforges JPW, Galbraith M, Ross PS (2015) Ingestion of microplastics by zooplankton in the Northeast Pacific Ocean. Arch Environ Contam Toxicol 69:320–330

Dey TK, Uddin ME, Jamal M (2021) Detection and removal of microplastics in wastewater: evolution and impact. Environ Sci Pollut Res 28:16925–16947

Ding L, Zhang S, Wang X, Yang X, Zhang C, Qi Y, Guo X (2020) The occurrence and distribution characteristics of microplastics in the agricultural soils of Shaanxi Province, in northwestern China. Sci Total Environ 720:137525. https://doi.org/10.1016/j.scitotenv.2020.137525

Ding R, Tong L, Zhang W (2021) Microplastics in freshwater environments: sources, fates and toxicity. Water Air Soil Pollut 232(5):1–19. https://doi.org/10.1007/s11270-021-05081-8

Drummond JD, Schneidewind U, Li A, Hoellein TJ, Krause S, Packman AI (2022) Microplastic accumulation in riverbed sediment via hyporheic exchange from headwaters to mainstems. Sci Adv 8(2):eabi9305

Duan J, Bolan N, Li Y, Ding S, Atugoda T, Vithanage M, Sarkar B, Tsang DC, Kirkham MB (2021) Weathering of microplastics and interaction with other coexisting constituents in terrestrial and aquatic environments. Water Res 196:117011

Dubaish F, Liebezeit G (2013) Suspended microplastics and black carbon particles in the Jade system, southern North Sea. Water Air Soil Pollut 224(2):1–8

Duis K, Coors A (2016) Microplastics in the aquatic and terrestrial environment: sources (with a specific focus on personal care products), fate and effects. Environ Sci Eur 28(1):1–25

Eckert EM, Di Cesare A, Kettner MT, Arias-Andres M, Fontaneto D, Grossart H, Corno G (2018) Microplastics increase the impact of treated wastewater on the freshwater microbial community. Environ Pollut 234:495–502

EIA (2015) Lost at sea: the urgent need to tackle marine litter. EIA, Washington

Eich A, Mildenberger T, Laforsch C, Weber M (2015) Biofilm and diatom succession on polyethylene (PE) and biodegradable plastic bags in two marine habitats: early signs of degradation in the pelagic and benthic zone? PloS One 10(9):e0137201

Elgarahy AM, Akhdhar A, Elwakeel KZ (2021) Microplastics prevalence, interactions, and remediation in the aquatic environment: a critical review. J Environ Chem Eng 9(5):106224

Eriksen M, Mason S, Wilson S, Box C, Zellers A, Edwards W, Farley H, Amato S (2013) Microplastic pollution in the surface waters of the Laurentian Great Lakes. Mar Pollut Bull 77(1–2):177–182

Esfandiari A, Mowla D (2021) Investigation of microplastic removal from greywater by coagulation and dissolved air flotation. Process Saf Environ Prot 151:341–354

Essel R, Engel L, Carus M, Ahrens RH (2015) Sources of microplastics relevant to marine protection in Germany. Text 64(2015):1219–1226

Ferreira P, Fonte E, Soares ME, Carvalho F, Guilhermino L (2016) Effects of multi-stressors on juveniles of the marine fish Pomatoschistus microps: gold nanoparticles, microplastics and temperature. Aquatic Toxicol 170:89–103

Funck M, Al-Azzawi MS, Yildirim A, Knoop O, Schmidt TC, Drewes JE, Tuerk J (2021) Release of microplastic particles to the aquatic environment via wastewater treatment plants: the impact of sand filters as tertiary treatment. Chem Eng J 426:130933

Gérigny O, Pedrotti ML, El Rakwe M, Brun M, Pavec M, Henry M, Galgani F (2022) Characterization of floating microplastic contamination in the bay of Marseille (French Mediterranean Sea) and its impact on zooplankton and mussels. Mar Pollut Bull 175:113353

Gil-Ortiz R, Naranjo MÁ, Ruiz-Navarro A, Caballero-Molada M, Atares S, García C, Vicente O (2020) New eco-friendly polymeric-coated urea fertilizers enhanced crop yield in wheat. Agronomy 10:438. https://doi.org/10.3390/agronomy10030438

Giustozzi F, Xuan DL, Enfrin M, Masood H, Audy R, Boom YJ (2021) Use of road-grade recycled plastics for sustainable asphalt pavements: towards the selection of road-grade plastics: an evaluation framework and preliminary experimental results (No. AP-R663-21)

Gohla J, Bračun S, Gretschel G, Koblmüller S, Wagner M, Pacher C (2021) Potassium Carbonate (K2CO3)–A cheap, non-toxic and high-density floating solution for microplastic isolation from beach sediments. Mar Pollut Bull 170:112618

Gola D, Tyagi PK, Arya A, Chauhan N, Agarwal M, Singh SK, Gola S (2021) The impact of microplastics on marine environment: a review. Environ Nanotechnol Monit Manag 16:100552

Gopinath A, Divyapriya G, Srivastava V, Laiju AR, Nidheesh PV, Kumar MS (2021) Conversion of sewage sludge into biochar: a potential resource in water and wastewater treatment. Environ Res 194:110656

Grogan AE, Mallin MA, Cahoon LB (2021) Investigation of polyethylene terephthalate (PET) drinking bottles as marine reservoirs for fecal bacteria and phytoplankton. Mar Pollut Bull 173:113052

Guo W, Khan MA, Ngo HH, Johir MAH, Nghiem LD, Ni BJ (2020) Anaerobic membrane bioreactors – an introduction. In: Current developments in biotechnology and bioengineering. Elsevier, pp 1–24

Guzzetti E, Sureda A, Tejada S, Faggio C (2018) Microplastic in marine organism: environmental and toxicological effects. Environ Toxicol Pharmacol 64:164–171

Hahladakis JN, Iacovidou E (2018) Closing the loop on plastic packaging materials: what is quality and how does it affect their circularity? Sci Total Environ 630:1394–1400

Hartman N, Hüfer T, Thompson RC, Hassellöv M, Verschoor A, Daugaard AE, Rist S, Karlsson TK, Brennholt N, Cole M, Maria P, Herrling M, Heß M, Ivleva NP, Amy L, Lusher AL, Wagner M (2019) Are we speaking the same language. Recommendations for a definition and categorization framework for plastic debris. Sci Total Environ 53:1039–1047. https://doi.org/10.1021/acs.est.8b05297

Hchaichi I, Bandini F, Spini G, Banni M, Cocconcelli PS, Puglisi E (2020) Enterococcus faecalis and Vibrio harveyi colonize low-density polyethylene and biodegradable plastics under marine conditions. FEMS Microbiol Lett 367(15):fnaa125

He S, Jia M, Xiang Y, Song B, Xiong W, Cao J, Zeng G (2022) Biofilm on microplastics in aqueous environment: Physicochemical properties and environmental implications. J Hazard Mat 424:127286

Hermabessiere L, Dehaut A, Paul-Pont I et al (2017) Occurrence and effects of plastic additives on marine environments and organisms: a review. Chemosphere 182:781e793. https://doi.org/10.1016/J.CHEMOSPHERE.2017.05.096

Hernandez LM, Yousefi N, Tufenkji N (2017) Are there nanoplastics in your personal care products. Environ Sci Technol Lett 4(7):280–285

Hidalgo-Ruz V, Gutow L, Thompson RC, Thiel M (2012) Microplastics in the marine environment: a review of the methods used for identification and quantification. Environ Sci Technol 46(6):3060–3075

Hidayaturrahman H, Lee TG (2019) A study on characteristics of microplastic in wastewater of South Korea: identification, quantification, and fate of microplastics during treatment process. Mar Pollut Bull 146:696–702

Hodkovicova N, Hollerova A, Svobodova Z, Faldyna M, Faggio C (2022) Effects of plastic particles on aquatic invertebrates and fish–a review. Environ Toxicol and Pharmacol 104013
Horton AA, Barnes DK (2020) Microplastic pollution in a rapidly changing world: implications for remote and vulnerable marine ecosystems. Sci Tot Environ 738:140349. https://doi.org/10.1016/j.scitotenv.2020.140349
Hou L, Kumar D, Yoo CG, Gitsov I, Majumder ELW (2021) Conversion and removal strategies for microplastics in wastewater treatment plants and landfills. Chem Eng J 406:126715
Hsieh L, He L, Zhang M, Lv W, Yang K, Tong M (2022) Addition of biochar as thin preamble layer into sand filtration columns could improve the microplastics removal from water. Water Res 221:118783
Huang Y, Liu Q, Jia W, Yan C, Wang J (2020) Agricultural plastic mulching as a source of microplastics in the terrestrial environment. Environ Pollut 260:114096. https://doi.org/10.1016/j.envpol.2020.114096
Huuskonen H, i Folguera JS, Kortet R, Akkanen J, Vainikka A, Janhunen M, Kekäläinen J (2020) Do whitefish (Coregonus lavaretus) larvae show adaptive variation in the avoidance of microplastic ingestion?. Environ Pollut 262:114353
Ivleva NP, Wiesheu AC, Niessner R (2017) Microplastic in aquatic ecosystems. Angew Chem Int Ed Engl 56(7):1720–1739
Jin Y, Xia J, Pan Z, Yang J, Wang W, Fu Z (2018) Polystyrene microplastics induce microbiota dysbiosis and inflammation in the gut of adult zebrafish. Environ Pollut 235:322–329
Ju S, Yi J, Lee J, Kim J, Lim C, Lee J et al (2023) High-efficiency microplastic sampling device improved using CFD analysis. Sustain For 15(5):3907
Kang JH, Kwon OY, Shim WJ (2015) Potential threat of microplastics to zooplanktivores in the surface waters of the Southern Sea of Korea. Arch Environ Contam Toxicol 69:340–351
Karaoğlu K, Gül S (2020) Characterization of microplastic pollution in tadpoles living in small water-bodies from Rize, the northeast of Turkey. Chemosphere 255:126915
Katare Y, Singh P, Sankhla MS, Singhal M, Jadhav EB, Parihar K, Nikalje BT, Trpathi A, Bhardwaj L (2021) Microplastics in aquatic environments: sources, ecotoxicity, detection and remediation. Biointerface Res Appl Chem 12:3407–3428
Klöckner P, Reemtsma T, Wagner S (2021) The diverse metal composition of plastic items and its implications. Sci Total Environ 764:142870
Klun B, Rozman U, Ogrizek M, Kalčíková G (2022) The first plastic produced, but the latest studied in microplastics research: the assessment of leaching, ecotoxicity and bioad
Kristanti RA, Hadibarata T, Wulandari NF, Sibero MT, Darmayati Y, Hatmanti A (2022) Overview of microplastics in the environment: type, source, potential effects and removal strategies. Bioprocess Biosyst Eng:1–13
Laskar N, Kumar U (2019) Plastics and microplastics: a threat to environment. Environ Technol Innov 14:100352
Lassen C, Hansen SF, Magnusson K, Norén F, Hartmann NIB, Jensen PR, Nielsen TG, Brinch A (2012) Microplastics: occurrence, effects and sources of releases to the environment in Denmark. Significance 2(2)
Lechner A, Keckeis H, Lumesberger-Loisl F, Zens B, Krusch R, Tritthart M, Glas M, Schludermann E (2014) The Danube so colourful: a potpourri of plastic litter outnumbers fish larvae in Europe's second largest river. Environ Pollut 188:177–181
Lee KE, Morad N, Teng TT, Poh BT (2012) Development, characterization and the application of hybrid materials in coagulation/flocculation of wastewater: A review. Chem Eng J 203:370–386
Leslie HA (2014) Review of microplastics in cosmetics. IVM Institute for environmental studies, vol 476, p 1–33
Li L, Xu G, Yu H, Xing J (2018) Dynamic membrane for micro-particle removal in wastewater treatment: performance and influencing factors. Sci Total Environ 627:332–340
Li C, Busquets R, Campos LC (2019) Assessment of microplastics in freshwater systems: a review. Sci Total Environ 707:135578

Li H, Lu X, Wang S, Zheng B, Xu Y (2021a) Vertical migration of microplastics along soil profile under different crop root systems. Environ Pollut 278:116833. https://doi.org/10.1016/j.envpol.2021.116833

Li C, Cui Q, Zhang M, Vogt RD, Lu X (2021b) A commonly available and easily assembled device for extraction of bio/non-degradable microplastics from soil by flotation in NaBr solution. Sci Total Environ 759:143482

Li B, Zhao J, Ge W, Li W, Yuan H (2022) Coagulation-flocculation performance and floc properties for microplastics removal by magnesium hydroxide and PAM. J Environ Chem Eng 10(2):107263

Li X, Zhang Y, Xu H, Sun Y, Gao B, Wu J (2023) Granular limestone amended sand filters for enhanced removal of nanoplastics from water: performance and mechanisms. Water Res 229:119443

Lin KH, Lin FW, Wu C, Chang YS (2019) Biostimulation of maize (Zea mays) and irrigation management improved crop growth and water use under controlled environment. Agronomy 9:559–573. https://doi.org/10.3390/agronomy9090559

Liu B, Gao Y, Yue Q, Guo K, Gao B (2023) The suitability and mechanism of polyaluminum-titanium chloride composite coagulant (PATC) for polystyrene microplastic removal: structural characterization and theoretical calculation. Water Res 232:119690

Lobelle D, Cunliffe M (2011) Early microbial biofilm formation on marine plastic debris. Mar Pollut Bull 62(1):197–200

Lu Y, Zhang Y, Deng Y, Jiang W, Zhao Y, Geng J, Ren H (2016) Uptake and accumulation of polystyrene microplastics in zebrafish (Danio rerio) and toxic effects in liver. Environ Sci Technol 50(7):4054–4060

Luo H, Liu C, He D, Xu J, Sun J, Li J, Pan X (2022) Environmental behaviors of microplastics in aquatic systems: A systematic review on degradation, adsorption, toxicity and biofilm under aging conditions. J Hazard Mat 423:126915

Ma B, Xue W, Hu C, Liu H, Qu J, Li L (2019) Characteristics of microplastic removal via coagulation and ultrafiltration during drinking water treatment. Chem Eng J 359:159–167

Maga D, Galafton C, Blömer J, Thonemann N, Özdamar A, Bertling J (2022) Methodology to address potential impacts of plastic emissions in life cycle assessment. Int J Life Cycle Assess 27(3):469–491

Malankowska M, Echaide-Gorriz C, Coronas J (2021) Microplastics in marine environment: a review on sources, classification, and potential remediation by membrane technology. Environ Sci Water Res Technol 7(2):243–258

Maliwan T, Pungrasmi W, Lohwacharin J (2021) Effects of microplastic accumulation on floc characteristics and fouling behavior in a membrane bioreactor. J Hazard Mater 411:124991

Marsono BD, Yuniarto A, Purnomo A, Soedjono ES (2022) Comparison performances of microfiltration and rapid sand filter operated in water treatment plant. IOP Conf. Ser. Earth Environ. Sci. 1111(1):012048

Matjasic T, Simcic T, Medvescek N, Bajt O, Dreo T, Mori N (2021) Critical evaluation of biodegradation studies on synthetic plastics through a systematic literature review. Sci Total Environ 752:141959. https://doi.org/10.1016/J.SCITOTENV.2020.141959

Mdlalose L, Chimuka L (2022) Mitigation approaches to prevent microplastics effects in the aquatic environment: exploration of microbeads from personal care and cosmetic products. Int J Environ Res 16(5):84

Mehra S, Sharma K, Sharma G, Singh M, Chadha P (2020) Sources, fate, and impact of microplastics in aquatic environment. Emerging contaminants

Meng F, Fan T, Yang X, Riksen M, Xu M, Geissen V (2020a) Effects of plastic mulching on the accumulation and distribution of macro and micro plastics in soils of two farming systems in Northwest China. Peer J 8:e10375. https://doi.org/10.7717/peerj.10375

Meng Y, Kelly FJ, Wright SL (2020b) Advances and challenges of microplastic pollution in freshwater ecosystems: a UK perspective. Environ Pollut 256:113445. https://doi.org/10.1016/j.envpol.2019.113445

Mintenig SM, Int-Veen I, Löder MG, Primpke S, Gerdts G (2017) Identification of microplastic in effluents of waste water treatment plants using focal plane array-based micro-Fourier-transform infrared imaging. Water Res 108:365–372

Miraj SS, Parveen N, Zedan HS (2021) Plastic microbeads: small yet mighty concerning. Int J Environ Health Res 31(7):788–804

Mishra S, Singh RP, Rout PK, Das AP (2022) Membrane bioreactor (MBR) as an advanced wastewater treatment technology for removal of synthetic microplastics. In: Development in wastewater treatment research and processes, pp 45–60

Monira S, Bhuiyan MA, Haque N, Shah K, Roychand R, Hai FI, Pramanik BK (2021) Understanding the fate and control of road dust-associated microplastics in stormwater. Process Saf Environ Prot 152:47–57

Murphy F, Quinn B (2018) The effects of microplastic on freshwater Hydra attenuata feeding, morphology and reproduction. Environ Pollut 234:487–494

Murphy F, Ewins C, Carbonnier F, Quinn B (2016) Wastewater treatment works (WwTW) as a source of microplastics in the aquatic environment. Environ Sci Technol 50(11):5800–5808

Na SH, Kim MJ, Kim JT, Jeong S, Lee S, Chung J, Kim EJ (2021) Microplastic removal in conventional drinking water treatment processes: performance, mechanism, and potential risk. Water Res 202:117417

Naik RA, Rowles LS, Hossain AI, Yen M, Aldossary RM, Apul OG, Conkle J, Saleh NB (2020) Microplastic particle versus fber generation during photo-transformation in simulated seawater. Sci Total Environ 736:139690. https://doi.org/10.1016/j.scitotenv.2020.139690

Napper IE, Bakir A, Rowland SJ, Thompson RC (2015) Characterisation, quantity and sorptive properties of microplastics extracted from cosmetics. Mar Pollut Bull 99(1–2):178–185

Napper IE, & Thompson RC (2016) Release of synthetic microplastic plastic fibres from domestic washing machines: Effects of fabric type and washing conditions. Mar Pollut Bull 112(1-2):39–45

Nelms SE, Galloway TS, Godley BJ, Jarvis DS, Lindeque PK (2018) Investigating microplastic trophic transfer in marine top predators. Environ Pollut 238:999–1007

Ngo PL, Pramanik BK, Shah K, Roychand R (2019) Pathway, classification and removal efficiency of microplastics in wastewater treatment plants. Environ Pollut 255:113326

Nhamo C, Kebede K, Innocent C, Isaac N, Willis G, Welldone M, Mamba, Francis OA (2021) Microplastics in the aquatic environment—the Occurrence, sources, ecological impacts, fate, and remediation challenges

Nobre CR, Santana MFM, Maluf A, Cortez FS, Cesar A, Pereira CDZ, Turra A (2015) Assessment of microplastic toxicity to embryonic development of the sea urchin Lytechinus variegatus (Echinodermata: Echinoidea). Mar Pollut Bull 92(1-2):99–104

O'Brien S, Okoffo ED, Rauert C, O'Brien JW, Ribeiro F, Burrows SD, Toapanta T, Wang X, Thomas KV (2021) Quantification of selected microplastics in Australian urban road dust. J Hazard Mater 416:125811

O'Connor D, Pan S, Shen Z, Song Y, Jin Y, Wu WM, Hou D (2019) Microplastics undergo accelerated vertical migration in sand soil due to small size and wet-dry cycles. Environ Pollut 249:527–534. https://doi.org/10.1016/j.envpol.2019.03.092

Olatunji O (2022) Plastic and polymer industry by region: production, consumption and waste management in the African continent. Springer Nature

Padervand M, Lichtfouse E, Robert D, Wang C (2020) Removal of microplastics from the environment. A review. Environ Chem Lett 18(3):807–828

Pagano M, Vazzana I, Gentile A, Caracappa G, Faggio C (2019) Hematological and biochemical parameters in Sea turtles (Caretta caretta) after stranding. Reg Stud Mar Sci 32:100832

Pan Y, Gao SH, Ge C, Gao Q, Huang S, Kang Y, Luo G, Zhang Z, Fan L, Zhu Y, Wang AJ (2022) Removing microplastics from aquatic environments: a critical review. Environ Sci Ecotechnol 100222

Peda C, Caccamo L, Fossi MC, Gai F, Andaloro F, Genovese L, Maricchiolo G (2016) Intestinal alterations in European sea bass Dicentrarchus labrax (Linnaeus, 1758) exposed to microplastics: preliminary results. Environ Pollut 212:251–256

Peng G, Xu P, Zhu B, Bai M, Li D (2018) Microplastics in freshwater river sediments in Shanghai, China: a case study of risk assessment in mega-cities. Environ Pollut 234:448–456

Pepper LR, Truscott L (2021) Preferred fiber and materials market report 2021

Philipp C, Unger B, Ehlers SM, Koop JH, Siebert U (2021) First evidence of retrospective findings of microplastics in harbour porpoises (Phocoena phocoena) from German waters. Front Mar Sci 508

Platjouw FM, Pozdnakova A (Eds.) (2023) The environmental rule of law for oceans: designing legal solutions. Cambridge University Press

Periyasamy AP, Tehrani-Bagha A (2022) A review of microplastic emission from textile materials and its reduction techniques. Polym Degrad Stab 199:109901

Priya A, Anusha G, Thanigaivel S, Karthick A, Mohanavel V, Velmurugan P et al (2022) Removing microplastics from wastewater using leading-edge treatment technologies: a solution to microplastic pollution – a review. Bioprocess Biosyst Eng:1–13

Raddadi N, Fava F (2019) Biodegradation of oil-based plastics in the environment: existing knowledge and needs of research and innovation. Sci Total Environ 679:148e158. https://doi.org/10.1016/J.SCITOTENV.2019.04.419

Rajala K, Grönfors O, Hesampour M, Mikola A (2020) Removal of microplastics from secondary wastewater treatment plant effluent by coagulation/flocculation with iron, aluminum and polyamine-based chemicals. Water Res 183:116045

Rajapakse N, Zargar M, Sen T, Khiadani M (2022) Effects of liquid phase physicochemical characteristics on microbubble dynamics in dissolved air flotation: a review. Sep Purif Technol 289:120772

Rakib M, Jahan R, Sarker A, Ram K, Uddin M, Walker TR, Chowdhury T, Uddin J, Khandaker MU, Rahman MM, Idris AM (2023) Microplastic toxicity in aquatic organisms and aquatic ecosystems: a review. Water Air Soil Pollut 234(1):1–28

Raza M, Lee JY (2019) Factors affecting spatial pattern of groundwater hydro chemical variables and nitrate in agricultural region of Korea. Episodes 42:135–148. https://doi.org/10.18814/epiiugs/2019/019011

Re V (2019) Shedding light on the invisible: addressing the potential for groundwater contamination by plastic microfibers. Hydrogeol J 27:2719–2727. https://doi.org/10.1007/s10040-019-01998-x

Rochman CM et al (2015). Scientific evidence supports a ban on microbeads

Rochman CM, Brookson C, Bikker J, Djuric N, Earn A, Bucci K, Athey S, Huntington A, McIlwraith H, Munno K, De FH, Kolomijeca A, Erdle L, Grbic J, Bayoumi MS, Borrelle SB, Wu T, Santoro S, Werbowski LM, Zhu X, Giles RK, Hamilton BM, Thaysen CT, Kaura A, Klasios N, Ead L, Kim J, Sherlock C, Ho A, Hung C (2019) Rethinking microplastics as a diverse contaminant suite. Environ Toxicol Chem 38(4):703–711

Rodrigues MO, Abrantes N, Gonçalves FJM, Nogueira H, Marques JC, Gonçalves AMM (2019) Impacts of plastic products used in daily life on the environment and human health: what is known? Environ Toxicol Pharmacol 72:103239

Rummel CD, Schäfer H, Jahnke A, Arp HPH, Schmitt-Jansen M (2022) Effects of leachates from UV-weathered microplastic on the microalgae Scenedesmus vacuolatus. Anal Bioanal Chem 414(4):1469–1479

Sarkar DJ, Sarkar SD, Das BK, Praharaj JK, Mahajan DK, Purokait B, Mohanty TR, Mohanty D, Gogoi P, Kumar S, Behera BK, Manna RK, Samanta S (2021) Microplastics removal efficiency of drinking water treatment plant with pulse clarifier. J Hazard Mater 413:125347

Sarma H, Hazarika RP, Kumar V, Roy A, Pandit S, Prasad R (2022) Microplastics in marine and aquatic habitats: sources, impact, and sustainable remediation approaches. Environ Sustain 5(1):39–49

Sembiring E, Fajar M, Handajani M (2021) Performance of rapid sand filter–single media to remove microplastics. Water Supply 21(5):2273–2284

Shahnawaz M, Sangale MK, Ade AB (2019) Microplastics. Bioremediation technology for plastic waste. Springer, Singapore, pp 11–19. https://doi.org/10.1007/978-981-13-7492-0_2

Shahul Hamid F, Bhatti MS, Anuar N, Anuar N, Mohan P, Periathamby A (2018) Worldwide distribution and abundance of microplastic: how dire is the situation? Waste Manag Res 36(10):873–897

Shamshina JL, Kelly A, Oldham T, Rogers RD (2020) Agricultural uses of chitin polymers. Environ Chem Lett 18:53–60. https://doi.org/10.1007/s10311-019-00934-5

Shang X, Lu J, Feng C, Ying Y, He Y, Fang S, Ju J (2020) Microplastic (1 and 5 μm) exposure disturbs lifespan and intestine function in the nematode Caenorhabditis elegans. Sci Total Environ 705:135837

Shen M, Zhang Y, Almatrafi E, Hu T, Zhou C, Song B, Zeng Z, Zeng G (2022) Efficient removal of microplastics from wastewater by an electrocoagulation process. Chem Eng J 428:131161

Shi X, Zhang X, Gao W, Zhang Y, He D (2022) Removal of microplastics from water by magnetic nano-Fe3O4. Sci Total Environ 802:149838

Siegfried M, Koelmans AA, Besseling E, Kroeze C (2017) Export of microplastics from land to sea. A modelling approach. Water Res 127:249–257

Silva AB, Costa MF, Duarte AC (2018) Biotechnology advances for dealing with environmental pollution by micro (nano) plastics: lessons on theory and practices. Curr Opin Environ Sci Health 1:30–35

Sooriyakumar P, Bolan N, Kumar M, Singh L, Yu Y, Li Y, Weralupitiya C, Vithanage M, Ramanayaka S, Sarkar B, Wang F, Gleeson DB, Zhang D, Kirkham MB, Rinklebe J, Siddique KH (2022) Biofilm formation and its implications on the properties and fate of microplastics in aquatic environments: a review. J Hazard Mater Adv 6:100077

Sridharan S, Kumar M, Bolan NS, Singh L, Kumar S, Kumar R, You S (2021) Are microplastics destabilizing the global network of terrestrial and aquatic ecosystem services. Environ Res 198:111243

Su L, Cai H, Kolandhasamy P, Wu C, Rochman CM, Shi H (2018) Using the Asian clam as an indicator of microplastic pollution in freshwater ecosystems. Environ Pollut 234:347–355

Su Y, Zhang Z, Wu D, Zhan L, Shi H, Xie B (2019) Occurrence of microplastics in landfll systems and their fate with landfll age. Water Res 164:114968. https://doi.org/10.1016/j.watres.2019.114968

Sun X, Li Q, Zhu M, Liang J, Zheng S, Zhao Y (2017) Ingestion of microplastics by natural zooplankton groups in the northern South China Sea. Mar Pollut Bull 115(1–2):217–224

Sun M, Chen W, Fan X, Tian C, Sun L, Xie H (2020) Cooperative recyclable magnetic microsubmarines for oil and microplastics removal from water. Appl Mater Today 20:100682

Sun J, Zheng H, Xiang H, Fan J, Jiang H (2022) The surface degradation and release of microplastics from plastic films studied by UV radiation and mechanical abrasion. Sci Total Environ 838:156369

Tadsuwan K, Babel S (2022) Microplastic abundance and removal via an ultrafiltration system coupled to a conventional municipal wastewater treatment plant in Thailand. J Environ Chem Eng 10(2):107142

Tang N, Liu X, Xing W (2020) Microplastics in wastewater treatment plants of Wuhan, Central China: abundance, removal, and potential source in household wastewater. Sci Total Environ 745:141026

Thinh TQ, Sang TT, Viet TQ, Dan NP, Strady E, Chung KL (2020) Preliminary assessment on the microplastic contamination in the atmospheric fallout in the Phuoc Hiep landfll, Cu Chi, Ho Chi Minh city. Vietnam J Sci Techno Eng 62:83–89. https://doi.org/10.31276/VJSTE.62(3).83-89

Tian W, Song P, Zhang H, Duan X, Wei Y, Wang H, Wang S (2022) Microplastic materials in the environment: problem and strategical solutions. Prog Mater Sci 101035

Tirkey A, Pandey M, Tiwari A, Sahu RL, Kukkar D, Dubey R, Kim K, Pandey SK (2022) Global distribution of microplastic contaminants in aquatic environments and their remediation strategies. Water Environ Res 94(12):e10819

Tursi A, Baratta M, Easton T, Chatzisymeon E, Chidichimo F, De Biase M, De Filpo G (2022) Microplastics in aquatic systems, a comprehensive review: origination, accumulation, impact, and removal technologies. RSC Adv 12(44):28318–28340

UNEP (2009) Marine litter: a global challenge; UNEP: Nairobi, Kenya, p 232

VanCauwenberghe L, Claessens M, Vandegehuchte MB, Janssen CR (2015) Microplastics are taken up by mussels (Mytilus edulis) and lugworms (Arenicola marina) living in natural habitats. Environ Pollut 199:10–17

Vivekanand AC, Mohapatra S, Tyagi VK (2021) Microplastics in aquatic environment: challenges and perspectives. Chemosphere 282:131151

Vo HC, Pham MH (2021) Ecotoxicological effects of microplastics on aquatic organisms: a review. Environ Sci Pollut Res 28:44716–44725

Vogelsang C, Lusher A, Dadkhah ME, Sundvor I, Umar M, Ranneklev SB, Eidsvoll D, Meland S (2019) Microplastics in road dust–characteristics, pathways and measures. NIVA-rapport

Vroom RJ, Koelmans AA, Besseling E, Halsband C (2017) Aging of microplastics promotes their ingestion by marine zooplankton. Environ Pollut 231:987–996

Wagner S, Hüfer T, Klöckner P, Wehrhahn M, Hofmann T, Reemtsma T (2018) Tire wear particles in the aquatic environment – a review on generation, analysis, occurrence, fate and effects. Water Res 139:83–100

Wang W, Gao H, Jin S, Li R, Na G (2019) The ecotoxicological effects of microplastics on aquatic food web, from primary producer to human: a review. Ecotoxicol Environ Saf 173:110–117

Wang F, Gao J, Zhai W, Liu D, Zhou Z, Wang P (2020) The influence of polyethylene microplastics on pesticide residue and degradation in the aquatic environment. J Hazar Mat 394:122517

Wang W, Ge J, Yu X (2020a) Bioavailability and toxicity of microplastics to fish species: a review. Ecotoxicol Environ Saf 189:109913

Wang Z, Sedighi M, Lea-Langton A (2020b) Filtration of microplastic spheres by biochar: removal efficiency and immobilisation mechanisms. Water Res 184:116165

Wang J, Sun C, Huang QX, Chi Y, Yan JH (2021a) Adsorption and thermal degradation of microplastics from aqueous solutions by Mg/Zn modified magnetic biochars. J Hazard Mater 419:126486

Wang Y, Li YN, Tian L, Ju L, Liu Y (2021b) The removal efficiency and mechanism of microplastic enhancement by positive modification dissolved air flotation. Water Environ Res 93(5):693–702

Wang Q, Li Y, Liu Y, Zhou Z, Hu W, Lin L, Wu Z (2022) Effects of microplastics accumulation on performance of membrane bioreactor for wastewater treatment. Chemosphere 287:131968

Watts AJ, Lewis C, Goodhead RM, Beckett SJ, Moger J, Tyler CR, Galloway TS (2014) Uptake and retention of microplastics by the shore crab Carcinus maenas. Environ Sci Technol 48(15):8823–8830

Watts AJ, Urbina MA, Goodhead R, Moger J, Lewis C, Galloway TS (2016) Effect of microplastic on the gills of the shore crab Carcinus maenas. Environ Sci Technol 50(10):5364–5369

Westphalen H, Abdelrasoul A (2017) Challenges and treatment of MPs in water. Water Chall Urb World. https://doi.org/10.5772/intechopen.71494

Wojnowska-Baryła I, Bernat K, Zaborowska M (2022) Plastic waste degradation in landfill conditions: the problem with microplastics, and their direct and indirect environmental effects. Int J Environ Res Public Health 19(20):13223

Wright SL, Thompson RC, Galloway TS (2013) The physical impacts of microplastics on marine organisms: a review. Environ Pollut 178:483–492

Wright SL, Ulke J, Font A, Chan KL, Kelly FJ (2020) Atmospheric microplastic deposition in an urban environment and an evaluation of transport. Environ Intl 136:105411. https://doi.org/10.1016/j.envint.2019.105411

Wurm FR, Spierling S, Endres HJ, Barner L (2020) Plastics and the environment – current status and challenges in Germany and Australia. Macromol Rapid Commun 41(18):2000351

Xiao X, Shi W, Ruan W (2019) Performance and microbial community structure of anaerobic membrane bioreactor for lipids-rich kitchen waste slurry treatment: mesophilic and thermophilic processes. Processes 7(12):879

Xiong X, Chen X, Zhang K, Mei Z, Hao Y, Zheng J, Wang D (2018) Microplastics in the intestinal tracts of East Asian finless porpoises (Neophocaena asiaeorientalis sunameri) from Yellow Sea and Bohai Sea of China. Mar Pollut Bull 136:55–60

Xue J, Peldszus S, Van Dyke MI, Huck PM (2021) Removal of polystyrene microplastic spheres by alum-based coagulation-flocculation-sedimentation (CFS) treatment of surface waters. Chem Eng J 422:130023

Yin L, Chen B, Xia B, Shi X, Qu K (2018) Polystyrene microplastics alter the behavior, energy reserve and nutritional composition of marine jacopever (Sebastes schlegelii). J Hazard Mater 360:97–105

Yu Y, Chen H, Hua X, Dang Y, Han Y, Yu Z, Li H (2020) Polystyrene microplastics (PS-MPs) toxicity induced oxidative stress and intestinal injury in nematode Caenorhabditis elegans. Sci Total Environ 726:138679

Yurtsever M (2019) Glitters as a source of primary microplastics: an approach to environmental responsibility and ethics. J Agric Environ Ethics 32:459–478. https://doi.org/10.1007/s10806-019-09785-0

Ziajahromi S, Neale PA, Rintoul L, Leusch FD (2017) Wastewater treatment plants as a pathway for microplastics: development of a new approach to sample wastewater-based microplastics. Water Res 112:93–99

Ziino G, Nalbone L, Giarratana F, Romano B, Cincotta F, Panebianco A (2021) Microplastics in vacuum packages of frozen and glazed icefish (Neosalanx spp.): a freshwater fish intended for human consumption. Ital J Food Saf 10(4)

Zimmermann L, Göttlich S, Oehlmann J, Wagner M, Völker C (2020) What are the drivers of microplastic toxicity? Comparing the toxicity of plastic chemicals and particles to Daphnia magna. Environ Pollut 267:115392

Zhang D, Zhang H, Shi C (2017) Investigation of aging performance of SBS modified asphalt with various aging methods. Constr Build Mater 145:445–451

Zhang J, Wang L, Kannan K (2019) Polyethylene terephthalate and polycarbonate microplastics in pet food and feces from the United States. Environ Sci Technol 53(20):12035–12042

Zhang B, Yang X, Chen L, Chao J, Teng J, Wang Q (2020) Microplastics in soils: a review of possible sources, analytical methods, and ecological impacts. J Chem Technol Biotechnol 95:2052–2068. https://doi.org/10.1002/jctb.6334

Zhang M, Yang J, Kang Z, Wu X, Tang L, Qiang Z, Zhang D, Pan X (2021a) Removal of micron-scale microplastic particles from different waters with efficient tool of surface-functionalized microbubbles. J Hazard Mater 404:124095

Zhang Y, Wu D, Su Y, Xie B (2021b) Occurrence, influence and removal strategies of mycotoxins, antibiotics and microplastics in anaerobic digestion treating food waste and co-digestive bio-solids: a critical review. Bioresour Technol 330:124987

Zhang Y, Zhao J, Liu Z, Tian S, Lu J, Mu R, Yuan H (2021c) Coagulation removal of microplastics from wastewater by magnetic magnesium hydroxide and PAM. J Water Process Eng 43:102250

Zhang K, Xu S, Zhang Y, Lo Y, Liu M, Ma Y, Chau HS, Cao Y, Xu X, Wu R, Lam PK (2022) A systematic study of microplastic occurrence in urban water networks of a metropolis. Water Res 223:118992

Zhao T, Tan L, Huang W, Wang J (2019) The interactions between micro polyvinyl chloride (mPVC) and marine dinoflagellate Karenia mikimotoi: The inhibition of growth, chlorophyll and photosynthetic efficiency. Environ Pollut 247:883–889

Zhou X, Zhao Y, Pang G, Jia X, Song Y, Guo A, Wang A, Zhang S, Ji M (2022) Microplastic abundance, characteristics and removal in large-scale multi-stage constructed wetlands for effluent polishing in northern China. Chem Eng J 430:132752

Zhou Y, Ashokkumar V, Amobonye A, Bhattacharjee G, Sirohi R, Singh V, Flora G, Kumar V, Pillai S, Zhang Z, Awasthi MK (2023) Current research trends on cosmetic microplastic pollution and its impacts on the ecosystem: a review. Environ Pollut 121106:121106

Zhu Y, Ma J, Zeng S, Li X, Lisak G, Chen F (2022) Advanced treatment of microplastics and antibiotic-containing wastewater using integrated modified dissolved air flotation and pulsed cavitation-impinging stream processes. J Hazard Mater Adv 7:100139

Chapter 4
Microplastics Pollution: Occurrence, Behaviour and Microbial Remediation Technologies

Anish Dua and Harmanpreet Kaur Gill

Abstract The particles that are having density greater than 5 mm are called as microplastics. They possess various physical and chemical properties like colour, density, chemical composition and size. Microplastics are great threat to the environment and cause adverse effects on the biological cells. Because of their small size, they get easily incorporated into bodies of living organisms and harm them at cellular levels. Microplastics possess hydrophobic nature, so they are able to attract different toxicants such as pharmaceuticals waste, heavy metals, plasticizers, etc. These contaminants enter the biological organisms and get transported through food chain and finally pass to humans. Mainly marine environment is responsible for microplastic pollution, but it was estimated that around 4–23 times more plastic is found in terrestrial environment. Research is going on to eradicate microplastic pollution, according to recent studies; microorganisms such as algae, bacteria and fungi are widely used in elimination of microplastics from surrounding environment. The enzymes present in these microbes are responsible for the degradation of waste related to microplatics. This chapter will discuss about the occurrence of microplastic in the surrounding environment along with their behaviour toward other living organisms residing in this environment. The different types of possible microbial remediation techniques for microplastics will also be discussed.

Keywords Microplastics · Pollution · Surface area · Microbial · Remediation

1 Introduction

The use of plastic is increased in enormous amount in the present scenario. This leads to extensive accumulation of plastic waste in our surrounding biota. According to recent studies, around 2.5 trillion pieces of plastic especially microplastic have

A. Dua · H. K. Gill (✉)
Aquatic Biology Laboratory, Department of Zoology Guru Nanak Dev University, Amritsar, Punjab, India

K. Khanna et al. (eds.), *Microplastics and Soil Microbiome*,
https://doi.org/10.1007/978-981-96-4978-5_4

been dumped into oceans every year. During the era of COVID-19, there observes alarming increase in the disposal of plastic waste including gloves, face masks, etc.; it was reported that by the end of 2050, it is estimated that there will be around 33 billion tonnes of increase in plastic waste. The oceans are the main target of plastic waste. When the plastic waste reaches the water bodies, photolysis takes place and degrades the plastic into microplastic, the most harmful form of plastic. Microplastic pollution is ubiquitous (Van Melkebeke et al. 2020). Microplastics are categorized into two classes. First one is primary microplastics. These are mainly used as artificial powder for moulding purpose. Second are secondary microplastics (Cera et al. 2022; Yu et al. 2018). The major sources of microplastic pollution include irrigation systems, landfills, agriculture fields and industrial effluents (Conley et al. 2019). Belzagui et al. (2019) explained that variety of sources, including atmospheric deposition, land-based sources, fertilizers, artificial turf, road and air transportation, textiles, coastal and tourism activities, commercial fishing, marine vessels, aquaculture and oilrigs are responsible for microplastic pollution. Microplastic then settled in the deep sea, the virgin polar regions, the ice sheet, living things and toxicants such as heavy metals, antibiotics and polynuclear aromatic hydrocarbons (PNAs) (Obbard 2018; Woodall et al. 2014; Yu et al. 2018). Microplastic waste gets further broken down into nanoplastic which is absorbed by living organisms and affects their behaviour pattern such as how they feed, digest, excrete and reproduce (Amelia et al. 2021; Westphalen and Abdelrasoul 2018).

As compared to large particles, there observed increased environmental concern about these tiny particles because of their deleterious environmental issues (Anik et al. 2021). The process of weathering broke these plastics debris into minute particles that are termed as microplastics/nanoparticles (Gigault et al. 2018). Not only the breakdown of microplastics but the disposal of industrial waste and consumer products are equally responsible for the generation of microplastics, and these are easily found in environmental compartment and even in tap water too (Biginagwa et al. 2016; Blumenröder et al. 2017; Dümichen et al. 2017). The aquatic and terrestrial organisms including zooplankton, crustaceans, birds, fishes, turtles and mammals ingest these microplastics, and it leads to serious consequences such as asphyxiation, starvation, injuries because of abrasion and restriction to movement mammals (Free et al. 2014; Naidoo et al. 2020; Hara et al. 2020; Alimba and Faggio 2019). This also leads to harmful effects on tissues and cellular level and also disrupts energy reallocation, sibling execution and successive reproduction (Karlsson et al. 2017; Leslie et al. 2017).

There are two main features of microplastics toxicity to organisms; first, the intake or accumulation of microplastics that inhibits the normal metabolism of animals and human beings that triggers their neurological and intestinal functions which can be lethal in serious cases (Zhu et al. 2019). Microplastics may also contain several additives such as UV stabilizers, biocides, colour pigments, flame retardants, etc. During stressful condition like strong air and water pressure, strong irradiation of UV light, natural weathering and microplastics releases these components into environments and causes damage to the natural biota (Lambert et al. 2017). The degradation of microplastics based on the intervention of biotic components such as fungi, bacteria and algae and also various types of advanced oxidation processes (AOPs) such as photocatalytic degradation, photodegradation and electrochemical degradation which help for their structural destruction via depolymerization into oligomers, dimers and monomers. The main factors responsible for the destruction of plastic polymers are enzymatic actions, temperature and specificity (Ukil et al. 2022). AOPs are the hot topic for the degradation of microplastics (Du et al. 2021). The persistent nature of microplastics makes them hard to eradicate them by the use of conventional approaches (Zhang et al. 2021). Presently, the most promising approach for the fate of MPs in recycling and utilizing is the chemical recycling approach that transforms the wastes into useful products which can be used as fuel (Miao et al. 2020).

2 Sources of MPs Contamination in Aquatic Environments

On the basis of sources, MPs are categorized into two types that are responsible for contamination of aquatic environments:

1. Primary microplastics.
2. Secondary microplastics.

   - *Primary microplastics*: The primary microparticles are the nano-sized plastics that come directly into aquatic environments from different effluents, for example, microbeads that are present in plastic pellets, cosmetic products, plastic fibres, etc.
   - *Secondary microplastics*: These are the types of microplastics that are formed by the fragmentation of bigger particles into smaller particles, for example, ultraviolent rays from sunlight (Katare et al. 2021).

3 Presence of Microplastics in the Surrounding Environment

- : *Occurrence of Microplastics in the Environment*

Microplastics are present across all environmental compartments. They are also observed in the most remote areas of the environment (e.g., Artic ice and deep-sea sediment) (Gong and Xie 2020; Eerkes-Medrano et al. 2015); the exact prevalence of microplastics in the environment is unknown due to several factors including the following:

- Microplastics are unevenly present in the environment because of human activity and environmental factors. The concentration and character of microplastics can be highly variable spatially and temporally in the environment.
- Global occurrence about microplastics research is most often siloed according to the environmental compartment (water, soil, air and living organisms [biota]),

shape (bead, fibre, fragment) or chemical composition. Widely cited research from the International Union for Conservation of Nature estimates that about two-thirds of microplastics in the global marine environment are from fragmentation of virgin plastics (tires, synthetic textiles, city dust, road markings, marine coatings, etc.) (Boucher and Friot 2017).

- Another commonly repeated estimate is that microplastics annually released to land are 4–23 times greater than microplastics released to oceans (Horton et al. 2017; Mendoza et al. 2021).

3.1 Occurrence of MPs in the Aquatic Bodies

Microplastic particles are present everywhere in the world such as in surface water. Microplastics are commonly present at the Italian, Singapore and Portuguese coast, at beaches of Hawaii and islands of the equatorial Western Atlantic as well as at shores of German and Greek islands. Environmental monitoring studies divide microplastics into different categories such as spheres, fibres, foams and fragments.

There are studies that report that in freshwater ecosystems like lakes and rivers, there observed different concentrations as compared to marine ecosystems.

A study conducted with lakeshore sediments collected from Lake Garda (Italy) showed high abundances of polyethylene (PE) and polystyrene (PS) microplastics, indicating the importance of buoyant microplastics for shore sediments. In addition, polymer particles of a higher density, such as PVC and PET were also identified in this study underlining the variety of microplastics present inshore sediments. Microplastics were also detected in the Laurentian Great Lakes in North America with concentrations ranging between 0 and 466,305 particles m^2. Higher abundances of microplastics have been detected in proximity to urban areas. The determination of microplastics in gastrointestinal tract of fish from Lake Victoria (Tanzania) shows first evidence for the microplastic pollution of African lakes. A monitoring study of Lake Taihu in China detected up to 6,000,000 particles km^2 in surface water samples. In bulk water samples, microplastics were present in concentrations between 3 and 26 particlesL^1, while the sediments of Lake Taihu contained microplastics in the range of 11–235 particles kg^1. The reported concentrations of microplastics in the sediments of Lake Taihu are lower compared to those detected in the sediments of the European lakes. Microplastics were quantified in the range of 997–4435 particles km^2, but it should be taken into account that the catchment area of Lake Hovsgol is less populated compared to the above-mentioned lakes.

3.2 Occurrence of MPs in Humans

Humans can be exposed to microplastics by inhalation, digestion and cutaneous contact (such as through hair follicles, sweat glands and damaged skin). Scientists believe that studies undercount particles in human tissue as a result of measurement constraints, which have consequences for comprehending impacts at the chemical, cellular, and tissue levels (Huang et al. 2023; Stapleton et al. 2023; Li et al. 2022). The Kutralam-Muniasamy et al. (2022) review revealed 20 biomonitoring studies, the majority of which concentrated on a single tissue type, about the prevalence of microplastics in 16 distinct types of human biological samples. According to several research on this study, people with sickness have higher microplastic concentrations than people in good health. To understand whether diseases are brought on by microplastics or whether diseases prevent the elimination of microplastics, more research is needed (Alberghini et al. 2023).

3.3 Microplastics Occurrence in Food

Microplastics have been studied in various foods because a major entry point of microplastics into humans is the ingestion of contaminated food (Pironti et al. 2021).

3.4 Occurrence of MPs in Flora and Fauna (Biota)

Investigations on the potential effects of microplastics on the flora and fauna existing in freshwater, marine and terrestrial ecosystems were spurred by the existence of microplastics in those settings (Gautam et al. 2022). In a sizable fraction of the examined biota, microplastics have been found.

3.5 Occurrence of MPs in Soil (Terrestrial Compartment)

Because soil is a primary carrier of microplastic pollutants to the aquatic environment, soil is regarded as a substantial sink for collecting microplastics. Another direct route for microplastics to enter the food chain is through it (Kim and An 2020; Li et al. 2021). According to a single source's estimate, the amount of microplastics released each year on land is 4–23 times more than the amount released into the oceans (Horton et al. 2017).

3.6 Occurrence of MPs in the Air (Atmospheric Compartment)

Airborne microplastic appears to be the least studied compartment. Because air is contaminated continuously by particles released by building materials, furniture, clothing and lab equipment, contamination of air samples is concerned when measuring microplastics (Prata et al. 2020). The few studies conducted in specific regions or cities in North America, Europe and Asia had short periods of monitoring, making estimates of concentration highly uncertain. Studies in several European and Asian countries found that indoor environments generally had higher concentrations of microplastics than outdoor environments, although the concentration of microplastics in indoor air samples depended heavily on where the sample was collected (Prata et al. 2020). Researchers postulate that the higher indoor concentrations might be due to the closed environment that traps and recirculates microplastics generated from common household products (e.g., carpeting, clothing, bedding, cleaning products and kitchenware) (Kacprzak and Tijing 2022). Additionally, study findings suggest that the primary sources of atmospheric particles in the western US are roads, the ocean and agricultural soil dust (Brahney et al. 2021).

4 Environmental Degradation of Microplastics

The strong resistance to environmental impacts of many synthetic polymers is one of the factors contributing to their great versatility. This characteristic causes synthetic polymers to enter the environment with very little degradation and a very long residence duration. Synthetic polymer degradation can generally be divided into biotic and abiotic processes, each of which uses a distinct mechanism and is influenced by a number of physical, chemical, or biological factors. Polymers are broken down into smaller molecules throughout the degradation process, such as oligomers, monomers, orchemically altered forms. They may even entirely mineralize. The most important processes for the degradation of synthetic polymers can be divided into:

- Physical degradation (abrasive forces, heating/cooling, freezing/thawing, wetting/drying).
- Photodegradation (usually by UV light).
- Chemical degradation (oxidation or hydrolysis).
- Biodegradation by organisms (bacteria, fungi, algae).

Apart from these, mechanical degradation is an important factor with regard to microplastics in the aquatic environment. In most cases, aging of the polymer by environmental influences, such as photodegradation or chemical degradation of additives, changes the polymer properties and leads to embrittlement of the polymer. The recalcitrant material is then shredded into smaller particles by friction forces occurring during the movement through different environmental. This degradation generally leads to smaller plastic particles, which can result in particles with sizes between 1 and 5000 μm. Such particles are classified as microplastics. If the particles fall into the range of microplastics, however, the mechanical degradation continues. Thus, it is extremely conceivable that even smaller particles, or so-called nanoplastics, will arise. When compared to the original macroplastics or microplastics, these nanoplastics may have different characteristics. Both times, the mechanical breakdown causes a reduction in particle size and, as a result, an increase in surface area of the polymer particles, which causes the degradation to proceed more quickly due to higher reactivity. Thermal degradation is not significant for freshwaters, since the temperature in aquatic systems is not high enough to initiate chemical changes in synthetic polymers. The degradation of synthetic polymers in the environment on a molecular basis is usually initiated by photooxidation (with UV radiation) or by hydrolysis and is eventually followed by chemical oxidation. The predominant mechanisms strongly depend on the type of polymer, as there are numerous different compositions of synthetic polymers produced (i.e., polyolefins, polyesters, and polyamides). After the initial reactions, the molecular weight of the polymer is decreased and the reacted groups become available for microbial degradation. Photooxidation is usually a fast process, but the degradation rate also depends on the extent of additives in a particular polymer that could prevent oxidation processes (i.e., antioxidants). Moreover, the photodegradation of plastics floating in the aquatic environment is slower compared to degradation in terrestrial exposure. According to experiments, PE and PS degraded faster on the water's surface than they did while partially or entirely immersed. This is probably due to the light's waning intensity and the slower rate of photooxidation. Because of this, many plastics can survive for decades or even centuries in the aquatic environment. Both aerobic and anaerobic settings can result in the biodegradation of synthetic polymers. Both full and partial mineralization of polymers into CO_2, H_2O, N_2, H_2, CH_4, salts, minerals, and biomass is possible. Partial or primary degradation of the polymer chain leads to stable or temporarily stable transformation products. Biodegradation is coupled to three essential criteria:

- Microorganisms must be capable to depolymerize the target substance and mineralize the monomeric compounds with the help of enzymes of an appropriate metabolic pathway.
- Temperature, pH, moisture, and salinity must provide conditions that are essential for biodegradation.
- The size of synthetic polymers is too large to penetrate the cell membranes of microorganisms, so the first step is the cleavage of side chains and the formation of smaller polymer units by extracellular enzymes.
- Mainly, at first, depolymerization step consists of an enzyme bases hydrolysis of amides and esters bonds. These smaller molecules can easily be absorbed and metabolized by microorganisms.
- It is widely documented in the literature how poly(-caprolactam) and water-soluble polyethylene glycol completely biodegrade. However, the majority of naturally occurring plastics are not water soluble, and many of the synthetic

polymers found in aquatic environments, like PET, PE, and polypropylene (PP), degrade very slowly or not at all. These polymers typically degrade by a combination of biotic and abiotic processes.
- Polyolefins, including PE and PP, are a class of chemicals with substantial industrial production volumes and are frequently found in environmental samples. Because the alkyl backbone of these polymers is inaccessible to microorganisms and requires an abiotic change, they are often not biodegradable. Although polyolefins' alkyl backbones have a strong resistance to hydrolysis, they are frequently vulnerable to oxidative destruction. The oxidative or photooxidative breakdown of the polymer is postponed until the antioxidants are consumed in order to avoid this by adding additives throughout the manufacture process (Li et al. 2021). The decomposition of polyolefins after the initial oxidation of their surfaces may take many weeks, but it produces microplastics as potential intermediates. These smaller and oxidized plastic fragments are more susceptible to microbial attack; e.g., biodegradation of PE is described for pre-oxidized fragments of the original material by *Pseudomonas* sp.

5 Bioremediation of MPs

Bioremediation is the process of turning polymers into minerals with the help of biotic elements like fungi, bacteria, and algae. It entails the development of biofilms on the surface of plastics, which leads to the depolymerization of plastics under the influence of particular enzymes, destroying their structural integrity. In addition to carbon dioxide, water, and by-products of the depolymerization, the products may include oligomers, dimers, and monomers (Zhang et al. 2021). It is possible for biotransformation to take place, producing persistent organic pollutants (POPs) that are harmful to the environment. The final by-products of biodegrading plastics are water and carbon dioxide under aerobic settings and carbon dioxide and methane under anaerobic conditions (Giacomucci et al. 2020). According to Alomar and Deudero (2017), MPs are consumed by marine species ranging from zooplanktons to top predators and decomposers, even when it endangers them. Since the majority of tests are carried out ex vivo, their harmful consequences in the natural environment are still unknown, necessitating additional research for adequate validation (Lenz et al. 2016; Gaur et al. 2022).

- Microalgae was found to be best fit for bioremediation of MPs over years, and its profound potential was observed. There are only few strains of algae species known to have the capacity for the breakdown of microplastic particles. In order to break down the plastic waste, lignin and extracellular polysaccharides on their surface were used (Nabeela et al. 2021).
- Khoironi and Anggoro (2019) showed that extent of degradation of polyethylene terephthalate (PET) was observed, followed by polyethylene (PE), indicating

that degradation can be defined with reference to microplastic polymer and their candidate species (Corinaldesi et al. 2021).

- Algae can be used as a source of engineered PETase by the help of biotechnological approach (Zurier and Goddard 2021). For about 60 days, *Bacillus* sp. And *Paenibacillus* sp. significantly decrease the size of polypropylene (PP) (Park et al. 2023).
- Under laboratory conditions, *Bacillus cereus* and *Bacillus gottheilii* decreased the size of different MPs components such as poly(methylene), polyethylene polypropylene (PP), and pvc in around 40 days, indicating a decrease in the weight and quantity of MPs (Auta et al. 2017).
- According to Zhang et al. (2021), *Aspergillus flavus*, a fungus, also contributed significantly to the degradation of plastics. The environmental friendliness and low cost of biodegradation have led to its acceptance on a global scale. However, there are a number of downsides, including time-consuming, challenges with on-site deployments, gaps particular to certain species, etc.

6 Conclusion

Since 2014, researchers have focused a lot of attention on the category of organic pollutants known as microplastics, which is cause for increased concern. It is crucial to create sustainable ways to decrease microplastics' negative impacts and lessen their presence in the ecosystem, while their negative effects keep growing. The varieties, shapes, sources, and worldwide response to microplastics are only a few of the topics covered in this paper. Although there are many aquatic bodies where microplastics can be found, land-based sources make up the majority of environmental pollution (80–90%). Several actions should be taken, like educational programmes to create awareness among people regarding microplastics and media sources like television programmes, journals and social media platforms, in order to raise public awareness of microplastic concerns and encourage the development of practical solutions. Microplastics have been discovered in a variety of human biological specimens, including faeces, sputum, saliva, blood, bronchoalveolar lavage fluid, placenta and other organs, indicating that these particles may have a negative impact on human health. In addition to possible health hazards like cancer, immunotoxicity, intestinal, pulmonary, cardiovascular and inflammatory illnesses, these effects can also have a adverse effect on pregnancy and the subjection of the mother to her offspring.

Microplastics must be made into useful by-products, better separated from other contaminants, and their environmental fate must be determined. When creating reuse and recycle techniques for the plastic waste generated by the medical industry, finding viable substitutes for single-use face masks is essential. Additionally, efforts should be made to enhance the effectiveness and quality of plastic substitutes, such as bioplastics, as well as to incorporate microplastic treatment technologies to maximize their removal effectiveness and decrease negative effects. The public's readiness to make the transition to an economy not dependent on plastic should also be taken into account when choosing a plan to minimize plastic consumption. These elements include infrastructure, economic conditions, the types of microplastics discharged, alternative options, and infrastructure.

References

Alberghini L, Truant A, Santonicola S, Colavita G, Giaccone V (2023) Microplastics in fish and fishery products and risks for human health: a review. Int J Environ Res Public Health 20(1):789

Alimba CG, Faggio C (2019) Microplasticsinthemarineenvironment: current trends in environmental pollution and mechanisms of toxicological profile. Environ Toxicol Pharmacol 68:61–74

Alomar C, Deudero S (2017) Evidence of microplastic ingestion in the shark Galeus melastomus Rafinesque, 1810 in the continental shelf off the western Mediterranean Sea. Environ Pollut 223:223–229

Amelia TSM, Khalik WMAWM, Ong MC, Shao YT, Pan HJ, Bhubalan K (2021) Marine microplastics as vectors of major ocean pollutants and its hazards to the marine ecosystem and humans. Prog Earth Planet Sci 8(1):1–26

Anik AH, Hossain S, Alam M, Sultan MB, Hasnine MT, Rahman MM (2021) Microplastics pollution: a comprehensive review on the sources, fates, effects, and potential remediation. Environ Nanotechnol Monit Manag 16:100530

Auta HS, Emenike CU, Fauziah SH (2017) Distribution and importance of microplastics in the marine environment: a review of the sources, fate, effects, and potential solutions. Environ Int 102:165–176

Belzagui F, Crespi M, Álvarez A, Gutiérrez-Bouzán C, Vilaseca M (2019) Microplastics' emissions: microfibers' detachment from textile garments. Environ Pollut 248:1028–1035

Biginagwa FJ, Mayoma BS, Shashoua Y, Syberg K, Khan FR (2016) First evidence of microplastics in the African Great Lakes: recovery from Lake Victoria Nile perch and Nile tilapia. J Great Lakes Res 42(1):146–149

Blumenröder J, Sechet P, Kakkonen JE, Hartl MG (2017) Microplastic contamination of intertidal sediments of Scapa flow, Orkney: a first assessment. Marinepollutionbulletin 124(1):112–120

Boucher J, Friot D (2017) Primary microplastics in the oceans: a global evaluation of sources, vol 10. Iucn, Gland

Brahney J, Mahowald N, Prank M, Cornwell G, Klimont Z, Matsui H, Prather KA (2021) Constraining the atmospheric limb of the plastic cycle. Proc Natl Acad Sci 118(16):e2020719118

Cera A, Sighicelli M, Sodo A, Lecce F, Menegoni P, Scalici M (2022) Microplastics distribution and possible ingestion by fish in lacustrine waters (Lake Bracciano, Italy). Environ Sci Pollut Res 29(45):68179–68190

Conley K, Clum A, Deepe J, Lane H, Beckingham B (2019) Wastewater treatment plants as a source of microplastics to an urban estuary: removal efficiencies and loading per capita over one year. Water research X 3:100030

Corinaldesi C, Canensi S, Dell'Anno A, Tangherlini M, DiCapua I, Varrella S, Danovaro R (2021) Multiple impacts of microplastics can threaten marine habitat-forming species. Commun Biol 4(1):431

Du H, Xie Y, Wang J (2021) Microplastic degradation methods and corresponding degradation mechanism: research status and future perspectives. J Hazard Mater 418:126377

Dümichen E, Eisentraut P, Bannick CG, Barthel AK, Senz R, Braun U (2017) Fast identification of microplastics in complex environmental samples by a thermal degradation method. Chemosphere 174:572–584

Eerkes-Medrano D, Thompson RC, Aldridge DC (2015) Microplastics in freshwater systems: a review of the emerging threats, identification of knowledge gaps and prioritisation of research needs. Water Res 75:63–82

Free CM, Jensen OP, Mason SA, Eriksen M, Williamson NJ, Boldgiv B (2014) High-levels of microplastic pollution in a large, remote, mountain lake. Mar Pollut Bull 85(1):156–163

Gaur VK, Gupta S, Sharma P, Gupta P, Varjani S, Srivastava JK, Bui XT (2022) Metabolic cascade for remediation of plastic waste: a case study on microplastic degradation. Curr Pollut Rep 8(1):30–50

Gautam R et al (2022) Evaluation of potential toxicity of polyethylene microplastics on human derived cell lines. Sci Total Environ 838:156089

Giacomucci L, Raddadi N, Soccio M, Lotti N, Fava F (2020) Biodegradation of polyvinyl chloride plastic films by enriched anaerobic marine consortia. Mar Environ Res 158:104949

Gigault J, Ter Halle A, Baudrimont M, Pascal PY, Gauffre F, Phi TL, Reynaud S (2018) Current opinion: what is a nanoplastic. Environ Pollut 235:1030–1034

Gong J, Xie P (2020) Research progress in sources, analytical methods, eco- environmental effects, and control measures of microplastics. Chemosphere 254:126790

Hara J, Frias J, Nash R (2020) Quantification of microplastic ingestion by the decapod crustacean Nephrops norvegicus from Irish waters. Mar Pollut Bull 152:110905

Horton AA, Walton A, Spurgeon DJ, Lahive E, Svendsen C (2017) Microplastics in freshwater and terrestrial environments: evaluating the current understanding to identify the knowledge gaps and future research priorities. Sci Total Environ 586:127–141

Huang H, Mohamed BA, Li LY (2023) Accumulation and fate of microplastics in soils after application of biosolids on land: a review. Environ Chem Lett 21:1–15

Kacprzak S, Tijing LD (2022) Microplastics in indoor environment: sources, mitigation and fate. J Environ Chem Eng 107359:107359

Karlsson TM, Vethaak AD, Almroth BC, Ariese F, van Velzen M, Hassellöv M, Leslie HA (2017) Screening for microplastics in sediment, water, marine invertebrates and fish: method development and microplastic accumulation. Mar Pollut Bull 122(1–2):403–408

Katare Y, Singh P, Sankhla MS, Singhal M, Jadhav EB, Parihar K, Bhardwaj L (2021) Microplastics in aquatic environments: sources, ecotoxicity, detection & remediation. Biointerface Res Appl Chem 12:3407–3428

Khoironi A, Anggoro S et al (2019) Evaluation of the interaction among microalgae spirulina sp, plastics polyethylene terephthalate and polypropylene in freshwater environment. J Ecol Eng 20(6):161–173

Kim SW, An YJ (2020) Edible size of polyethylene microplastics and their effects on springtail behavior. Environ Pollut 266:115255

Kutralam-Muniasamy G, Pérez-Guevara F, Shruti VC (2022) A critical synthesis of current peer-reviewed literature on the environmental and human health impacts of COVID-19 PPE litter: new findings and next steps. J Hazard Mater 422:126945

Lambert S, Scherer C, Wagner M (2017) Ecotoxicitytestingofmicroplastics: considering the heterogeneity of physicochemical properties. Integr Environ Assess Manag 13(3):470–475

Lenz R, Enders K, Nielsen TG (2016) Microplastic exposure studies should be environmentally realistic. Proc Natl Acad Sci 113(29):E4121–E4122

Leslie HA, Brandsma SH, Van Velzen MJM, Vethaak AD (2017) Microplastics en route: field measurements in the Dutch river delta and Amsterdam canals, wastewater treatment plants, North Sea sediments and biota. Environ Int 101:133–142

Li J, Ouyang Z, Liu P, Zhao X, Wu R, Zhang C, Guo X (2021) Distribution and characteristics of microplastics in the basin of Chishui River in Renhuai, China. Sci Total Environ 773:145591

Li J, Yu S, Yu Y, Xu M (2022) Effects of microplastics on higher plants: a review. Bull Environ Contam Toxicol 109(2):241–265

Mendoza LMR, Vargas DL, Balcer M (2021) Microplasticsoccurrenceand fate in the environment. Curr Opin Green Sustain Chem 32:100523

Miao F, Liu Y, Gao M, Yu X, Xiao P, Wang M, Wang X (2020) Degradation of polyvinylchloride microplastics via an electro-Fenton-like system with a TiO2/graphite cathode. J Hazard Mater 399:123023
Nabeela K, Thorat MN, Backer SN, Ramachandran AM, Thomas RT, Preethikumar G, Pillai S (2021) Hydrophilic 3D interconnected network of bacterial Nanocellulose/black titania photothermal foams as an efficient interfacial solar evaporator. ACS Appl Bio Mater 4(5):4373–4383
Naidoo T, Thompson RC, Rajkaran A (2020) Quantification and characterisation of microplastics ingested by selected juvenile fish species associated with mangroves in Kwa Zulu-Natal, South Africa. Environ Pollut 257:113635
Obbard RW (2018) Microplastics in polar regions: the role of long range transport. Curr Opin Environ Sci Health 1:24–29
Park YJ, Kim YJ, Yu HH, Lee NK, Paik HD (2023) Cell-free supernatants of *Bacillus subtilis* and *bacillus polyfermenticus* inhibit listeria monocytogenes biofilm formation. Food Control 144:109387
Pironti C, Ricciardi M, Motta O, Miele Y, Proto A, Montano L (2021) Microplastics in the environment: intake through the food web, human exposure and toxicological effects. Toxics 9(9):224
Prata JC, da Costa JP, Lopes I, Duarte AC, Rocha-Santos T (2020) Environmental exposure to microplastics: An overview on possible human health effects. Sci Total Environ 702:134455
Stapleton MJ, Ansari AJ, Hai FI (2023) Antibiotic sorption onto microplastics in water: acritical review of the factors, mechanisms and implications. Water Res 119790:119790
Ukil M, Roy S, Choudhury AR, Ganesh PS (2022) Role of microplastics as attachment Media for the Growth of microorganisms. In: Microplastics pollution in aquatic media: occurrence, detection, and removal, pp 313–325
Van Melkebeke M, Janssen C, De Meester S (2020) Characteristics and sinkingbehavioroftypicalmicroplasticsincludingthepotentialeffect of biofouling: implications for remediation. Environ Sci Technol 54(14):8668–8680
Westphalen H, Abdelrasoul A (2018) Challenges and treatment of microplastics in water. Water Challengesofan UrbanizingWorld 5:71–82
Woodall LC, Sanchez-Vidal A, Canals M, Paterson GL, Coppock R, Sleight V, Thompson RC (2014) The deep sea is a major sink for microplastic debris. R Soc Open Sci 1(4):140317
Yu P, Liu Z, Wu D, Chen M, Lv W, Zhao Y (2018) Accumulation of polystyrene microplastics in juvenile Eriocheir sinensis and oxidative stress effects in the liver. Aquat Toxicol 200:28–36
Zhang Y, Gao T, Kang S, Allen S, Luo X, Allen D (2021) Microplastics in glaciers of the Tibetan plateau: evidence for the long-range transport of microplastics. Sci Total Environ 758:143634
Zhu F, Zhu C, Wang C, Gu C (2019) Occurrence and ecological impacts of microplastics in soil systems: a review. Bull Environ Contam Toxicol 102:741–749
Zurier HS, Goddard JM (2021) Biodegradation of microplastics in food and agriculture. Curr Opin Food Sci 37:37–44

Chapter 5
Microplastic: Interaction with Agroecosystem and Microbial Remediation

Deepak Kumar, Roohi Sharma, Raman Tikoria, Mohd Ali, Parkirti, Vikram, Rajesh Kumari Manhas, Harish Changotra, and Puja Ohri

Abstract Due to their ubiquity, resistance and protracted persistence in the ecosystem, microplastic pollution has emerged as a major environmental hazard on the global scale. For the time being, the majority of investigation has been concentrated on freshwater and marine environments, but there hasn't been much investigation on terrestrial ecosystems, especially agroecosystem and very little is known about how plants absorb and store microplastics. Microplastics' detrimental impacts on the physical, chemical and biological characteristics of soil are gradually becoming clear. However, little research has been done on how microplastics affect plant growth and yield. By impacting the rhizosphere's microbial ecology and upsetting the agroecosystem's homeostasis, soil contamination with microplastics might change plants' functioning. Additionally, it might enter the plant's system through water and nutrient absorption pathways and influence plant physiology. Microplastic accumulation in plants can negatively affect plant growth and development. Since plants are the primary living component of terrestrial ecosystems, understanding how microplastics and plants interact is crucial. Furthermore, it is decisive to create innovative methods for microplastic remediation. Among the several techniques now in use, microbial remediation has the potential to degrade microplastics in a sustainable manner. Scientists have thus far focused on microorganisms including bacteria, fungi and algae as a means of remediating microplastic. Considering above-mentioned aspects, the current chapter focuses on sources and effects of

D. Kumar · R. Sharma · M. Ali · Parkirti · Vikram · P. Ohri (✉)
Department of Zoology, Guru Nanak Dev University, Amritsar, Punjab, India

R. Tikoria
Department of Zoology, Guru Nanak Dev University, Amritsar, Punjab, India

Department of Zoology, Central University of Punjab, Ghudha, Bathinda, Punjab, India

R. K. Manhas
Department of Microbiology, Guru Nanak Dev University, Amritsar, Punjab, India

H. Changotra
Department of Molecular Biology and Biochemistry, Guru Nanak Dev University, Amritsar, Punjab, India

K. Khanna et al. (eds.), *Microplastics and Soil Microbiome*,
https://doi.org/10.1007/978-981-96-4978-5_5

microplastic in agroecosystem, uptake and toxic effects of microplastic on plants and its microbial remediation.

Keywords Rhizosphere · Remediation · Microbes · Agroecosystem · Toxicity · Plant physiology

1 Introduction

A global concern for the future has been raised by the ubiquitous presence of microplastics (˂5 mm) in our ecosystem. Microplastics are 4–23 times more common in soil ecosystems than in aquatic habitats, and they enter ecosystems either directly or indirectly through a number of processes (Horton et al. 2017). The amount of plastic waste entering the ocean alone each year is estimated to be 8 million metric tonnes. This volume allows microplastics to actively contaminate agriculturally important ecosystems (air, soil, and water), which qualifies them as "pollutants of importance and agents of global change" (Okeke et al. 2022). Microplastics, which are agents of global change, alter the composition of impacted ecosystems, which has an impact on the production and growth of accustomed fauna or flora. In order to ensure their own survival and well-being, humans must be willing to gently alter their natural ecosystems in order to increase the yields of advantageous products. Through a variety of connections and interactions with surroundings, climatic factors, water, domesticated plants and species at various trophic levels, agroecosystems provide benefits to humans (Zabala et al. 2021, Liu et al. 2022a, b, c, d). Agroecosystems are thus ecosystems that people have modified to provide benefits and are negatively impacted by various human activities (Yang et al. 2021a, b, c, d; Sharma et al. 2023a, b). Due to growing anthropogenic activities, the use of specific agricultural products and containers introduces levels of environmental pollutants and contaminants, including pesticides, fertilizers, plant growth regulators, and, recently, microplastics (Rafique et al. 2020; Zheng et al. 2020; Kumar and Ohri 2023). Indirectly and directly, microplastics in agricultural environments have an impact on crop yields, health of plants and animals that are part of the food chain and, invariably, the health of people who are the food's final consumers (Blackburn and Green 2022; Okeke et al. 2022; Zolotova et al. 2022).

Unlike the microplastics found in the ocean, little research has been done on the microplastics found in terrestrial systems. The level of microplastics in terrestrial ecosystems will likely rise with time due to soil's reduced capacity for movement. Microplastics can build up in plants and have a deleterious impact on plant growth, according to recent studies (Khalid et al. 2020; Yu et al. 2021). Understanding the interplay between microplastics and plant communities is essential, since plants are the fundamental living element of terrestrial ecosystems. Microplastic buildup in plants has the potential to have direct ecological effects and consequences for food safety and agricultural sustainability (Uwamungu et al. 2022). They can accumulate in these frameworks either by being adsorbed onto the root surface or by being absorbed by roots, fruits and vegetables. Different microplastic kinds, shapes, sizes and concentrations have different effects on plants. Consequently, different types of microplastics have different effects on plant growth, as seen in different situations with lettuce, wheat, welsh onion and common bean (Qi et al. 2018a, b; Lian et al. 2019; Li et al. 2020a, b, c, Li et al. 2022; Biba et al. 2023a, b; Roy et al. 2023).

Considering that microplastics have significant negative impact on the agroecosystem, finding ways to reduce microplastic levels has become a hot topic. Numerous studies have looked at the distribution, intake, fate, behaviour, concentration and impact of microplastics in recent years (Qi et al. 2020; Okeke et al. 2023). The effectiveness of techniques created for the eradication and correction of microplastics is still unknown, though. Furthermore, a thorough summary of the content concerning the decline of microplastics has not yet been completed. Researchers have been astonished to investigate that despite microplastics can persist in the environment and withstand degradation; some microbes can degrade them, which has led them to explore ways to reduce microplastic pollution (Ghosh et al. 2013; Debroy et al. 2022). Microorganisms have the capacity to break down a wide range of substances, including microplastics, and have an inherent ability to adjust to almost any environment (Kotova et al. 2021). In order to facilitate native bioremediation and improve the cleaning of ecosystems without having a negative impact on the environment, the use of microbes that can degrade microplastics will boost biodegradation without disturbing the environment (Abraham et al. 2017; Yuan et al. 2020). However, few viable microorganisms have now been isolated, the interactions between microbes and microplastics have not yet been clarified, and there is still a dearth of information regarding microplastic removal. Determining how to acquire more useful bacteria, enhance their MP-degrading activity and develop an improved grasp of how microbes metabolize and utilize microplastics require a summary and analysis of the existing state of knowledge (Yuan et al. 2020). Considering this, the current chapter summarizes different sources of microplastics, interaction of microplastic with agroecosystem particularly, with the uptake and toxicity of microplastic to plants and its microbial remediation strategies.

2 Sources of Microplastics

Plastics have grown into an indispensable and flexible component with vast variety of attributes, chemical nature and usage over the past six decades (Gregory 2009). The challenge of their degradation would likely to be worsen as microplastic production, and demand is being escalated despite some restrictions, leading to the likelihood that plastic will permeate more extensively in the form of microplastics and nanoplastics (Klemeš et al. 2020; Bernard 2014). Synthetic textiles, tires, personal care items, etc. have been listed as prospective microplastics sources. It enters the environment in a variety of ways via point sources like mulching, biosolids application, reclaimed water usage and treatment facilities of wastewater and various soil amendment and conditioners and through non-point sources such as manufacturing and processing sites, industrial emissions and effluents, textile industry, domestic waste and biomedical sectors (Kang et al. 2022; Modak and Basu 2022).

The lengthy period of storage, inappropriate usage and dumping of the substance lead to an accumulation of plastic waste in the natural environment. Yet, a significant danger to the natural world originates from the discharge of plastic particles resulting from the industrial production of these materials as up to 13% of the overall carbon budget on the globe will be employed in plastics manufacture by 2050 (Sharma et al. 2023a, b; Gazal and Gheewala 2020; Hale et al. 2020). Additionally, during the disintegration process, chemical additives as well as hazardous substances in the plastics are liberated (Alabi et al. 2019). These substances are resistant to natural deterioration and can amass readily in lithosphere and hydrosphere (Hahladakis et al. 2018). Plastic pollution negatively impacts soil creatures and vegetation, particularly in the lithosphere (Allouzi et al. 2021). It is impossible to separate the widespread use of throwaway plastic goods from the substantial soil contamination caused by microplastics and nanoplastics (Azeem et al. 2021). Primary microplastics are those that are produced and discharged into the surrounding atmosphere in micron (nano) size range (such as from cosmetics, shampoos), while secondary microplastics originate from the splitting of bigger materials made of plastic (Zhao et al. 2022a, b).

By 2050, it's predicted that 450–500 million tonnes of trash made of plastic would be assembled (Jebaranjitham et al. 2022). Besides this, it is estimated that more than 300 million tonnes of microplastics will be generated, whether primary or secondary, that will keep contaminating the agroecosystems (Guo et al. 2020) (Fig. 5.1).

2.1 Packaging

The biggest supplier of plastic in the marketplace is packaging, and as the rate of plastic resin manufacture grows, consumers sector are accumulating greater quantities of plastic constantly (Jacobsen et al. 2022; Kan and Miller 2022). Plastic has been utilized for encasing even more frequently, which has eased daily life. According to latest figures, more than 98.6 billion bags made of polyethylene are disposed of as unrenewable solid waste over the course of a year (Steensgaard et al. 2017). Following its usage, the bags end up going straight into the environment or discarded as garbage (Barnes et al. 2009). Rapid growth in the economy, which accounted for 26% of the total volume in 2015, according to the International Energy Agency (IEA)'s forecast (IEA 2015), will even double the share of plastic packaging in coming years. This share could increase to 318 million tonnes per year by 2050, which would be more than the entire global plastics industries (Alabi et al. 2019).

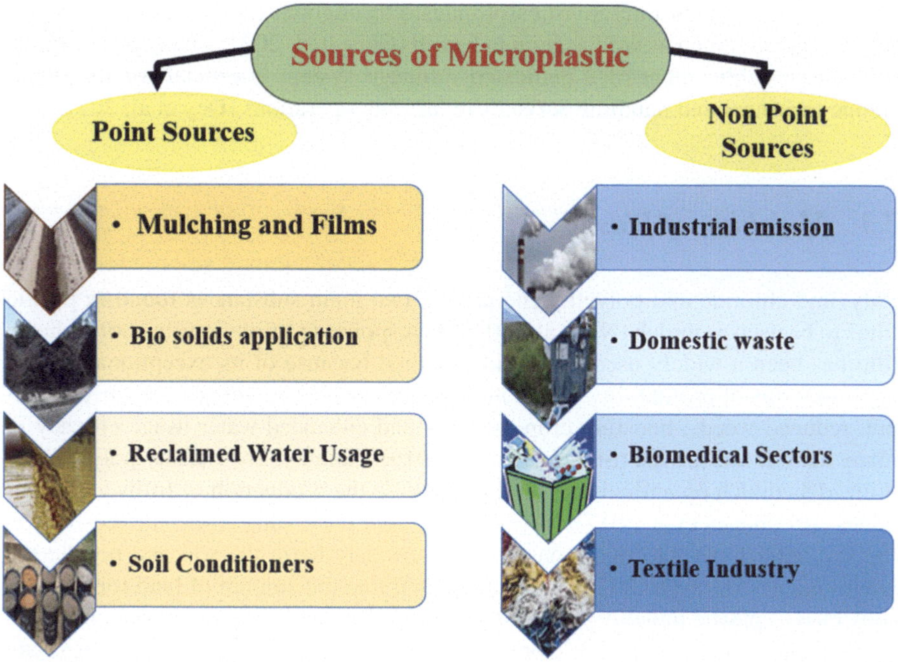

Fig. 5.1 Different sources of microplastic

2.2 *Sludge as Wastewater*

Clothing fabric fibres, personal grooming item microbeads, cleaners and debris from plastics are the most common contributors to microplastics and nanoplastics in municipal wastewater treatments (Ziajahromi et al. 2021). An earlier investigation found that the primary constituents of plastics in wastewater samples were acrylic polyester (67%) and fibres made of acrylic (17%) and polyamide (16%). It was discovered that the proportion of plastic in relation to actual waste material was comparable to the makeup of textiles (5% acrylic and 78% polyester). Over 700,000 fibres could be discharged from a 6 kg wash load on average. In the long run, it was determined that sewage microplastics come mainly from laundering (Browne et al. 2011; Napper and Thompson 2016). Regular wastewater treatment can discharge 300 million pieces of plastic into adjacent waterways on a daily basis (Hu et al. 2020). According to Li et al.'s research (2018a, b), the concentration of microplastics in sludge from wastewater treatment plants ranges from 1500 to 56,400 particles kg^{-1}. Also, sludge is most commonly used as fertilizer on land for agriculture (Seleiman et al. 2020). Similar to that, an additional study revealed that fertilizer made from organic materials contains up to 895 microplastics particles per kg (Weithmann et al. 2018). Hence, sludge fertilizers may transport microplastics into soils, where they become part of the soil and cause severe poisoning of farmlands and could affect how microplastics propagate throughout the ecosystem affecting soil's flora and fauna as well as human health (Chia et al. 2021; Zhang et al. 2020a, b). Thus, establishment of a dedicated drainage system is crucial for the better sludge treatment and handling services in farming operations (Dey et al. 2021).

2.3 Mulching of Plastic

Polyvinyl chloride and polyethylene are the two main substances found in plastic films to be used as mulch (Ng et al. 2018). In response to its obvious benefits, plastic film has been a widely used cultivation method because of its exceptional advantages in terms of profitability which include greater fruit suppleness, improved output, reduced weeds, boosting crop maturity and enhanced water usage efficacy in many nations worldwide (Steinmetz et al. 2016). Notably, low-density polythene is utilized as mulch on agricultural land to increase the production of fruits and vegetables (Brodhagen et al. 2017). Nevertheless, after harvesting crops, the leftovers in the soil split down into microplastics and seriously harm the soil and pollute the environment (Maraveas 2020). It is anticipated that the amount of land for farming subjected to plastic mulching was 13% in 2012, 1.455 million tonnes mulched land in 2015 and it will rise by 7.1% or even higher annually in China, where there are roughly 20 million hectares of such land accounting 60% mulching worldwide (Sun et al. 2020a, b; Liu et al. 2014, 2018). The accumulation of plastic film remains can alter the particle morphology, respiration and permeability of the soil, which may result in impacting the activity of enzymes and microbial functional diversity of soil as a whole (Jin et al. 2022; Hodson et al. 2017). In a study, it was found that 15 days following the treatment of soil with microplastic amendment (2000 particles per kg soil), the activity levels of enzyme urease and catalase were dramatically elevated inferencing that these changes will greatly impact plant community biomass and organization (Huang et al. 2019). Considering its small size (0.01–0.03 mm), once plastic is embedded in soil, it is logistically exceedingly challenging to reuse or eliminate from the site (Zhu et al. 2019). The field-based leftover films might gradually disintegrate into microplastics and nanoplastics when exposed to a mixture of UV light and physiochemical, chemical and biological factors that cause microplastic and nanoplastic environmental pollution in the soil (Sajjad et al. 2022; Briassoulis et al. 2015).

2.4 Atmospheric Dispersion

Microplastics and nanoplastics can enter the surface soil through deposition from atmosphere (Yang et al. 2021a, b, c, d). Microplastics are increasingly using airborne transmission to relocate to far-flung sites, especially those that are challenging to access (Sridharan et al. 2021). Low-density polyethylene microplastics are blown into the soil, and they have been located in far-off locations such as the arctic area, the Qinghai Tibet Plateau and a section of the Mariana Trench, Paris, Shangai, Hamburg and Dongguan, and their concentration in sea ice is considerable (Liu et al. 2019a, b; Horton et al. 2017). Paris urban regions experienced an atmospheric precipitation of plastic between 2 and 35 particles per cubic metre per day (Dris et al. 2016). According to Allen et al. (2019), air mass trajectory estimation indicated that faraway and little-populated locations could be influenced by atmospheric movement of microplastics up to 95 km. Microplastics have also been found in soil from distant, uninhabited or unsettling high mountain places (Scheurer and Bigalke 2018).

Plastics can be found in soil utilized for agriculture from a variety of sources, but the proportion of microplastics and nanoplastics from those sources fluctuates, and the investigation of the source is still in the beginning stages, so it is uncertain what the precise source is for the soil used for diverse applications on the land.

3 Microplastic and Agroecosystem

Plastics have a wide range of benefits, such as chemical and mechanical properties, but the pollution brought on by their widespread use has become a major issue in recent years. With the development of agriculture, there are more direct and indirect sources of microplastics in agroecosystems. Composts formed from home rubbish that has not been sorted or combined municipal solid waste, as well as source-separated garden organic waste, are sources of plastic contamination in agroecosystem (Boucher and Friot 2017). Exposure to microplastics has both direct and indirect effects in addition to raising temperature and oxygen availability. Following this, microplastics become dispersed on the food chain as a result of their ubiquity, size, source volume, chemical components and migration vertically through the soil profile and horizontally along the soil surface in the agroecosystems (Panigrahi et al. 2019) causing plastic contamination to spread throughout a variety of habitats, such as subsurface soil, groundwater, aquatic environments, etc. (Briassoulis et al. 2015). Soil amendments, the use of sewage sludge and wastewater to irrigate agricultural soils, surface runoff, landfills for waste, atmospheric deposition of fibres and fragments, mulching, nets, irrigation pipes, packaging bags, seed and fertilizer containers and contemporary agricultural practise are the main sources of plastics in agroecosystems (Steinmetz et al. 2016; Li et al. 2018a, b; Liu et al. 2019a, b; Zhang et al. 2020b). In the environment, microplastic particles come in a variety of shapes, sizes and forms, including fibres, beads, pellets, granules, foam, microfibrils and films (Bläsing and Amelung 2018). In recent years, the use of plastic mulch in agriculture has increased at a rate of 5–10% annually, and it is thought to be one of the major sources of microplastics in agricultural soil (He et al. 2018). Plastic particles from mulch waste and animal consumption (including sheep and, unexpectedly, people) could contaminate the soil. These would consume food that contained plastic and eggest MP particles, which would cause plastic waste to spread throughout the ecosystem (Beriot et al. 2021; Zhang et al. 2019; Zhang et al. 2021a, b, c). The ecological environment is under a significant deal of pressure, particularly for agricultural ecosystems, as a result of microplastic pollution in plant communities (de Souza Machado et al. 2018). Microplastics in the agroecosystem not only disrupt plant nutrient cycles but also decrease soil microbial biomass, activity and functional diversity (Iqbal et al. 2023). Microplastic buildup in crops could potentially be harmful to people's health. In agricultural fields with high microplastic pressure, it is more likely that microplastics will have a deleterious impact on plant communities.

3.1 Microplastics Flexibility of Agroecosystems

It is likely that lands serve as a permanent reservoir for microplastic waste (de Souza Machado et al. 2018). The basic reasons, such as particle size and form (Lebreton and Andrady 2019), as well as the mechanisms, such as soil matrix redistribution and transfer to groundwater, runoff, water and wind erosion (Nizzetto et al. 2016), are poorly understood. The size, shape and pore size distribution of the profile as well as the microplastics themselves control and limit the leaching of microplastics in the soil profile. Microplastic particles may be transported faster and deeper into the soil layers through preferential diffusion through macropores like fissures or bio-pores (Yu et al. 2019). Plastic particles are easily swept up by the wind and taken into the atmosphere, especially in areas subject to wind erosion, which has an effect on the air quality. Wind-borne microplastic migration may result in exposure of people and the environment far from the source (Lwanga et al. 2022). Further research is necessary to fully understand the mobility of microplastics from farmlands to land-based avian nests, because microplastics in agricultural soils can be transferred by either vertebrates (terrestrial) or invertebrates. Other classes, such as reptiles and amphibians, should also be the subject of further study. Figure 5.2 illustrates microplastics flow in agroecosystem.

4 Microplastics and Plants

Soil is the primary pathway for microplastics to penetrate plant communities in terrestrial habitats. In addition to soil, the introduction of surface runoff and agricultural irrigation water, the residual decomposition of agricultural mulching film and plastic waste, the use of civil sludge on land and the sediment of microplastics from the atmosphere are the birthplace of microplastics establishment in plant communities. The airborne microplastics among them have the potential to directly damage plant roots and aboveground parts. Many studies have been done on the mechanisms that allow the uptake and aggregation of microplastics in specific plants. Fluorescent labels were added to microplastics in leafy greens, wheat and other plant species,

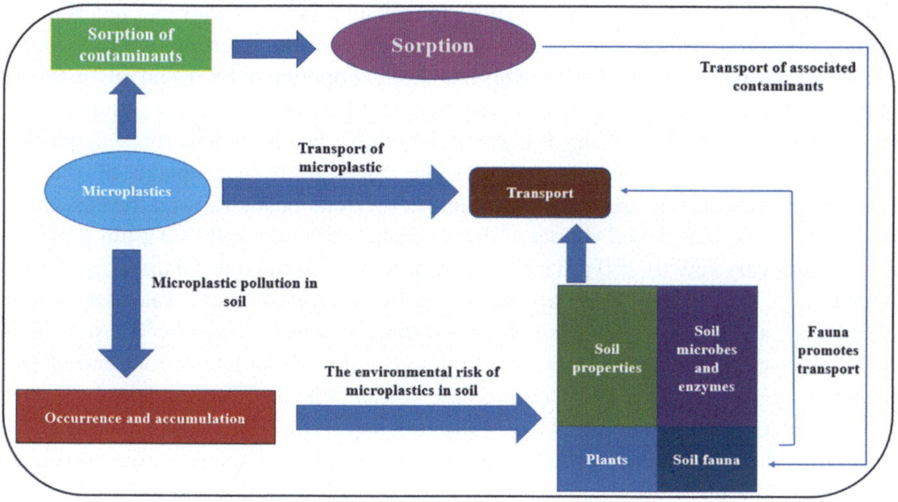

Fig. 5.2 Microplastics in agroecosystem

like *Arabidopsis thaliana* (Jiang et al. 2019b; Sun et al. 2020a, b). It was discovered that on a micron scale, these components were caught and adsorbed onto root surface due to the mucus secreted by the root cap, and they could not penetrate the root cortex or reach the stele through (Li et al. 2022). It is observed that microplastics can alter the physicochemical characteristics of the soil, such as its pH, proportion of organic matter, water-stable structures and other variables, leaving the soil less suitable for plant growth and affecting the vertical organization of plant communities (Boots et al. 2019; Fu et al. 2019). The terrestrial plants have unique root, stem, xylem, plasma membrane and vacuole architecture that make it easier for plants to soak up microplastics from the soil. Only nanoplastics can enter a plant's cell membrane; microplastics, due to their tiny size and high molecular structure, are unable to do so (He et al. 2018). However, microplastics and nanoplastics have an impact on plants by altering the microbial community in the rhizosphere and by affecting the physiological process of plants (de Souza Machado et al. 2019).

4.1 Uptake and Transport of Microplastics in Plants

It is frequently believed that plants won't be able to uptake microplastics because of their large size and high molecular mass, which precludes them from entering the cellulose-rich cell wall of plants. However, when they are reduced to their nanoforms, several reports have indicated that they can be absorbed by plants (Ebere et al. 2019; Kumari et al. 2022). Consequently, nanoplastics can infiltrate plant cells (Li et al. 2020b). According to an experiment on *Nicotiana tabacum*, it was demonstrated that nanoplastics can directly enter into the cell walls of plants, where it was observed that tobacco plants did not assimilate nano-polystyrene beads having100 nm size, but beads having 20–40 nm were readily taken up by the tobacco plants (Bandmann et al. 2012). Another study conducted by Li et al. (2020a) depicted the uptake of polystyrene beads having different sizes. They observed that in a hoagland solution having 0.2 μm polystyrene beads, wheat roots displayed distinct concentration-dependent fluorescence indicating the uptake of plastic. Lettuce and wheat roots exposed to 2.0 μm polystyrene beads showed fluorescence, as did the roots, leaves and shoots of these plants when exposed to 0.2 μm polystyrene beads. Very few for 5.0 μm and almost none for 7.0 μm and 10 μm polystyrene microbeads, luminescence signals were seen in the epidermis and vascular system of lettuce and wheat roots. Further, polystyrene beads having size 0.2 μm were found throughout the apical meristem and across the whole lateral root cap of lettuce and wheat roots after connecting to the border cells indicating that polystyrene beads were trapped in the mucilage of the root cap, promoting their penetration of the cell wall and allowing diffusion into the highly porous apical meristem tissue. Negatively charged mucilage and exudates, which serve as the first line of defence in plants, have been found to prevent positively charged metal nanoplastics from adhering to the outside of the cell wall. Li et al. (2020a) further observed that, unlike the 0.2 μm polystyrene beads, the 2.0 μm polystyrene beads only entered the stele once a secondary root that had just broken through the main root's epidermis was present, which is in accordance with the finding that 2.0 μm microbeads were absent from the whole wheat roots without secondaries. According to scanning electron microscopy photographs of the roots of wheat and lettuce, the 0.2 μm polystyrene beads largely accumulated in the xylem and on the cell walls of the cortical tissue. This suggests that the beads travelled through the intercellular space using the apoplastic transport mechanism, whereas micrometre-sized beads were observed in the spaces between the epidermal cells of the primary and secondary wheat roots. Further, Giorgetti et al. (2020) allowed the *Allium cepa* seeds to germinate for 72 h in the presence of polystyrene beads (50 nm size at concentrations of 0.01, 0.1, and 1 g/L) in order to study the mechanisms of action of nanoplastics on plants. Transmission electron microscopy showed more electron dense structures in the cytoplasm of nano-polystyrene-treated cells in 1 g/L than in 0.1 g/L, suggesting that the lipid bodies had not yet mobilized. Numerous vacuoles, tiny mitochondria, dictyosomes and rough endoplasmic reticulum were found in these bodies. Both the cytoplasm and vacuoles contained the internalized nano-polystyrene. They frequently took the form of aggregates of two to five particles and were cylindrical or parallelepiped in shape. Their diameters ranged from 25 nm to 130 nm. Rarely, the nucleus included nano-polystyrene that were about 25 nm in size (Giorgetti et al. 2020; Matthews et al. 2021). The presence of nano-PS in the nucleus suggests that a smaller nano-subfraction of this material may also pass through the nuclear membrane and disrupt the structure and function of chromatin (Giorgetti et al. 2020). However, no study has yet shown how microplastics are absorbed and transported throughout the entire plant. Understanding the microplastic and nanoplastic transport and absorption mechanisms in plants will require in-depth research.

Aggregates of microplastics and nanoplastics can be absorbed or adsorbed by plants (Azeem et al. 2021; Xu et al. 2022). The absorption and transport of plastic particles by plants are significantly influenced by transpiration pull. Plastic particles penetrate the epidermal tissue of primary and secondary roots of wheat and are facilitated through the pericycle and then transported into the xylem (Li et al. 2020a). These particles might migrate to the plant's aerial portion inside the central cylinder via the xylem. Further, the vascular system moved plastic particles from the root to the shoot via the transpiration stream. Confocal pictures demonstrated that the vascular system of the stem was the primary location of plastic luminescence signals. Microplastics and nanoplastics can enter the vasculature responsible for transporting water by passing through tiny extracellular channels. The stem, leaves and perhaps fruits can swiftly absorb the water transport system that supports nanoplastics. Lian et al. (2020a, b) recent research on wheat crops indicated plastic nanoplastic absorption and translocation. Confocal microscopy with 3D laser scanning made it obvious that polystyrene nanoplastics are taken up by the root tips, whereas the fluorescent signal of polystyrene nanoplastics in leaves was blocked by autofluorescence of plant tissues. Similar to that, scanning electron microscopy also found polystyrene nanoplastics absorbed in the root and shoot xylem. The xylem channel may be used to convey polystyrene nanoplastics (100 nm) from root to shoot, which may then enter the grain. Through epifluorescence and confocal microscopy, Bosker et al. (2019) discovered clusters of plastic particles in the leaf and epidermis of *Lepidium sativum* after 48 and 72 h of microplastic exposure. The scanning electron microscopy detected the polystyrene beads (0.2 μm) in the intercellular space of the lettuce leaf veins as well as on the outside of the cell walls in the stem xylem of wheat. The cucumber plant's stem and leaves can accumulate polystyrene nanoplastics and microplastics (Li et al. 2021). 500 and 700 nm microplastic particles were found in the first fruit interspace tissue of a calyx cross-section. Based on the findings, they also proposed that polystyrene nanoplastics could build up in cucumber plant blooms and fruits. Through the stomata via foliar spray is another potential route for plastic infiltration into the plant leaf. According to Sun et al. (2021), stomatal uptake is one potential route through which nanoplastics enter the leaves and then go to the vasculature. The maize plant showed a high frequency of micro-fluorescent polystyrene nanoplastics in the stems and a low frequency in the roots. Fluorescence microscope imaging analysis also showed a high frequency of polystyrene nanoplastics in and around the vascular system of the stem. These findings demonstrate nanoplastic transport through the vascular bundle from the leaves to the stems and then from the stems to the roots. The polystyrene nanoplastics bind to the plant's stomata, pierce through the phloem and reach the lettuce plant's roots (Lian et al. 2021). Figure 5.3 illustrates some pathways of microplastics uptake by plants. The processes of microplastic and nanoplastic absorption and transportation in plant shoots need to be further studied.

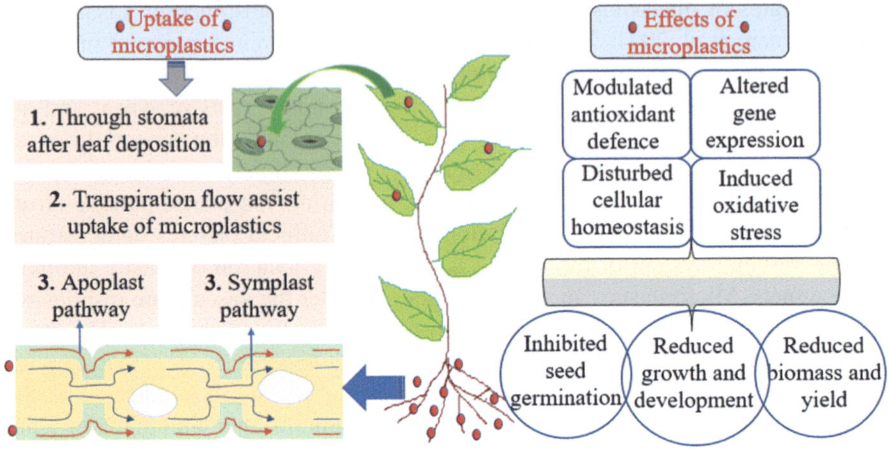

Fig. 5.3 Uptake and effects of microplastics in plants

4.2 Effects of Microplastic on Plants

Microplastics and their effect on plants are recent topic of concern (Table 5.1), because plants play a vital role in our ecosystem by providing oxygen, fixing carbon, conserving water, protecting soil and mainly being the source of food and fodder for other organisms. Excess use of plastic in our day-to-day life had adversely affect the whole agroecosystem by contaminating it in its minute form such as microplastics. Microplastics are taken up by the plants and are translocated to other parts as shown in a study on rice plant where treatment of fluorescently labelled polystyrene microspheres was given to experimental plant for 14–40 days. The results showed that there was accumulation of microplastic in vascular system, intercellular regions and in cell walls. The traces of microplastics were also observed in roots, stems and leaves of the plants. These microplastics got accumulated with time and pose certain toxic effects on various physiological and biochemical processes of plants (Azeem et al. 2021; Li et al. 2022; Liu et al. 2022a, b, c, d). The accumulation of microplastics in plant shows various effects like they can block their cell wall pores, cell junctions and pores of seed. Microplastic with uneven sharp edges and rough surface can damage the roots. It indirectly causes reactive oxygen species production, exert cytotoxic effect, reduce antioxidant enzyme, etc. (Li et al. 2022). Microplastics can affect plant directly or indirectly as they consist of toxic chemicals, metals, additives, etc., or they act as carriers of other toxic substances from the environment. Moreover, they can also change the microenvironmental properties of the soil around the plant by altering the bioavailability and bioaccumulation of the toxic substances in vicinity of plants and alters the soil properties of soil also. As degradation of plastic is difficult, it remains in soil for long and is usually taken up by plants through roots as well as though foliar surface as they have great adsorption capacity (Wang et al. 2022). Moreover, the aging and the

Table 5.1 Effects of different microplastics on plants

Microplastic	Concentration	Plant species	Reported effect	Study
Polystyrene, polymethyl methacrylate	0.01, 0.1 and 1 g L^{-1}	*Allium cepa*	Induced production of reactive oxygen species and elevated catalase and peroxidase activities	Biba et al. (2023a, b)
Poly (styrene-co-methyl methacrylate)	50, 100 mg/L	*Lemna minuta*	Induced phytotoxic effect in roots	Ceschin et al. (2023)
Polystyrene	10, 50 and 100 mg/L	*Cicer* sp.	Inhibited growth of seedlings and increased oxidative stress	Dey et al. (2023)
Polystyrene beads (small and large)	50 mg/L	*Zea mays*	Induced accumulation of microplastic in crop plant as toxin contaminant and altered the absorption properties of different parts of roots	Li et al. (2023)
Polyethylene	0.25, 0.50, 0.75 and 1.0%	*Vigna mungo* and *Solanum lycopersicum*	Reduced seedling germination	Sahasa et al. (2023)
Polyethylene	0.5% and 5% w/w	*Zea mays*	Reduced yield	Sun et al. (2023)
Polystyrene (fluorescence)	0, 10, 20, 30, 40 and 50 mg/L	*Lactuca sativa*	Hypertonic injury, downregulated ion homeostasis and upregulated stress related genes	Wang et al. (2023a, b, c)
Polyvinyl chloride	0, 10, 100 and 1000 mg/L	*Spirodela polyrhiza*	Decreased adventitious root elongation, decreased anthocyanin content	Wang et al. (2023a, b, c)
Polystyrene, polyethylene and polyvinyl chloride	0.05 g/kg and 0.5 g/kg	*Cucumis sativus*	Reduced fresh and dry weight, decreased water use efficiency and induced mechanical damage	Zhuang et al. (2023)

(continued)

Table 5.1 (continued)

Microplastic	Concentration	Plant species	Reported effect	Study
Polyethylene	50 and 150 mg/L	*Triticum aestivum*	Altered metabolic responses and phytotoxicity	Bao et al. (2022)
Polyethylene terephthalate, polyvinyl chloride, polypropylene and polyethylene	0.02 and 0.1%	*Cucurbita pepo*	Reduced root and shoot fresh weight, chlorophyll content and leaf size	Colzi et al. (2022)
Polymethyl methacrylate	0.05 g/L	*Brassica campestris*	Decreased chlorophyll content and induced structural damage of cells	Dong et al. (2022)
Polystyrene	0, 10, 50, 100 and 500 mg/L	*Trifolium repens*, *Impatiens balsamina* and *Orychophragmus violaceus*	Decreased seed germination tendency, root hair density, catalase and superoxide dismutase activities	Gou et al. (2022)
Polyamide, polyethylene, polystyrene and polyethylene terephthalate	100 mg/L	*Pisum sativum*	Increased peroxide content, catalase and superoxide dismutase activities	Kang et al. (2023)
Polystyrene and polyvinyl chloride	0, 1,5 and 3.0 mgL^{-1}	*Oryza sativa*	Reduced photosynthetic rate, growth and biomass	Ma et al. (2022a, b)
Polystyrene	1.0 and 5.0 mg/L	*Triticum aestivum*	Affected growth and chlorophyll content	Ma et al. (2022a, b)
Polystyrene	68 microgram per litre	*Cymodocea nodosa*	Degenerated adventitious roots, decreased number of leaves and increased oxidative damage	Menicagli et al. (2022)
Polystyrene	0.05, 0.25, 1.25 and 6 mg L^{-1}	*Egeria densa*	Increased hydrogen peroxide content and antioxidant activity	Senavirathna et al. (2022)
Polystyrene, polypropylene and polyethylene	0, 10, 100, 500 and 1000 mg/L	*Lycopersicon esculentum*	Inhibited seedling germination and growth and produced oxidative stress	Shi et al. (2022)

(continued)

Table 5.1 (continued)

Microplastic	Concentration	Plant species	Reported effect	Study
Polyethylene	10, 50 and 100 mg/L	*Lens culinaris*	Reduced internal activities of a plant during seed germination, caused pore blockage and stunted growth	de Silva et al. (2022)
Polyethylene	0, 25 and 500 mg/L	*Oryza sativa*	Reduced growth and altered biochemical and physiological attributes	Tan et al. (2022)
Polyethylene	50, 250 and 500 mg/L	*Oryza sativa*	Inhibited root growth and decreased mineral uptake	Tang et al. (2022)
Low-density polyethylene	0, 10, 100 and 1000 mg/L	*Nicotiana tabacum*	Inhibited growth, increased level of malondialdehyde and superoxide dismutase and reduced biomass	Zhang et al. (2022)
Low-density polyethylene	0.135, 0.27, 0.81 and 1.35 mg·kg^{-1}	*Solanum nigrum*	Inhibited growth and decreased chlorophyll content	Zou et al. (2022)
Polycarbonate	0.03% to 6.7% (w/w)	*Lepidium sativum*	Inhibited germination, decreased chlorophyll content and reduced seedling growth	Pflugmacher et al. (2021)
Polyethylene	0, 10, 50, 100, 200 and 500 mg/L	*Glycine max* and *Vigna radiata*	Inhibited root and shoot length and decreased germination	Wang et al. (2021)
Polystyrene	0, 0.1 and 10 mg/L	*Oryza sativa*	Oxidative stress increases, decreased root length, inhibited nutrient uptake	Wu et al. (2021)
Polystyrene	0.1, 10 and 1000 mg/L	*Oryza sativa*	Induced production of reactive oxygen species and upregulated genes for flavanol and flavonoid biosynthesis	Zhang et al. (2021a, b, c)

(continued)

Table 5.1 (continued)

Microplastic	Concentration	Plant species	Reported effect	Study
Polystyrene and polytetrafluoroethylene	0.2 g/L	*Oryza sativa*	Inhibited photosynthesis, reduced biomass and induced mechanical damage	Dong et al. (2020)
Polyvinyl chloride	0.5, 1 and 2%	*Lactuca sativa*	Increased activity of superoxide dismutase and reduced light absorption capacity and electron transfer	Li et al. (Li et al. 2020a, b, c)
Fluorescent polystyrene	15, 70 and 140 mg/L	*Utricularia vulgaris*	Inhibited growth, resulted accumulation of microplastic in bladder, high oxidative damage and ecotoxicity	Yu et al. (2020)
Polyethylene and polyvinylchloride	50 mg/L	*Utricularia aurea*	Accumulation of microplastic in plant and inhibited growth	Zhou et al. (2020)
Fluoro-max Green fluorescent polymer microspheres	10^3, 10^4, 10^5, 10^6 and 10^7 particles m/L	*Lepidium sativum*	Delayed germination and inhibited root growth	Bosker et al. (2019)
Polystyrene	10, 50, 100 mg/L	*Vicia faba*	Generated oxidative stress, genotoxicity showed toxic effects on root and shoot and blocked cell connecting pores	Jiang et al. (2019a, b)
Polystyrene	0, 10, 100, 200 mg/L	*Triticum aestivum*	Inhibited growth, decreased photosynthetic pigments and decreased activities of superoxide dismutase and catalase	Liao et al. (2019)

environmental factors affect phototoxicity of plastic present in soil as pollutant as observed in an experiment in which the exposure of microplastic was given to *Lolium multiflorum* (Italian ryegrass). The observations of various growth parameters including root length, shoot length and germination were adversely affected by microplastic and the leachates (Esterhuizen et al. 2022). In another study, tomato plant was treated with sewage slurry containing microplastic which decreased biomass and production of tomato fruit (Hernández-Arenas et al. 2021). Microplastics not only affect the terrestrial plant but also affect the aquatic flora due to its accumulation in water bodies. Municipal wastewater treatment plants are the main microplastic receptors of domestic and industrial wastewater which finally opens into large water bodies and contaminate the flora present there (Yuan et al. 2022; Ge et al. 2021). The organisms living in aquatic environment at all trophic levels in both fresh and marine environment are prone to the harmful exposure of microplastics, especially the organisms of first trophic level living on the surface of water as well as in the sediments that perform photosynthesis (Ceschin et al. 2023). The effect of microplastics on aquatic plants includes effect on growth parameters, pigment content and stress indicators as observed in marine microalgae *Chlorella vulgaris* (Su et al. 2022). When microplastics interacts with other pollutants, it shows combined effect on submerged plants as observed in a study on *Myriophyllum verticillatum* where plant was treated with microplastics in combination with an antibiotic resulting in their adverse effects on plant. The observed effects included increased antioxidant enzyme activity, decreased photosynthetic pigments and reduced rate of uptake of essential elements like nitrogen and phosphorus from the provided environment (Hong et al. 2022). In aquatic ecosystem, microalgae are the main food producer. Microplastics induce genotoxicity and ecotoxicity in plants. They also induce cytotoxicity by inhibiting genes like *cytokinase 2* gene and by overproducing reactive oxygen species in plants (Ullah et al. 2021).

Microplastics have various effects on plants as discussed below:

Photosynthesis in plants: The accumulation of microplastics in plants can affect the photosynthesis process of a plant. The main reason behind this can be the overproduction of reactive oxygen species. Reactive oxygen species blocks chlorophyll synthesis and damages cell structure. In an experiment, the effect of microplastic polyethylene along with other pollutants was observed on lettuce resulting in decrease in stomatal conductance and rate of photosynthesis along with decrease in Rubisco activity. The mechanism behind hindrance of photosynthesis was the inhibition of photoelectron flow (Gao et al. 2019). In aquatic plants also, the effect of polyvinyl chloride and polypropylene was seen in *C. pyrenoidosa* which decreased chlorophyll a content (Wu et al. 2019). In another study, effect of microplastic polyacrylonitrile polymer was observed on *C. pyrenoidosa*, and the results exhibited both decreased chlorophyll a and chlorophyll b content with increasing levels of polyacrylonitrile polymer (Lin et al. 2020).

Germination of seeds: Microplastics can affect seed germination by blocking pores physically. This was observed in a study where fluorescent microscopy revealed accumulation of microplastics as a film coating on *Lepidium sativum* seeds treated with plastic particles at different concentrations (Bosker et al. 2019). Similarly, in new and aged microplastic, polycarbonate treatment was given to the seeds of *L. sativum*, and it was observed that germination rate was reduced and higher amount of toxic chemicals were secreted by more aged microplastics (Pflugmacher et al. 2021). Another study examined the effect of different microplastics on *L. sativum* which reported decrease in percent seed germination (Pignattelli et al. 2020). In blackgram, also the rate of seed germination declined when it was treated with different concentrations of microplastics in different types of soils (Sahasa et al. 2022). Similarly in *Nelumbo nucifera*, the seed germination was inhibited when microplastic treatment was given in the absence of lake sediments (Esterhuizen and Kim 2022), whereas in some cases, microplastics do not affect seed germination as observed in an experiment on wheat and rice (Lian et al. 2020a, b; Zhang et al. 2021a, b, c).

Growth of plant: Microplastics affect the growth of plants in different possible ways. It may affect microbes around plants which in turn produce secondary metabolites with harmful effects on plant growth. It can also produce volatile compounds in rhizosphere of plants that in turn may inhibit growth of respective plant by altering its hormone levels. Also, during plastic degradation, some metabolites affect plant growth (Yang et al. 2021a, b, c, d). In a study on *Amaranthus mangostanus*, the effect of microplastics like polyethylene, polystyrene and polypropylene showed inhibition of shoot growth and decrease in root length as compared to control (Wang et al. 2023a, b, c). In another study, the plant of Romain lettuce was treated for 30 days with polyethylene microplastic at different concentrations like 50, 250 and 1000 mg/kg. The results showed decrease in fresh weight of plant, root length and plant height (Lei and Engeseth 2022). An experiment on plant *Plantago lanceolata* was done to check the effect of microplastic ethylene propylene diene monomer, and the effect was observed on growth and survival of the plant under the influence of microplastic. The roots were more adversely affected than shoot as indicated by lower specific root length and root weight ratio (van Kleunen et al. 2020).

Plant biomass: Microplastics also have impact on the biomass of the affected root, shoot and bud (Zantis et al. 2023). A study on wheat had shown decrease in total biomass of the plant in the presence of microplastic (Qi et al. 2018a, b). In another study, biomass of strawberry plant decreased when treated with microplastic along with other contaminants like heavy metal (Pinto-Poblete et al. 2022). It was also observed in some plants that as compared to shoot biomass, root biomass was much more in weight. Similar observation was made in another experiment on macrophyte, and the reason behind this was the blocking of roots, resulting in decline in nutrient acquisition and improper root functioning. To tackle this issue, root grows more in length to reach out to the nutrients available in vicinity of soil (van Weert et al. 2019).

Altered gene expression: Microplastics affect the gene expression in some plants and show cytotoxic effects. The cytotoxicity induced by microplastic was observed in an experiment on *Allium cepa* when the plant was treated with different concentrations of micropolystyrene (MPS); the reason behind this was the reduced mitotic index that hinders the normal cell cycle. There was decrease in *cdc 2* expression also, which is an important cell cycle regulator (Maity et al. 2020). Microplastics were also found to upregulate the genes in rice crop, related to antioxidant activity of roots like flavanol biosynthesis and flavonoid. It was also responsible for downregulating the genes functional in nitrogen linoleic acid metabolism pathway (Zhang et al. 2021a, b, c). Microplastics can also regulate the gene expression of genes involved in transport of metal ion, as observed in microplastic treated wheat plant in which the absorption rate of micronutrients was lower as compared to control plant (Lian et al. 2019). In another study, the treatment of microplastic polystyrene on rice plant showed inhibition of a protein encoding gene in TCA cycle (Wu et al. 2022).

Mechanical damage to plants: Microplastic with rough uneven edges can damage plant roots physically and affect their functioning. In a study on duckweed, it was found that root length was decreased because of sharp edges and rough surface (Rozman et al. 2021). While studying the effect of microplastic on lettuce, the scanning electron microscopy revealed the damaged root parts and adhered particles of polystyrene to the root surface due to which root pores were blocked. Also, in the presence of microplastic, the membrane was found separated from cell wall followed by cell damage (Gao et al. 2021). In this way, microplastics hinder root growth and activity by damaging it or clogging its surface.

Bioavailability of nutrients and other pollutants: The adhered microplastic on root surface reduces the absorption rate of nutrients available in vicinity of the root (Gao et al. 2021). Microplastic also alters the microbial community structure and the available nutrients around the plant soil. Moreover, being hydrophobic, they absorb heavy metals and induce phototoxicity in plants (Kumar et al. 2022). In a study on maize plant, the cadmium accumulation was elevated when treated with microplastic (Zhao et al. 2023). In another study, microplastic enhanced metal uptake by 14.6% and 13.5% in root and shoot respectively (Huang et al. 2023). Microplastic can also accumulate heavy metals in aquatic plants as shown in a study on submerged macrophyte *P. crispus* (Ogo et al. 2022). The study on *Brassica napus* also exhibited that the presence of microplastic facilitates the transportation of toxic heavy metals leading to their bioaccumulation in plants (Jia et al. 2022). These heavy metals further affect plants like decreased growth rate, decrease in yield, altered functions of enzymes and change in gene expressions (Pinto-Poblete et al. 2022).

5 Microbial Remediation

Microplastic pollutants can be removed practically, thoroughly and affordably through microbial degradation. The microbial breakdown of micro-nanoplastics involves numerous biochemical reaction pathways. To protect our environment, we must fully understand the different microplastic degrading methods based on their chemical structures (Pham et al. 2021; Vaid et al. 2021). Enhancing the biodegradation rate of pollutants in nutrient-limited environments requires two key processes: bio-stimulation (the input of limiting nutrients to stimulate microbial growth) and bio-augmentation (the inoculation of living cells that are able to breakdown) (Ubalua 2011). Numerous factors influence the microbial breakdown of polymers, changing their physicochemical characteristics. Various biotic and abiotic environmental factors have an impact on the degradation process (Hakkarainen and Albertsson 2004; Restrepo-Flórez et al. 2014). Despite the difficulty in understanding how microplastic degrade, there are potential microbial plastic remediation processes which take place in three successive steps: biodeterioration, bio-fragmentation and mineralization (Tiwari et al. 2020).

Microorganisms use two key processes called intracellular and extracellular degradation to depolymerize large plastic particles (Lin et al. 2022; Zhang et al. 2021a, b, c). Microbes build up on the surface of microplastics due to intracellular breakdown, which hydrolyses the plastic into short chains. Microbes release extracellular enzymes such hydrolases during extracellular degradation, which breaks down complex polymers into simpler pieces that can then be metabolized by various metabolic pathways (Pathak 2017; Wilkes and Aristilde 2017). Additionally, through aerobic or anaerobic metabolism, microbes transform these short chains into end products (H_2O, CH_4 or CO_2,) in a process known as mineralization (Megonigal et al. 2004; Mohan 2011). Biological natural attenuation refers to the process by which these final products are used as sources of carbon and energy (Eskander and Saleh 2017; Tiwari et al. 2020). Various microbes including fungus, bacteria and algae have been reported to remediate microplastic contamination (Table 5.2).

5.1 Fungus

Fungi comprises a wide group of organisms that constitute saprotrophs, opportunistic or obligatory parasites (Aguilar-Trigueros et al. 2023). They are extremely adaptable and may flourish in a variety of aquatic and terrestrial ecosystems and in a wide range of climatic conditions (Madouh and Quoreshi 2023). Several studies have reported that fungi produce a wide variety of extracellular enzymes and natural biosurfactants like hydrophobics that can break down complex polymers into simple monomers, making them a source of electrons and carbons for microorganisms, facilitating the degradation and mineralization of complex pollutants (Varma et al. 2023; Anand et al. 2023; Dashora et al. 2023; Venkatesan et al. 2022). *Zalerion maritimum*, *Aspergillus niger*, *Cladosporium* and *Penicillium simplicissimum* are the main genera involved to the breakdown of several types of polymers, including polyethylene, polypropylene and polyethylene terephthalate (Oliveira et al. 2020). They utilize microplastic as their prime source of carbon after being degraded by extracellular enzyme and reduce their hydrophobicity and encourage the formation of various chemical bonds, including carboxyl, ester and carbonyl functional groups. *A. fumigatus*, *A. tubingensis*, *Cladosporium pseudocladosporioides*,

Table 5.2 Microbial remediation of microplastics

Microplastic	Microorganism used	Time for degradation	Effects	Study
Polystyrene	*Lichtheimia ramosa*	30 days	Rise in fungal biomass by consuming the carbon structure from polystyrene	Bhanot and Panwar (2023)
Low-density polyethylene and high-density polyethylene	*Cephalosporium*	56 days	Weight loss24.53 ± 0.73% and 18.22 ± 0.31%	Chaudhary et al. (2023)
Polyethylene	*Cladosporium halotolerans*	15 days	Contacted with plastic materials via its hyphae and produced enzymes necessary for the breakdown of polyethylene	Di Napoli et al. (2023)
Polystyrene	*Aspergillus nidulans*	60 days	Converted waste streams containing post-consumer polyethylene into carboxylic acid	Rabot et al. (2023)
Polystyrene	*Phanerochaete chrysosporium*	35 days	Weight loss by 19.17%	Wu et al. (2023)
Low-density polyethylene	*A. flavus, A. niger* and *brown rot and white rot fungi*	30 days	Weight loss by 22.9, 16.1, 18.4 and 22.7%	Maroof et al. (2022)
Polyurethane, polystyrene and polyethylene	*Aspergillus* and *Penicillium*	90 days	Weight loss	Oviedo-Anchundia et al. (2021)
Polystyrene	*Vibrio* and *Exiguobacterium*	48 hours	The bacteria formed the biofilms (white flocs) over the microplastics and thereby assisted in sinking of the microplastics in the seawater	Zhao et al. (2022a, b)
Polyethylene	*A. niger, A. flavus* and *A. oryzae*	55 days	Weight loss by 26.15%	Dsouza et al. (2021)
Polyethylene terephthalate	*Pseudomonas aestusnigri*	2 weeks	A new polyester hydrolase secreted by bacteria broke down the microplastics	Bollinger et al. (2020)
Polypropylene	*Aspergillus* sp. *Penicillium* sp.	30 days	Mass loss	De Oliveira et al. (2020)

(continued)

Table 5.2 (continued)

Microplastic	Microorganism used	Time for degradation	Effects	Study
Polyethylene terephthalate (terephthalic acid)	*Chlamydomonas reinhardtii*	5 days	*C. reinhardtii* produced plastic-degrading enzymes that were engineered to express polyethylene terephthalatease, and its cell lysate was combined with polyethylene terephthalate that produced terephthalic acid, which is the totally decomposed form of the polyethylene terephthalate. Dents and holes on the surface of the polyethylene terephthalate film were formed	Kim et al. (2020)
Polychlorinated biphenyls	Anaerobic bacteria	10 weeks	The composition of sorbed polychlorinated biphenyls—Microplastics combinations and consequently, the toxicity related to polychlorinated biphenyls—Polluted microplastics change as a result of microbial colonization of contaminated microplastics. Colonies dehalogenated the polychlorinated biphenyls	Rosato et al. (2020)
Polyethylene	*A. flavus*	28 days	Mass and molecular weight loss	Zhang et al. (2020a, b)
Polyhydroxybutyrate films	*Pseudomonas* sp.	21 hours	Polyhydroxybutyrate-degrading depolymerize degraded the films producing polyhydroxybutyrate monomers	Di et al. (2019)
Polyurethane	*Penicillium* sp.	2 months	Weight and mass loss	Magnin et al. (2019)
Polyethylene terephthalate (terephthalic acid)	*Phaeodactylum tricornutum*	10 days	The microbial cell factory secreted polyethylene terephthalatease that had activity against polyethylene terephthalate and almost degraded the microplastics	Moog et al. (2019)
Micro-polyethylene	*Bacillus* and *Paenibacillus* sp.	2 months	The quantity and kinds of organic matters eluted out from polyethylene microplastics were significantly lower in the early decay phase, but they rose in the later period, indicating that they undertook biological deterioration over time by bacterial consortia	Park and Kim (2019)

Polyethylene	*A. terreus*, *A. sydowii*	60 days	Weight loss and reduction in tensile strength	Sangale et al. (2019)
Polyhydroxybutyrate	*Microbacteriumparaoxydans*		Extracellular polyhydroxybutyrate-depolymerase produced by the bacteria isolated from the plastic-contaminated sites degraded the PHB efficiently under variable conditions	Sayyed et al. (2019)
Polycaprolactone and polyvinyl chloride	*A. brasiliensis*	28 days	Significant mass loss and colour alterations	Vivi et al. (2019)
Polypropylene	*Rhodococcus* sp.	40 days	The microbe from mangrove sediment reduced the weight of the polypropylene by 6.4%	Auta et al. (2018)
Polyurethane	*Cladosporium*	14 days	Visible clearance zone (or "halo") caused by the enzymatic breakdown of plastic in growing cultures	Brunner et al. (2018)
Low-density polyethylene	*B. simplex* and *Bacillus* sp.	21 days	Bacteria isolated from *Lumbricus terrestris* gut proved very efficacious in soil restitution where 60% of microplastics was degraded by the microbes producing nanoplastics	Lwanga et al. (2018)
Polyethylene	*A. oryzae*	30 days	Weight loss	Muhonja et al. (2018)
Low-density polyethylene	*B. cereus* and *Brevibacillus borstelensis*	16 weeks	Efficient biological degradation of low density polyethylene	Muhonja et al. (2018)
Low-density polyethylene	*Trichoderma viride*, *A. nomius*	45 days	Decreased weight by 5.13 and 6.63%	Munir et al. (2018)
Low-density polyethylene	Cyanobacteria (*Phormidium lucidum*) and (*Oscillatoria subbrevis*)	42 days	Low-density polyethylene biodegrades effectively in the absence of pretreatment or pro-oxidative chemicals by spreading the polyethylene surfaces	Sarmah and Rout (2018)

(continued)

Table 5.2 (continued)

Microplastic	Microorganism used	Time for degradation	Effects	Study
Low-density polyethylene, high-density polyethylene and polypropylene	*Brevibacillus* spp. and *Aneurinibacillus* sp.	4 months and 20 days	The bacterial communities utilized in waste disposal landfills and treatment plants for sewage decomposed low-density polyethylene, high-density polyethylene and polypropylene, respectively, by 58.9, 48 and 58% at 50 °C	Skariyachan et al. (2018)
High-density polyethylene	Souda and Agios consortium	60 days	Two microbial communities degraded the high-density polyethylene (pretreated with UV-radiation) efficiently. Souda consortia decreased the microplastics weight by 18% and Agios by 8%	Tsiota et al. (2018)
Polyethylene, polyethylene terephthalate, polypropylene, polystyrene	*B. gottheilii* and *Bacillus cereus*	40 days	Following 40 days, *B. cereus* indicated that the weight reduction percentage of the microplastics were 1.6, 6.6 and 7.4% for polyethylene, polyethylene terephthalate and polystyrene, respectively. However, for *B. gottheilii*, the percentage change was 6.2, 3.0, 3.6 and 5.8%, respectively	Auta et al. (2017)
Polyester polyurethane	*Monascus sanguineus*, *Monascus ruber* and *Monascus* sp.	14 days	*Monascus ruber* produced maximum esterase concentration followed by *M. sanguineus* to evaluate polyurethane biodegradation	El-Morsy et al. (2017)
Low-density polyethylene	Cyanobacterium, *Anabaena spiroides*, *Navicula pupula* and *Scenedesmus dimorphus*	45 days	*Anabaena spiroides*, a blue-green cyanobacterium alga, demonstrated the greatest rate of low-density polyethylene hydrolysis (8.18%), accompanied by the diatom *Navicula pupula* (4.44%) and the green microalga *Scenedesmus dimorphus* (3.74%)	Kumar et al. (2017)
Polyethylene sheet	*Penicillium oxalicum* and *Penicillium chrysogenum*	90 days	Reduced weight and mass of pellet	Ojha et al. (2017)
Polyethylene pellets	*Zalerion maritimum*	28 days	Decrease in both mass and size	Paço et al. (2017)

Polyethylene films	*Lysinibacillus* and *Salinibacterium* genera	6 months	The microplastics after bacterial therapy had a wider molecular mass dispersion as well as reduced average molar mass, showing biodegradation	Syranidou et al. (2017)
Polyvinyl alcohol	*Pseudomonas* sp. and *P. vesicularis*	45 days	Polyvinyl alcohol oxidase and esterase enzyme produced by the consortia helped in the degradation of the polyvinyl alcohol with greater efficiency	Wilkes and Aristilde (2017)
High-impact polystyrene	*Pseudomonas* and *Bacillus* spp.	30 days	The ability of *Pseudomonas* and *Bacillus* strains to break down brominated high-impact polystyrene was observed. *Bacillus* spp. (degradation of 23% in 30 days) with high-impact polystyrene emulsion exhibited 94% decline in turbidity in 4 days of incubation, while *Pseudomonas* spp. had 94% reduction in turbidity	Mohan et al. (2016)
Polyethylene	*Staphylococcus* sp., *Pseudomonas* sp. and *Bacillus* sp.	40 days	*Bacillus* sp. (P1C) demonstrated highest degradation by 42.5% weight loss in 40 micron polythene, followed by *Staphylococcus* sp. (P1A) 20%, *Pseudomonas* sp. (P1B) 7.5% and consortium of 15 bacteria isolates (PID) by 5%	Singh et al. (2016)
Polyethylene	*B. subtilis*	30 days	UV radiation exposure of polymer films makes them more accessible to microbes as food, permitting a relatively fast rate of degradability. *B. subtilis* and its biosurfactant (when given to pretreated polyethylene films) showed to be the most efficacious, causing a weight reduction % of 9.26	Vimala and Mathew (2016)
Polyethylene terephthalate	*Ideonella sakaiensis*	6 weeks	Employing two enzymes (polyethylene terephthalatease and MHETase) to lyse polyethylene terephthalate and a major reaction intermediate, bacteria isolated outside of the bottle-recycling facility decomposed plastic, generating basic components for growth	Yoshida et al. (2016)

(continued)

Table 5.2 (continued)

Microplastic	Microorganism used	Time for degradation	Effects	Study
Polyethylene bag)	*Desulfotomaculum nigrificans* and *P. alcaligenes*	30 days	Polyethylene was destroyed by soil bacteria found in a field of soil that had been polluted with plastic. *Desulfotomaculum nigrificans* (16.2%) was shown to be less potent in degrading polythene bags than *P. alcaligenes* (20.1%)	Begum et al. (2015)
Polystyrene	*Exiguobacterium* sp. and *Rhodococcus ruber*	28 days	Polystyrene-degrading gut bacteria isolated from mealworm-degraded polystyrene in the gut itself	Yang et al. (2015a, b)
Polyethylene	*Enterobacter asburiae* and *Bacillus* sp.	28 days	Polyethylene-degrading bacteria species in the intestine of waxworms that consume plastic offered encouraging proof that polyethylene is being biodegraded in the environment	Yang et al. (2014)
Low-molecular-weight ethylene	*Chelatococcus* sp.		Compost-derived thermophile bacterium contributed to the low-molecular-weight ethylene degradation. Alkene-specific FTIR peaks showed greater intensity, demonstrating that dehydrogenations took place concurrently with microbially driven oxidation	Jeon and Kim (2013)
Low-density polyethylene	*Bacillus* spp., *Pseudomonas* spp., *Streptococcus* spp., *Staphylococcus* spp. and *Micrococcus* spp.	3 weeks	When compared to other isolated bacteria, *B. amylolyticus* was more beneficial for low-density polyethylene degradation, while *B. subtilis* had lower capacity to digest plastic	Prabhat et al. (2013)
Low-density polyethylene	*Microbacteriumparaoxydans* and *P. Aeruginosa*	60 days	*M. paraoxydans* decomposed 61.0% of low-density polyethylene while *P. aeruginosa* degraded 50.5%, according to FTIR-ATR spectroscopy measurements of bioremediation employing different amounts of low-density polyethylene	Rajandas et al. (2012)
Low-density polyethylene	*P. syringae*, *P. syringae* and *P. putida*	120 days	Different *Pseudomonas* sp. decomposed low-density polyethylene with great efficacy	Sakharkhar et al. (2012)

Polyhydroxybutyrate hydroxyvalerate	*Acidovorax temperans*	4 days	Bacteria isolated from sludge degraded the strain-producing eco-friendly monomers via enzymatic activity	Wang et al. (2012)
Polyhydroxyalkanoates	*Bacillus* spp., *E. cloacae* and *Gracilibacillus* spp.	160 days	Decreased molecular mass of polyhydroxyalkanoates	Volova et al. (2011)
Dibutyl phthalate	*Enterobacter* sp. T5	3 days	Bacteria flourished in the presence of dimethyl phthalate, diethyl phthalate or dioctyl phthalate solution as its only carbon and energy source, indicating that it is resistant to the harmful effects of phthalic acid diester suggesting that the microbe would be a possibility for PAE-containing waste remediation	Fang et al. (2010)
Bisphenol-A	*Chlorella vulgaris*	7 days	After approximately 7 hours, in the absence of estrogenic actions, Bisphenol-A was quickly broken down by algae and its amounts were under limits of detection	Gulnaz and Dincer (2009)
High-density polyethylene and low-density polyethylene	*B. cereus*, *B. pumilis* and *Arthrobacter*	14 days	At 400 °C, high-density polyethylene-treated consortia (22.41% weight loss) disintegrated more quickly than low-density polyethylene (21.70% weight loss)	Satlewal et al. (2008)
Nylon 66 and nylon 6	*B. cereus* and *B. sphaericus*	3 months	Nylon 66 and 6 was degraded by the bacteria producing carbon sources as by-products. The average molecular weights of nylon 66 and nylon 6 fell by 42 and 31%, respectively, and the weight fell by 7 and 2%	Sudhakar et al. (2007)
Bisphenol-A	*C. fusca* var. *vacuolata*	7 days	The estrogenic activity disappeared and deterioration of Bisphenol-A occurred in the presence of *C. fusca* in landfill leachates	Hirooka et al. (2005)
Polyethylene	*Chlorella* sp.	–	In comparison to the composites built from unaltered PE, the mechanical characteristics and thermoplasticity of the layer formed of *chlorella* sp. and altered polyethylene were higher and also more appropriate	Otuski et al. (2004)

Fusarium solani, *Penicillium chrysogenum* and isolates of *Pestalotiopsis microspora* were among the fungi that showed similar degradation of polyurethane (Ali et al. 2021; Anand et al. 2023). It has been observed that serine hydrolase often plays a crucial part in the breakdown of polyurethane. *A. tubingensis* VRKPT1 and *A. flavus* VRKPT2 were found to degrade high-density polyethylene from marine coastal and resulted 6.02 and 8.51% weight loss, respectively (Yuan et al. 2020; Anand et al. 2023). It is known that *A. niger* will degrade microplastics more quickly when they have been pretreated with chemicals like nitric acid and sodium hydroxide, such as polyethylene. The breakdown of low-density polyethylene caused by *A. niger* and *Penicillium pinophilum* needed physical pretreatment operations, such as thermo-oxidization at 80 °C for 15 days, demonstrating 0.57 and 0.37% following incubation over 30 months (Taghavi et al. 2021; Anand et al. 2023). However, in order to combat a wide range of complex molecules, some of which are harmful and polluting substances, fungi have developed a number of exceptional tactics. These methods consist of a potent enzymatic system, the capacity for adsorption and the generation of organic biosurfactants (i.e., hydrophobins), which allow them to employ against polymers (Sánchez 2020; Sharma et al. 2022).

5.2 Bacteria

Bacteria are cosmopolitan in nature where the majority can be found in soil, water and environment (Yu et al. 2022). Due to the diversity of their metabolic routes, bacteria are the microorganisms that are used in bioremediation the most frequently (Coelho et al. 2015). The capacity of many bacteria to break down environmental toxins is one of their primary traits (Kaur et al. 2021). The use of pure bacterial cultures, bacterial consortiums and bacterial biofilms under laboratory settings has been the main focus of numerous investigations on the bacteria-mediated breakdown of microplastics (Mishra et al. 2022; Anand et al. 2023). Through enriched culturing, these colonies were separated from a range of environments including sediments, sludge, sewerage, soil and the aquatic ecosystems (Liu et al. 2022a, b, c, d; Wei et al. 2022).

Muhonja and his colleagues (2018) exposed pure bacterial cultures taken from a garbage dump to 30- and 40-m sheets of low-density polyethylene. Microbes from the genera of *Bacillus*, *Brevibacillus*, *Cellulosimicrobium*, *Lysinibacillus*, *Ochrobactrum* and *Pseudomonas* were used in the study. After 16 weeks of growth, *B. cereus* and *Brevibacillus borstelensis* demonstrated the greatest biodegradations of 35.7 and 20.4%, accordingly. By using FTIR spectroscopy, bioremediation was verified. The study showed that aldehyde, ether, ketone and carboxyl functional groups were present. *P. aeruginosa* and *Microbacterium paraoxydans* were found to have even superior low-density polyethylene biodegradation. The achieved weight losses were 50.5 and 61.0%, respectively, after just 2 months of incubation. But they had prior accustomed the bacteria to low-density polyethylene powder for 2 months, and this appears to have major impact on the degradability (Rajandas et al. 2012).

In a similar manner, mesophilic diverse bacterial culture strains, primarily *Bacillus* sp. and *Paenibacillus* sp., were employed to expedite the biodegradation of micro-sized polyethylene from 40 to 600 μm. Following 60 days, the findings indicated that the average particle diameter had decreased to 22.8%, and the dry weight of the particles had fallen to 14.7% according to the field-emission SEM analysis (Park and Kim 2019). In a related work, Auta et al. (2018) reported that *Rhodococcus* sp. strain 36 and *Bacillus* sp. strain 27obtained from mangrove ecosystem can be used to disintegrate polyethylene (Auta et al. 2018). Numerous studies of bacterial consortia of *Enterobacter*, *Pseudomonas*, *Rhodococcus* sp., *Bacillus* sp. degrading polypropylene and polyvinyl chloride have been reported (Skariyachan et al. 2021; Auta et al. 2018). The synergistic biodegradation of PP by *B. flexus* with *P. azotoformans* and *B. flexus* with *B. subtilis* was examined. After 12 months of prolonged exposure, they observed a small weight reduction of the both tested plastic which rose exponentially in situations when polyethylene was prepared thermally or by UV irradiation (Aravinthan et al. 2016). Similarly, the polyvinylchloride was mineralized by *Micrococcus* sp. consortium to 8.87% (Patil and Bagde 2012). *B. flexus* and *P. citronellolis* are also reported for the same (Giacomucci et al. 2019). To the extent possible, these studies don't offer precise details regarding the enzymes involved in the biodegradation process. Although pretreated plastics are the focus of majority of the papers, there is some data given on the bioremediation of unprocessed plastics too (Othman et al. 2021; Miri et al. 2022).

Application of microbes as bioplastics is the modern tactic to combat the use of plastics (Mal et al. 2022). Among all, most common is polyhydroxyalkanoate (Gill 2014). The bacterial consortium can deliver maximum mineralization in around 15 days using the plastic sheet as its only carbon source. Thus, for the biodegradation of polymers, using biofilm-forming bacteria is also a promising option (Puri et al. 2022). A growing body of research suggests that microplastics' fate is affected by biofilms by altering and deteriorating their physical characteristics (Rummel et al. 2017). In a study, by oxidizing the polymer backbone, the laccase enzyme secreted by *Rhodococcus ruber* resulted polyethylene to depolymerize to form a biofilm on the polyethylene surface after 8 weeks of contact and destroyed 7.5% of the polymer (Santo et al. 2013).

5.3 Algae

With the benefits of a quick growing cycle, efficient photosynthesis, eco-friendliness, convenience of usage and biomass yield, algae have created prospects for environmental purification (Song et al. 2020; Yong et al. 2021). Algae are non-toxic, manageable organisms. When examining the impact of microplastics, algae are frequently utilized as test organisms. Algae are also employed to biodegrade microplastics (Shahi Khalaf Ansar et al. 2022; Barone et al. 2020). These microplastics can be found in sediment as well as beneath the surface water in a water body (Fan et al. 2022). By secreting extracellular polymeric molecules, they can colonize the surface plastics, which may ensure efficient dissolution. Plastics and other polymers can encourage the formation of extracellular polymeric substances (Shiu et al. 2020). Algal bioremediation is the best method for eliminating microplastics, since it can break down complicated polymers and does not contain endotoxins (Yang et al. 2021a; b, c, d).

The use of phycoremediation, a strategy to remove or reduce harmful chemicals using algae requires an array of criteria, including the kind of contaminant and algal species, aging of plastics and exposure time of material in the surroundings (Fu et al. 2019; Ugya et al. 2021; Manzi et al. 2022). In a study, *Anabaena spiroides*, a blue-green alga (Cyanobacterium), demonstrated the highest rate of low-density polyethylene degradation via surface degradation (8.18%), followed by diatom, *Navicula pupula* (4.44%) and green alga *Scenedesmus dimorphus* (3.74%). They were found to be most frequently on polyethylene bags that were taken from the metropolitan watercourses, and their impacts on polyethylene degradation were investigated where *A. spiroides* demonstrated maximum polyethylene breakdown (Kumar et al. 2017). Similarly, *Phormidium lucidum* and *Oscillatoria subbrevis* have shown a great potential for degrading microplastics, because they can utilize low-density polyethylene as a source of energy and carbon (Sarmah and Rout 2018). Polyethylene and polypropylene can also be broken down by *Spirulina* sp. (blue-green algae). New functional groups, such as the hydroxyl, carboxylic acid and carbonyl groups, are formed on the surface of the particles as a result of this microbial degradation (Hadiyanto et al. 2021). In addition, it is possible to genetically modify microalgae to create a microbial cell factory that can produce and secrete enzymes that degrade plastic. For instance, when the green microalga *C. reinhardtii* was modified to express polyethylene terephthalatease and its cell lysate was combined with polyethylene terephthalate, the outcome was terephthalic acid (which is the totally decomposed form of the polyethylene terephthalate), along dents and holes on the top of the polyethylene terephthalate film (Kim et al. 2020).

Priority consideration of cyanobacteria capacity to remove microplastics or nanoplastics from nature has not yet been made (Debroy et al. 2022). Also, as microplastics degrade, they release a number of hazardous chemicals, including plasticizers, polychlorinated biphenyls, dichlorodiphenyltrichloroethane and heavy metals (*Skeletonema costatum*) (Campanale et al. 2020). The microalgal cell membrane may be attacked by the secreted additives, resulting in cell injury and growth inhibition (Capolupo et al. 2020; Zhu et al. 2020). It is therefore, necessary to conduct more research on algal bioremediation. It's noteworthy that microalgae can degrade microplastics by producing toxins or enzymes and using the carbon-rich plastic polymers as food (Hu et al. 2022). Yet, development into algae-based deterioration is still naive and is not likely to see mainstream commercialization. Technological advancement and ongoing bioplastics research and development are still required (Chia et al. 2020; Nagi et al. 2021).

6 Conclusion and Future Perspectives

Despite the fact that study on microplastics began many years ago and that scientific understanding has advanced greatly in recent years, the research of so many aspects of the microplastic life cycle continue to be a serious issue. We presented some fresh implications of microplastic interactions with plants in this chapter. These interactions, in particular the uptake and storage of microplastics by vascular plants, can have an assortment of ecological implications on terrestrial ecosystems. It can, first and primarily, lead to the transfer of microplastics at various trophic levels. As a result, the presence of microplastics raises two important issues: threats to food security and permanence in various environmental matrices. Scientists have not yet tackled a pollution issue that is as complex and ambiguous as the studies conducted on microplastics. Therefore, the only way to successfully and efficiently address such a complex subject is through interdisciplinary collaboration among scientists. Additionally, while tackling microplastics pollution, a global perspective must be taken into account, and it must not be limited to the ocean and its impact on marine creatures. It must consider every imaginable environment, including the species that exists inside them and that might interact with microplastics. Also, it is vitally necessary to conduct studies on the screening of efficient microbial strains and other better removal techniques in order to minimize the environmental dangers caused by microplastics to a manageable level.

References

Abraham J, Ghosh E, Mukherjee P, Gajendiran A (2017) Microbial degradation of low density polyethylene. Environ Prog Sustain Energy 36(1):147–154

Aguilar-Trigueros CA, Krah FS, Cornwell WK, Zanne AE, Abrego N, Anderson IC, Andrew CJ, Baldrian P, Bässler C, Bissett A, Chaudhary VB (2023) Symbiotic status alters fungal eco-evolutionary offspring trajectories. Ecol Lett 26(9):1523–1534

Alabi OA, Ologbonjaye KI, Awosolu O, Alalade OE (2019) Public and environmental health effects of plastic wastes disposal: a review. J Toxicol Risk Assess 5(021):1–13

Ali SS, Elsamahy T, Al-Tohamy R, Zhu D, Mahmoud YAG, Koutra E, Metwally MA, Kornaros M, Sun J (2021) Plastic wastes biodegradation: mechanisms, challenges and future prospects. Sci Total Environ 780:146590

Allen S, Allen D, Phoenix VR, Le Roux G, Durántez Jiménez P, Simonneau A, Binet S, Galop D (2019) Atmospheric transport and deposition of microplastics in a remote mountain catchment. Nat Geosci 12(5):339–344

Allouzi MMA, Tang DYY, Chew KW, Rinklebe J, Bolan N, Allouzi SMA, Show PL (2021) Micro (nano) plastic pollution: the ecological influence on soil-plant system and human health. Sci Total Environ 788:147815

Anand U, Dey S, Bontempi E, Ducoli S, Vethaak AD, Dey A, Federici S (2023) Biotechnological methods to remove microplastics: a review. Environ Chem Lett 21:1–24

Aravinthan A, Arkatkar A, Juwarkar AA, Doble M (2016) Synergistic growth of *bacillus* and *pseudomonas* and its degradation potential on pretreated polypropylene. Prep Biochem Biotechnol 46(2):109–115

Auta HS, Emenike CU, Fauziah SH (2017) Screening of bacillus strains isolated from mangrove ecosystems in peninsular Malaysia for microplastic degradation. Environ Pollut 231:1552–1559

Auta HS, Emenike CU, Jayanthi B, Fauziah SH (2018) Growth kinetics and biodeterioration of polypropylene microplastics by *bacillus* sp. and *Rhodococcus* sp. isolated from mangrove sediment. Mar Pollut Bull 127:15–21

Azeem I, Adeel M, Ahmad MA, Shakoor N, Jiangcuo GD, Azeem K, Ishfaq M, Shakoor A, Ayaz M, Xu M, Rui Y (2021) Uptake and accumulation of nano/microplastics in plants: a critical review. Nano 11(11):2935

Bandmann V, Müller JD, Köhler T, Homann U (2012) Uptake of fluorescent nano beads into BY2-cells involves clathrin-dependent and clathrin-independent endocytosis. FEBS Lett 586(20):3626–3632

Bao Y, Pan C, Li D, Guo A, Dai F (2022) Stress response to oxytetracycline and microplastic-polyethylene in wheat (*Triticum aestivum* L.) during seed germination and seedling growth stages. Sci Total Environ 806:150553

Barnes DK, Galgani F, Thompson RC, Barlaz M (2009) Accumulation and fragmentation of plastic debris in global environments. Philos Trans R Soc B Biol Sci 364(1526):1985–1998

Barone GD, Ferizović D, Biundo A, Lindblad P (2020) Hints at the applicability of microalgae and cyanobacteria for the biodegradation of plastics. Sustain For 12(24):10449

Begum MA, Varalakshmi B, Umamagheswari K (2015) Biodegradation of polythene bag using bacteria isolated from soil. Int J Curr Microbiol App Sci 4(11):674–680

Beriot N, Peek J, Zornoza R, Geissen V, Lwanga EH (2021) Low density-microplastics detected in sheep faeces and soil: a case study from the intensive vegetable farming in Southeast Spain. Sci Total Environ 755:142653

Bernard M (2014) Industrial potential of polyhydroxyalkanoate bioplastic: a brief review. USURJ 1(1)

Bhanot V, Panwar J (2023) Unveiling the potential of Lichtheimiaramosa AJP11 for myco-transformation of polystyrene sulfonate and its driving molecular mechanism. J Environ Manag 325:116579

Biba R, Cvjetko P, Jakopčić M, Komazec B, Tkalec M, Dimitrov N et al (2023a) Phytotoxic effects of polystyrene and Polymethyl methacrylate microplastics on Allium cepa roots. Plan Theory 12(4):747

Biba R, Cvjetko P, Jakopčić M, Komazec B, Tkalec M, Dimitrov N, Begovic T, Balen B (2023b) Phytotoxic effects of polystyrene and polymethyl methacrylate microplastics on *Allium cepa* roots. Plan Theory 12(4):747

Blackburn K, Green D (2022) The potential effects of microplastics on human health: what is known and what is unknown. Ambio 51(3):518–530

Bläsing M, Amelung W (2018) Plastics in soil: analytical methods and possible sources. Sci Total Environ 612:422–435

Bollinger A, Thies S, Knieps-Grünhagen E, Gertzen C, Kobus S, Höppner A, Ferrer M, Gohlke H, Smits SH, Jaeger KE (2020) A novel polyester hydrolase from the marine bacterium pseudomonas aestusnigri–structural and functional insights. Front Microbiol 11:114

Boots B, Russell CW, Green DS (2019) Effects of microplastics in soil ecosystems: above and below ground. Environ Sci Technol 53(19):11496–11506

Bosker T, Bouwman LJ, Brun NR, Behrens P, Vijver MG (2019) Microplastics accumulate on pores in seed capsule and delay germination and root growth of the terrestrial vascular plant Lepidium sativum. Chemosphere 226:774–781

Boucher J, Friot D (2017) Primary microplastics in the oceans: a global evaluation of sources, vol 10. IUCN, Gland

Briassoulis D, Babou E, Hiskakis M, Kyrikou I (2015) Analysis of long-term degradation behaviour of polyethylene mulching films with pro-oxidants under real cultivation and soil burial conditions. Environ Sci Pollut Res 22:2584–2598

Brodhagen M, Goldberger JR, Hayes DG, Inglis DA, Marsh TL, Miles C (2017) Policy considerations for limiting unintended residual plastic in agricultural soils. Environ Sci Policy 69:81–84

Browne MA, Crump P, Niven SJ, Teuten E, Tonkin A, Galloway T, Thompson R (2011) Accumulation of microplastic on shorelines worldwide: sources and sinks. Environ Sci Technol 45(21):9175–9179

Brunner I, Fischer M, Rüthi J, Stierli B, Frey B (2018) Ability of fungi isolated from plastic debris floating in the shoreline of a lake to degrade plastics. PLoS One 13(8):e0202047

Campanale C, Massarelli C, Savino I, Locaputo V, Uricchio VF (2020) A detailed review study on potential effects of microplastics and additives of concern on human health. Int J Environ Res Public Health 17(4):1212

Capolupo M, Sørensen L, Jayasena KDR, Booth AM, Fabbri E (2020) Chemical composition and ecotoxicity of plastic and car tire rubber leachates to aquatic organisms. Water Res 169:115270

Ceschin S, Mariani F, Di Lernia D, Venditti I, Pelella E, Iannelli MA (2023) Effects of microplastic contamination on the aquatic plant *Lemnaminuta* (least duckweed). Plan Theory 12(1):207

Chaudhary AK, Chitriv SP, Chaitanya K, Vijayakumar RP (2023) Influence of ultraviolet and chemical treatment on the biodegradation of low-density polyethylene and high-density polyethylene by *Cephalosporium* strain. Environ Monit Assess 195(3):395

Chia WY, Tang DYY, Khoo KS, Lup ANK, Chew KW (2020) Nature's fight against plastic pollution: algae for plastic biodegradation and bioplastics production. Environ Sci Ecotechnol 4:100065

Chia RW, Lee JY, Kim H, Jang J (2021) Microplastic pollution in soil and groundwater: a review. Environ Chem Lett 19(6):4211–4224

Coelho LM, Rezende HC, Coelho LM, De Sousa PA, Melo DF, Coelho NM (2015) Bioremediation of polluted waters using microorganisms. In: Advances in bioremediation of wastewater and polluted soil, vol 10, p 60770

Colzi I, Renna L, Bianchi E, Castellani MB, Coppi A, Pignattelli S, Loppi S, Gonnelli C (2022) Impact of microplastics on growth, photosynthesis and essential elements in Cucurbita pepo L. J Hazard Mater 423:127238

Dashora K, Gattupalli M, Tripathi GD, Javed Z, Singh S, Tuohy M et al (2023) Fungal assisted valorisation of polymeric lignin: mechanism, enzymes and perspectives. Catalysts 13(1):149

de Oliveira TA, Barbosa R, Mesquita AB, Ferreira JH, de Carvalho LH, Alves TS (2020) Fungal degradation of reprocessed PP/PBAT/thermoplastic starch blends. J Mater Res Technol 9(2):2338–2349

De Silva YSK, Rajagopalan UM, Kadono H, Li D (2022) Effects of microplastics on lentil (Lens culinaris) seed germination and seedling growth. Chemosphere 303:135162

de Souza Machado AA, Kloas W, Zarfl C, Hempel S, Rillig MC (2018) Microplastics as an emerging threat to terrestrial ecosystems. Glob Chang Biol 24(4):1405–1416

de Souza Machado AA, Lau CW, Kloas W, Bergmann J, Bachelier JB, Faltin E, Becker R, Görlich AS, Rillig MC (2019) Microplastics can change soil properties and affect plant performance. Environ Sci Technol 53(10):6044–6052

Debroy A, George N, Mukherjee G (2022) Role of biofilms in the degradation of microplastics in aquatic environments. J Chem Technol Biotechnol 97(12):3271–3282

Dey TK, Uddin ME, Jamal M (2021) Detection and removal of microplastics in wastewater: evolution and impact. Environ Sci Pollut Res 28:16925–16947

Dey S, Guha T, Barman F, Natarajan L, Kundu R, Mukherjee A, Paul S (2023) Surface functionalization and size of polystyrene microplastics concomitantly regulate growth, photosynthesis and anti-oxidant status of Cicer arietinum L. Plant Physiol Biochem 194:41–51

Di Napoli M, Silvestri B, Castagliuolo G, Carpentieri A, Luciani G, Di Maro A, Sorbo S, Pezzella A, Zanfardino A, Varcamonti M (2023) High density polyethylene (HDPE) biodegradation by the fungus *Cladosporium halotolerans*. FEMS Microbiol Ecol 99(2):fiac148

Di Y, Xia H, Jiao Y, Zhang X, Fang Q, Li F, Chen S (2019) Biodegradation of polyhydroxybutyrate by *Pseudomonas* sp DSDY0501 and purification and characterization of polyhydroxybutyrate depolymerase. 3 Biotech 9:1–9

Dong Y, Gao M, Song Z, Qiu W (2020) Microplastic particles increase arsenic toxicity to rice seedlings. Environ Pollut 259:113892

Dong R, Liu R, Xu Y, Liu W, Sun Y (2022) Effect of foliar and root exposure to polymethyl methacrylate microplastics on biochemistry, ultrastructure, and arsenic accumulation in Brassica campestris L. Environ Res 215:114402

Dris R, Gasperi J, Saad M, Mirande C, Tassin B (2016) Synthetic fibers in atmospheric fallout: a source of microplastics in the environment? Mar Pollut Bull 104(1–2):290–293

DSouza GC, Sheriff RS, Ullanat V, Shrikrishna A, Joshi AV, Hiremath L, Entoori K (2021) Fungal biodegradation of low-density polyethylene using consortium of aspergillus species under controlled conditions. Heliyon 7(5):e07008

Ebere EC, Wirnkor VA, Ngozi VE (2019) Uptake of microplastics by plant: a reason to worry or to be happy? World Scient News 131:256–267

El-Morsy EM, Hassan HM, Ahmed E (2017) Biodegradative activities of fungal isolates from plastic contaminated soils. Mycosphere 8(8):1071–1087

Eskander S, Saleh HED (2017) Biodegradation: process mechanism. Environ Sci and Eng 8(8):1–31

Esterhuizen M, Kim YJ (2022) Effects of polypropylene, polyvinyl chloride, polyethylene terephthalate, polyurethane, high-density polyethylene, and polystyrene microplastic on Nelumbo nucifera (lotus) in water and sediment. Environ Sci Pollut Res 29(12):17580–17590

Esterhuizen M, Vikfors S, Penttinen OP, Kim YJ, Pflugmacher S (2022) Lolium multiflorum germination and growth affected by virgin, naturally, and artificially aged high-density polyethylene microplastic and leachates. Front Environ Sci 10:2174

Fan J, Zou L, Duan T, Qin L, Qi Z, Sun J (2022) Occurrence and distribution of microplastics in surface water and sediments in China's inland water systems: a critical review. J Clean Prod 331:129968

Fang CR, Yao J, Zheng YG, Jiang CJ, Hu LF, Wu YY, Shen DS (2010) Dibutyl phthalate degradation by Enterobacter sp. T5 isolated from municipal solid waste in landfill bioreactor. Int Biodeterior Biodegrad 64(6):442–446

Fu D, Zhang Q, Fan Z, Qi H, Wang Z, Peng L (2019) Aged microplastics polyvinyl chloride interact with copper and cause oxidative stress towards microalgae Chlorella vulgaris. Aquat Toxicol 216:105319

Gao M, Liu Y, Song Z (2019) Effects of polyethylene microplastic on the phytotoxicity of di-n-butyl phthalate in lettuce (*Lactuca sativa* L. var. ramosaHort). Chemosphere 237:124482

Gao M, Xu Y, Liu Y, Wang S, Wang C, Dong Y, Song Z (2021) Effect of polystyrene on di-butyl phthalate (DBP) bioavailability and DBP-induced phytotoxicity in lettuce. Environ Pollut 268:115870

Gazal AA, Gheewala SH (2020) Plastics, microplastics and other polymer materials – a threat to the environment. J Sustain Energy Environ 11:113–122

Ge J, Li H, Liu P, Zhang Z, Ouyang Z, Guo X (2021) Review of the toxic effect of microplastics on terrestrial and aquatic plants. Sci Total Environ 791:148333

Ghosh SK, Pal S, Ray S (2013) Study of microbes having potentiality for biodegradation of plastics. Environ Sci Pollut Res 20:4339–4355

Giacomucci L, Raddadi N, Soccio M, Lotti N, Fava F (2019) Polyvinyl chloride biodegradation by *pseudomonas citronellolis* and *Bacillus flexus*. New Biotechnol 52:35–41

Gill M (2014) Bioplastic: a better alternative to plastics. Int J Res Appl Nat Soc Sci 2:115–120

Giorgetti L, Spanò C, Muccifora S, Bottega S, Barbieri F, Bellani L, Castiglione MR (2020) Exploring the interaction between polystyrene nanoplastics and *Allium cepa* during germination: internalization in root cells, induction of toxicity and oxidative stress. Plant Physiol Biochem 149:170–177

Gregory MR (2009) Environmental implications of plastic debris in marine settings – entanglement, ingestion, smothering, hangers-on, hitch-hiking and alien invasions. Philos Trans R Soc B Biol Sci 364(1526):2013–2025

Gulnaz O, Dincer S (2009) Biodegradation of Bisphenol a by *Chlorella vulgaris* and Aeromonas Hydrophilia. J Appl Biol Sci 3(2):79–84

Guo J-J, Huang X-P, Xiang L, Wang Y-Z, Li Y-W, Li H, Cai Q-Y, Mo C-H, Wong M-H (2020) Source, migration and toxicology of microplastics in soil. Environ Int 137:105263
Guo M, Zhao F, Tian L, Ni K, Lu Y, Borah P (2022) Effects of polystyrene microplastics on the seed germination of herbaceous ornamental plants. Sci Total Environ 809:151100
Hadiyanto H, Khoironi A, Dianratri I, Suherman S, Muhammad F, Vaidyanathan S (2021) Interactions between polyethylene and polypropylene microplastics and *spirulina* sp. microalgae in aquatic systems. Heliyon 7(8):e07676
Hahladakis JN, Velis CA, Weber R, Iacovidou E, Purnell P (2018) An overview of chemical additives present in plastics: migration, release, fate and environmental impact during their use, disposal and recycling. J Hazard Mater 344:179–199
Hakkarainen M, Albertsson AC (2004) Environmental degradation of polyethylene. In: Long term properties of polyolefins, pp 177–200
Hale RC, Seeley ME, La Guardia MJ, Mai L, Zeng EY (2020) A global perspective on microplastics. J Geophys Res Oceans 125(1):e2018JC014719
He D, Luo Y, Lu S, Liu M, Song Y, Lei L (2018) Microplastics in soils: analytical methods, pollution characteristics and ecological risks. TrAC Trends Anal Chem 109:163–172
Hernández-Arenas R, Beltrán-Sanahuja A, Navarro-Quirant P, Sanz-Lazaro C (2021) The effect of sewage sludge containing microplastics on growth and fruit development of tomato plants. Environ Pollut 268:115779
Hirooka T, Nagase H, Uchida K, Hiroshige Y, Ehara Y, Nishikawa JI et al (2005) Biodegradation of bisphenol A and disappearance of its estrogenic activity by the green alga *Chlorella fusca* var. *vacuolata*. Environ Toxicol Chem Int J 24(8):1896–1901
Hodson ME, Duffus-Hodson CA, Clark A, Prendergast-Miller MT, Thorpe KL (2017) Plastic bag derived-microplastics as a vector for metal exposure in terrestrial invertebrates. Environ Sci Technol 51(8):4714–4721
Hong J, Huang X, Wang Z, Luo X, Huang S, Zheng Z (2022) Combined toxic effects of enrofloxacin and microplastics on submerged plants and epiphytic biofilms in high nitrogen and phosphorus waters. Chemosphere 308:136099
Horton AA, Walton A, Spurgeon DJ, Lahive E, Svendsen C (2017) Microplastics in freshwater and terrestrial environments: evaluating the current understanding to identify the knowledge gaps and future research priorities. Sci Total Environ 586:127–141
Hu D, Zhang Y, Shen M (2020) Investigation on microplastic pollution of Dongting Lake and its affiliated rivers. Mar Pollut Bull 160:111555
Hu K, Yang Y, Wang Y, Duan X, Wang S (2022) Catalytic carbon and hydrogen cycles in plastics chemistry. Chem Catalysis 2(4):724–761
Huang Y, Zhao Y, Wang J, Zhang M, Jia W, Qin X (2019) LDPE microplastic films alter microbial community composition and enzymatic activities in soil. Environ Pollut 254:112983
Huang F, Hu J, Chen L, Wang Z, Sun S, Zhang W, Jiang H, Luo Y, Zeng Y, Fang L (2023) Microplastics may increase the environmental risks of Cd via promoting Cd uptake by plants: a meta-analysis. J Hazard Mater 448:130887
IEA (2015) World Energy Outlook 2015, IEA, Paris https://www.iea.org/reports/world-energy-outlook-2015, License: CC BY 4.0 Accessed 13 Apr 2023
Iqbal B, Zhao T, Yin W, Zhao X, Xie Q, Khan KY, Zhao X, Nazar M, Li G, Du D (2023) Impacts of soil microplastics on crops: a review. Appl Soil Ecol 181:104680
Jacobsen LF, Pedersen S, Thøgersen J (2022) Drivers of and barriers to consumers' plastic packaging waste avoidance and recycling–a systematic literature review. Waste Manag 141:63–78
Jebaranjitham JN, Christyraj JDS, Prasannan A, Rajagopalan K, Chelladurai KS, Gnanaraja JKJS (2022) Current scenario of solid waste management techniques and challenges in Covid-19-a review. Heliyon 8:e09855
Jeon HJ, Kim MN (2013) Isolation of a thermophilic bacterium capable of low-molecular-weight polyethylene degradation. Biodegradation 24:89–98
Jia H, Wu D, Yu Y, Han S, Sun L, Li M (2022) Impact of microplastics on bioaccumulation of heavy metals in rape (*Brassica napus* L.). Chemosphere 288:132576

Jiang C, Yin L, Li Z, Wen X, Luo X, Hu S et al (2019a) Microplastic pollution in the rivers of the Tibet plateau. Environ Pollut 249:91–98

Jiang X, Chen H, Liao Y, Ye Z, Li M, Klobučar G (2019b) Ecotoxicity and genotoxicity of polystyrene microplastics on higher plant *Vicia faba*. Environ Pollut 250:831–838

Jin T, Tang J, Lyu H, Wang L, Gillmore AB, Schaeffer SM (2022) Activities of microplastics (MPs) in agricultural soil: a review of MPs pollution from the perspective of agricultural ecosystems. J Agric Food Chem 70(14):4182–4201

Kan M, Miller SA (2022) Environmental impacts of plastic packaging of food products. Resour Conserv Recycl 180:106156

Kang H, Park S, Lee B, Kim I, Kim S (2022) Concentration of microplastics in road dust as a function of the drying period – a case study in G City, Korea. Sustain For 14(5):3006

Kang M, Liu Y, Wang H, Weng Y, Gong D, Bai X (2023) Physiological toxicity and antioxidant mechanism of Photoaging microplastics on Pisum sativum L seedlings. Toxics 11(3):242

Kaur R, Kumari A, Sharma G, Singh D, Kaur R (2021) Biodegradation of endocrine disrupting chemicals benzyl butyl phthalate and dimethyl phthalate by *bacillus marisflavi* RR014. J Appl Microbiol 131(3):1274–1288

Khalid N, Aqeel M, Noman A (2020) Microplastics could be a threat to plants in terrestrial systems directly or indirectly. Environ Pollut 267:115653

Kim JW, Park SB, Tran QG, Cho DH, Choi DY, Lee YJ, Kim HS (2020) Functional expression of polyethylene terephthalate-degrading enzyme (PETase) in green microalgae. Microb Cell Factories 19(1):1–9

Klemeš JJ, Van Fan Y, Tan RR, Jiang P (2020) Minimising the present and future plastic waste, energy and environmental footprints related to COVID-19. Renew Sust Energ Rev 127:109883

Kotova IB, Taktarova YV, Tsavkelova EA, Egorova MA, Bubnov IA, Malakhova DV et al (2021) Microbial degradation of plastics and approaches to make it more efficient. Microbiology 90:671–701

Kumar D, Ohri P (2023) Say "NO" to plant stresses: unravelling the role of nitric oxide under abiotic and biotic stress. Nitric Oxide 130(1):36–57

Kumar RV, Kanna GR, Elumalai S (2017) Biodegradation of polyethylene by green photosynthetic microalgae. J Bioremed Biodegr 8(381):2

Kumar R, Ivy N, Bhattacharya S, Dey A, Sharma P (2022) Coupled effects of microplastics and heavy metals on plants: uptake, bioaccumulation, and environmental health perspectives. Sci Total Environ 836:155619

Kumari A, Rajput VD, Mandzhieva SS, Rajput S, Minkina T, Kaur R, Sushkova S, Kumari P, Ranjan A, Kalinitchenko VP, Glinushkin AP (2022) Microplastic pollution: an emerging threat to terrestrial plants and insights into its remediation strategies. Plan Theory 11(3):340

Lebreton L, Andrady A (2019) Future scenarios of global plastic waste generation and disposal. Palgrave Commun 5(1):1–11

Lei C, Engeseth NJ (2022) Comparison of growth and quality between hydroponically grown and soil-grown lettuce under the stress of microplastics. ACS ES T Water 2(7):1182–1194

Li J, Liu H, Chen JP (2018a) Microplastics in freshwater systems: a review on occurrence, environmental effects, and methods for microplastics detection. Water Res 137:362–374

Li X, Chen L, Mei Q, Dong B, Dai X, Ding G, Zeng EY (2018b) Microplastics in sewage sludge from the wastewater treatment plants in China. Water Res 142:75–85

Li L, Luo Y, Li R, Zhou Q, Peijnenburg WJ, Yin N et al (2020a) Effective uptake of submicrometre plastics by crop plants via a crack-entry mode. Nat Sustain 3(11):929–937

Li L, Yang J, Zhou Q, Peijnenburg WJ, Luo Y (2020b) Uptake of microplastics and their effects on plants. In: Microplastics in terrestrial environments: emerging contaminants and major challenges, pp 279–298

Li Z, Li Q, Li R, Zhao Y, Geng J, Wang G (2020c) Physiological responses of lettuce (*Lactuca sativa* L.) to microplastic pollution. Environ Sci Pollut Res 27:30306–30314

Li Z, Li Q, Li R, Zhou J, Wang G (2021) The distribution and impact of polystyrene nanoplastics on cucumber plants. Environ Sci Pollut Res 28:16042–16053

Li J, Yu S, Yu Y, Xu M (2022) Effects of microplastics on higher plants: a review. Bull Environ Contam Toxicol 109(2):241–265

Li H, Chang X, Zhang J, Wang Y, Zhong R, Wang L, Wei J, Wang Y (2023) Uptake and distribution of microplastics of different particle sizes in maize (*Zea mays*) seedling roots. Chemosphere 313:137491

Lian J, Shen M, Liu W (2019) Effects of microplastics on wheat seed germination and seedling growth. J Agro Environ Sci 38(4):737–745

Lian J, Wu J, Xiong H, Zeb A, Yang T, Su X, Su L, Liu W (2020a) Impact of polystyrene nanoplastics (PSNPs) on seed germination and seedling growth of wheat (*Triticum aestivum* L.). J Hazard Mater 385:121620

Lian J, Wu J, Zeb A, Zheng S, Ma T, Peng F et al (2020b) Do polystyrene nanoplastics affect the toxicity of cadmium to wheat (*Triticum aestivum* L.)? Environ Pollut 263:114498

Lian J, Liu W, Meng L, Wu J, Chao L, Zeb A, Sun Y (2021) Foliar-applied polystyrene nanoplastics (PSNPs) reduce the growth and nutritional quality of lettuce (*Lactuca sativa* L.). Environ Pollut 280:116978

Liao YC, Nazygul J, Li M, Wang XL, Jiang LJ (2019) Effects of microplastics on the growth, physiology, and biochemical characteristics of wheat (*Triticum aestivum*). Huanjingkexue 40(10):4661–4667

Lin W, Su F, Lin M, Jin M, Li Y, Ding K, Chen Q, Qian Q, Sun X (2020) Effect of microplastics PAN polymer and/or Cu2+ pollution on the growth of *Chlorella pyrenoidosa*. Environ Pollut 265:114985

Lin Z, Jin T, Zou T, Xu L, Xi B, Xu D et al (2022) Current progress on plastic/microplastic degradation: fact influences and mechanism. Environ Pollut:119159

Liu EK, He WQ, Yan CR (2014) 'White revolution'to 'white pollution' – agricultural plastic film mulch in China. Environ Res Lett 9(9):091001

Liu F, Xu B, He Y, Brookes PC, Tang C, Xu J (2018) Differences in transport behavior of natural soil colloids of contrasting sizes from nanometer to micron and the environmental implications. Sci Total Environ 634:802–810

Liu C, Li J, Zhang Y, Wang L, Deng J, Gao Y et al (2019a) Widespread distribution of PET and PC microplastics in dust in urban China and their estimated human exposure. Environ Int 128:116–124

Liu K, Wang X, Wei N, Song Z, Li D (2019b) Accurate quantification and transport estimation of suspended atmospheric microplastics in megacities: implications for human health. Environ Int 132:105127

Liu L, Xu M, Ye Y, Zhang B (2022a) On the degradation of (micro) plastics: degradation methods, influencing factors, environmental impacts. Sci Total Environ 806:151312

Liu Q, Sun X, Wu W, Liu Z, Fang G, Yang P (2022b) Agroecosystem services: a review of concepts, indicators, assessment methods and future research perspectives. Ecol Indic 142:109218

Liu S, Shang E, Liu J, Wang Y, Bolan N, Kirkham MB, Li Y (2022c) What have we known so far for fluorescence staining and quantification of microplastics: a tutorial review. Front Environ Sci Eng 16:1–14

Liu Y, Guo R, Zhang S, Sun Y, Wang F (2022d) Uptake and translocation of nano/microplastics by rice seedlings: evidence from a hydroponic experiment. J Hazard Mater 421:126700

Lwanga EH, Thapa B, Yang X, Gertsen H, Salánki T, Geissen V, Garbeva P (2018) Decay of low-density polyethylene by bacteria extracted from earthworm's guts: a potential for soil restoration. Sci Total Environ 624:753–757

Lwanga EH, Beriot N, Corradini F, Silva V, Yang X, Baartman J, Rezaei M, van Schaik L, Riksen M, Geissen V (2022) Review of microplastic sources, transport pathways and correlations with other soil stressors: a journey from agricultural sites into the environment. Chem Biol Technol Agric 9:20

Ma J, Aqeel M, Khalid N, Nazir A, Alzuaibr FM, Al-Mushhin AA, Hakimi O, Iqbal MF, Chen F, Alamri S, Hashem M, Noman A (2022a) Effects of microplastics on growth and metabolism of rice (Oryza sativa L.). Chemosphere 307:135749

Ma J, Chen F, Zhu Y, Li X, Yu H, Sun Y (2022b) Joint effects of microplastics and ciprofloxacin on their toxicity and fates in wheat: a hydroponic study. Chemosphere 303:135023

Madouh TA, Quoreshi AM (2023) The function of arbuscular mycorrhizal fungi associated with drought stress resistance in native plants of arid desert ecosystems: a review. Diversity 15(3):391

Magnin A, Hoornaert L, Pollet E, Laurichesse S, Phalip V, Avérous L (2019) Isolation and characterization of different promising fungi for biological waste management of polyurethanes. Microb Biotechnol 12(3):544–555

Maity S, Chatterjee A, Guchhait R, De S, Pramanick K (2020) Cytogenotoxic potential of a hazardous material, polystyrene microparticles on Allium cepa L. J Hazard Mater 385:121560

Mal N, Satpati G, Raghunathan S, Davoodbasha M (2022) Current strategies on algae-based biopolymer production and scale-up. Chemosphere 289:133178

Manzi HP, Abou-Shanab RA, Jeon BH, Wang J, Salama ES (2022) Algae: a frontline photosynthetic organism in the microplastic catastrophe. Trends Plant Sci

Maraveas C (2020) Environmental sustainability of plastic in agriculture. Agriculture 10(8):310

Maroof L, Iqbal M, Farman S, Faisal S (2022) Biodegradation of low-density polyethylene (LDPE) bags by fungi isolated from waste disposal soil. Appl Environ Soil Sci 2022. https://doi.org/10.1155/2022/8286344

Matthews S, Mai L, Jeong CB, Lee JS, Zeng EY, Xu EG (2021) Key mechanisms of micro-and nanoplastic (MNP) toxicity across taxonomic groups. Comp Biochem Physiol Part C Toxicol Pharmacol 247:109056

Megonigal JP, Hines ME, Visscher PT (2004) Anaerobic metabolism: linkages to trace gases and aerobic processes. Treatise Geochem 8:317–424

Menicagli V, Castiglione MR, Balestri E, Giorgetti L, Bottega S, Sorce C, Spano C, Lardicci C (2022) Early evidence of the impacts of microplastic and nanoplastic pollution on the growth and physiology of the seagrass *Cymodocea nodosa*. Sci Total Environ 838:156514

Miri S, Saini R, Davoodi SM, Pulicharla R, Brar SK, Magdouli S (2022) Biodegradation of microplastics: better late than never. Chemosphere 286:131670

Mishra S, Swain S, Sahoo M, Mishra S, Das AP (2022) Microbial colonization and degradation of microplastics in aquatic ecosystem: a review. Geomicrobiol J 39(3–5):259–269

Modak A, Basu S (2022) Microplastic-an imposing commination to the aquatic ecosystem and its removal strategies in wastewater treatment plants: a systematic review. J Sustain Environ Manag 1(2):265–274

Mohan K (2011) Microbial deterioration and degradation of polymeric materials. J Biochem Technol 2(4):210–215

Mohan AJ, Sekhar VC, Bhaskar T, Nampoothiri KM (2016) Microbial assisted high impact polystyrene (HIPS) degradation. Bioresour Technol 213:204–207

Moog D, Schmitt J, Senger J, Zarzycki J, Rexer KH, Linne U et al (2019) Using a marine microalga as a chassis for polyethylene terephthalate (PET) degradation. Microb Cell Factories 18(1):1–15

Muhonja CN, Makonde H, Magoma G, Imbuga M (2018) Biodegradability of polyethylene by bacteria and fungi from Dandora dumpsite Nairobi-Kenya. PLoS One 13(7):e0198446

Munir E, Harefa RSM, Priyani N, Suryanto D (2018) Plastic degrading fungi *Trichoderma viride* and *aspergillus nomius* isolated from local landfill soil in Medan. IOP Conf Ser Earth Environ Sci 126(1):012145. https://doi.org/10.1088/1755-1315/126/1/01214

Nagi GK, Minhas AK, Gaur S, Jain P, Mandal S (2021) Integration of algal biofuels with bioremediation coupled industrial commodities towards cost-effectiveness. Front Energy Res 9:735141

Napper IE, Thompson RC (2016) Release of synthetic microplastic plastic fibres from domestic washing machines: effects of fabric type and washing conditions. Mar Pollut Bull 112(1–2):39–45

Ng EL, Lwanga EH, Eldridge SM, Johnston P, Hu HW, Geissen V, Chen D (2018) An overview of microplastic and nanoplastic pollution in agroecosystems. Sci Total Environ 627:1377–1388

Nizzetto L, Futter M, Langaas S (2016) Are agricultural soils dumps for microplastics of urban origin? Environ Sci Technol 50(20):10777–10779

Ogo HA, Tang N, Li X, Gao X, Xing W (2022) Combined toxicity of microplastic and lead on submerged macrophytes. Chemosphere 295:133956

Ojha N, Pradhan N, Singh S, Barla A, Shrivastava A, Khatua P, Rai V, Bose S (2017) Evaluation of HDPE and LDPE degradation by fungus, implemented by statistical optimization. Sci Rep 7(1):39515

Okeke ES, Okoye CO, Atakpa EO, Ita RE, Nyaruaba R, Mgbechidinma CL, Akan OD (2022) Microplastics in agroecosystems-impacts on ecosystem functions and food chain. Resour Conserv Recycl 177:105961

Okeke ES, Chukwudozie KI, Addey CI, Okoro JO, Ezeorba TPC, Atakpa EO, Okoye CO, Nwuche CO (2023) Micro and nanoplastics ravaging our agroecosystem: a review of occurrence, fate, ecological impacts, detection, remediation, and prospects. Heliyon 9:e13296. https://doi.org/10.1016/j.heliyon.2023.e13296

Oliveira J, Belchior A, da Silva VD, Rotter A, Petrovski Ž, Almeida PL, Lourenço ND, Gaudêncio SP (2020) Marine environmental plastic pollution: mitigation by microorganism degradation and recycling valorization. Front Mar Sci 7:567126

Othman AR, Hasan HA, Muhamad MH, Ismail NI, Abdullah SRS (2021) Microbial degradation of microplastics by enzymatic processes: a review. Environ Chem Lett 19:3057–3073

Otsuki T, Zhang F, Kabeya H, Hirotsu T (2004) Synthesis and tensile properties of a novel composite of chlorella and polyethylene. J Appl Polym Sci 92(2):812–816

Oviedo-Anchundia R, del Castillo DS, Naranjo-MorÃ J, Francois N et al (2021) Analysis of the degradation of polyethylene, polystyrene and polyurethane mediated by three filamentous fungi isolated from the Antarctica. Afr J Biotechnol 20(2):66–76. https://doi.org/10.5897/AJB2020.17200

Paço A, Duarte K, da Costa JP, Santos PS, Pereira R, Pereira ME, Freitas AC, Duarte AC, Rocha-Santos TA (2017) Biodegradation of polyethylene microplastics by the marine fungus *Zalerion maritimum*. Sci Total Environ 586:10–15. https://doi.org/10.1016/j.scitotenv.2017.02.017

Panigrahi S, Velraj P, Subba Rao T (2019) Functional microbial diversity in contaminated environment and application in bioremediation. In: Das S, Dash HR (eds) Microbial diversity in the genomic era. Academic Press, London, pp 359–385

Park SY, Kim CG (2019) Biodegradation of micro-polyethylene particles by bacterial colonization of a mixed microbial consortium isolated from a landfill site. Chemosphere 222:527–533

Pathak VM (2017) Review on the current status of polymer degradation: a microbial approach. Bioresour Bioprocess 4:1–31

Patil R, Bagde US (2012) Isolation of polyvinyl chloride degrading bacterial strains from environmental samples using enrichment culture technique. Afr J Biotechnol 11(31):7947–7956

Pflugmacher S, Tallinen S, Kim YJ, Kim S, Esterhuizen M (2021) Ageing affects microplastic toxicity over time: effects of aged polycarbonate on germination, growth, and oxidative stress of *Lepidium sativum*. Sci Total Environ 790:148166

Pham TH, Do HT, Thi LAP, Singh P, Raizada P, Wu JCS, Nguyen VH (2021) Global challenges in microplastics: from fundamental understanding to advanced degradations toward sustainable strategies. Chemosphere 267:129275

Pignattelli S, Broccoli A, Renzi M (2020) Physiological responses of garden cress (*Lepidium sativum*) to different types of microplastics. Sci Total Environ 727:138609

Pinto-Poblete A, Retamal-Salgado J, López MD, Zapata N, Sierra-Almeida A, Schoebitz M (2022) Combined effect of microplastics and Cd alters the enzymatic activity of soil and the productivity of strawberry plants. Plan Theory 11(4):536

Prabhat S, Bhattacharyya S, Vishal V, Kalyan RK, Vijai K, Pandey KN, Singh M (2013) Studies on isolation and identification of active microorganisms during degradation of polyethylene/starch film. Int Res J Environ Sci 2(9):83–85

Puri A, Bajaj A, Singh Y, Lal R (2022) Harnessing taxonomically diverse and metabolically versatile genus *Paracoccus* for bioplastic synthesis and xenobiotic biodegradation. J Appl Microbiol 132(6):4208–4224

Qi Y, Yang X, Pelaez AM, Lwanga EH, Beriot N, Gertsen H et al (2018a) Macro-and microplastics in soil-plant system: effects of plastic mulch film residues on wheat (*Triticum aestivum*) growth. Sci Total Environ 645:1048–1056

Qi Y, Yang X, Pelaez AM, Lwanga EH, Beriot N, Gertsen H, Garbeva P, Geissen V (2018b) Macro-and micro-plastics in soil-plant system: effects of plastic mulch film residues on wheat (*Triticum aestivum*) growth. Sci Total Environ 645:1048–1056

Qi R, Jones DL, Li Z, Liu Q, Yan C (2020) Behavior of microplastics and plastic film residues in the soil environment: a critical review. Sci Total Environ 703:134722

Rabot C, Chen Y, Lin SY, Miller B, Chiang YM, Oakley CE, Oakley BR, Wang CC, Williams TJ (2023) Polystyrene upcycling into fungal natural products and a biocontrol agent. J Am Chem Soc 145(9):5222–5230

Rafique A, Irfan M, Mumtaz M, Qadir A (2020) Spatial distribution of microplastics in soil with context to human activities: a case study from the urban center. Environ Monit Assess 192:1–13

Rajandas H, Parimannan S, Sathasivam K, Ravichandran M, Yin LS (2012) A novel FTIR-ATR spectroscopy based technique for the estimation of low-density polyethylene biodegradation. Polym Test 31(8):1094–1099

Restrepo-Flórez JM, Bassi A, Thompson MR (2014) Microbial degradation and deterioration of polyethylene–a review. Int Biodeterior Biodegrad 88:83–90

Rosato A, Barone M, Negroni A, Brigidi P, Fava F, Xu P et al (2020) Microbial colonization of different microplastic types and biotransformation of sorbed PCBs by a marine anaerobic bacterial community. Sci Total Environ 705:135790

Roy T, Dey TK, Jamal M (2023) Microplastic/nanoplastic toxicity in plants: an imminent concern. Environ Monit Assess 195(1):27

Rozman U, Turk T, Skalar T, Zupančič M, Korošin NČ, Marinšek M, Olivero-Verbel J, Kalčíková G (2021) An extensive characterization of various environmentally relevant microplastics–material properties, leaching and ecotoxicity testing. Sci Total Environ 773:145576

Rummel CD, Jahnke A, Gorokhova E, Kühnel D, Schmitt-Jansen M (2017) Impacts of biofilm formation on the fate and potential effects of microplastic in the aquatic environment. Environ Sci Technol Lett 4(7):258–267

Sahasa RGK, Dhevagi P, Poornima R, Alagirisamy B, Moorthy PS, Karthikeyan S (2022) Influence of microplastics on seedling growth of blackgram under different soil types. Int J Environ Climate Change 12(11):642–649

Sahasa RGK, Dhevagi P, Poornima R, Ramya A, Moorthy PS, Alagirisamy B, Karthikeyan S (2023) Effect of polyethylene microplastics on seed germination of blackgram (*Vigna mungo* L.) and tomato (*Solanum lycopersicum* L.). Environ Adv 11:100349

Sajjad M, Huang Q, Khan S, Khan MA, Yin L, Wang J et al (2022) Microplastics in the soil environment: a critical review. Environ Technol Innov:102408

Sakharkar KR, Kyaw BM, Champakalakshmi R, Sakharkar MK, Lim CS (2012) Biodegradation of low density polythene (LDPE) by pseudomonas species. Indian J Microbiol 52:411–419

Sánchez C (2020) Fungal potential for the degradation of petroleum-based polymers: an overview of macro-and microplastics biodegradation. Biotechnol Adv 40:107501

Sangale MK, Shahnawaz M, Ade AB (2019) Potential of fungi isolated from the dumping sites mangrove rhizosphere soil to degrade polythene. Sci Rep 9:5390

Santo M, Weitsman R, Sivan A (2013) The role of the copper-binding enzyme–laccase–in the biodegradation of polyethylene by the actinomycete *Rhodococcusruber*. Int Biodeterior Biodegrad 84:204–210

Sarmah P, Rout J (2018) Efficient biodegradation of low-density polyethylene by cyanobacteria isolated from submerged polyethylene surface in domestic sewage water. Environ Sci Pollut Res 25:33508–33520

Satlewal A, Soni R, Zaidi MGH, Shouche Y, Goel R (2008) Comparative biodegradation of HDPE and LDPE using an indigenously developed microbial consortium. J Microbiol Biotechnol 18(3):477–482

Sayyed RZ, Wani SJ, Alyousef AA, Alqasim A, Syed A, El-Enshasy HA (2019) Purification and kinetics of the PHB depolymerase of *Microbacteriumparaoxydans* RZS6 isolated from a dumping yard. PLoS One 14(6):e0212324

Scheurer M, Bigalke M (2018) Microplastics in Swiss floodplain soils. Environ Sci Technol 52(6):3591–3598

Seleiman MF, Santanen A, Mäkelä PS (2020) Recycling sludge on cropland as fertilizer–advantages and risks. Resour Conserv Recycl 155:104647

Senavirathna MDHJ, Zhaozhi L, Fujino T (2022) Root adsorption of microplastic particles affects the submerged freshwater macrophyte *Egeria densa*. Water Air Soil Pollut 233(3):80

Shahi Khalaf Ansar B, Kavusi E, Dehghanian Z, Pandey J, Asgari Lajayer B, Price GW, Astatkie T (2022) Removal of organic and inorganic contaminants from the air, soil, and water by algae. Environ Sci Pollut Res 30:1–29

Sharma P, Singh SP, Parakh SK, Tong YW (2022) Health hazards of hexavalent chromium (Cr (VI)) and its microbial reduction. Bioengineered 13(3):4923–4938. https://doi.org/10.1080/21655979.2022.2037273

Sharma N, Khanna K, Jasrotia S, Kumar D, Bhardwaj R, Ohri P (2023a) Metabolites and chemical agents in the plant roots: an overview of their use in plant-parasitic nematode management. Nematology 25(3):243–257

Sharma S, Sharma V, Chatterjee S (2023b) Contribution of plastic and microplastic to global climate change and their conjoining impacts on the environment-a review. Sci Total Environ 875:162627

Shi R, Liu W, Lian Y, Wang Q, Zeb A, Tang J (2022) Phytotoxicity of polystyrene, polyethylene and polypropylene microplastics on tomato (*Lycopersicon esculentum* L.). J Environ Manag 317:115441

Shiu RF, Vazquez CI, Chiang CY, Chiu MH, Chen CS, Ni CW, Gong GC, Quigg A, Santschi PH, Chin WC (2020) Nano-and microplastics trigger secretion of protein-rich extracellular polymeric substances from phytoplankton. Sci Total Environ 748:141469

Singh G, Singh AK, Bhatt K (2016) Biodegradation of polythenes by bacteria isolated from soil. Int J Res Dev Pharm Life Sci 5(2):2056–2062

Skariyachan S, Patil AA, Shankar A, Manjunath M, Bachappanavar N, Kiran S (2018) Enhanced polymer degradation of polyethylene and polypropylene by novel thermophilic consortia of *Brevibacillus* spp. and *Aneurinibacillus* sp. screened from waste management landfills and sewage treatment plants. Polym Degrad Stab 149:52–68

Skariyachan S, Taskeen N, Kishore AP, Krishna BV, Naidu G (2021) Novel consortia of enterobacter and pseudomonas formulated from cow dung exhibited enhanced biodegradation of polyethylene and polypropylene. J Environ Manag 284:112030

Song C, Hu X, Liu Z, Li S, Kitamura Y (2020) Combination of brewery wastewater purification and CO_2 fixation with potential value-added ingredients production via different microalgae strains cultivation. J Clean Prod 268:122332

Sridharan S, Kumar M, Singh L, Bolan NS, Saha M (2021) Microplastics as an emerging source of particulate air pollution: a critical review. J Hazard Mater 418:126245

Steensgaard IM, Syberg K, Rist S, Hartmann NB, Boldrin A, Hansen SF (2017) From macro-to microplastics-analysis of EU regulation along the life cycle of plastic bags. Environ Pollut 224:289–299

Steinmetz Z, Wollmann C, Schaefer M, Buchmann C, David J, Tröger J, Muñoz K, Frör O, Schaumann GE (2016) Plastic mulching in agriculture. Trading short-term agronomic benefits for long-term soil degradation? Sci Total Environ 550:690–705

Su Y, Cheng Z, Hou Y, Lin S, Gao L, Wang Z, Bao R, Peng L (2022) Biodegradable and conventional microplastics posed similar toxicity to marine algae Chlorella vulgaris. Aquat Toxicol 244:106097

Sudhakar M, Priyadarshini C, Doble M, Murthy PS, Venkatesan R (2007) Marine bacteria mediated degradation of nylon 66 and 6. Int Biodeterior Biodegrad 60(3):144–151
Sun D, Li H, Wang E, He W, Hao W, Yan C, Li Y, Mei X, Zhang Y, Sun Z, Zhang F (2020a) An overview of the use of plastic-film mulching in China to increase crop yield and water-use efficiency. Natl Sci Rev 7(10):1523–1526
Sun Q, Ren SY, Ni HG (2020b) Incidence of microplastics in personal care products: an appreciable part of plastic pollution. Sci Total Environ 742:140218
Sun H, Lei C, Xu J, Li R (2021) Foliar uptake and leaf-to-root translocation of nanoplastics with different coating charge in maize plants. J Hazard Mater 416:125854
Sun H, Li Z, Wen J, Zhou Q, Gong Y, Zhao X, Mao H (2023) Co-exposure of maize to polyethylene microplastics and ZnO nanoparticles: impact on growth, fate, and interaction. Sci Total Environ 876:162705
Syranidou E, Karkanorachaki K, Amorotti F, Repouskou E, Kroll K, Kolvenbach B, Corvini PF, Fava F, Kalogerakis N (2017) Development of tailored indigenous marine consortia for the degradation of naturally weathered polyethylene films. PLoS One 12(8):e0183984
Taghavi N, Udugama IA, Zhuang WQ, Baroutian S (2021) Challenges in biodegradation of non-degradable thermoplastic waste: from environmental impact to operational readiness. Biotechnol Adv 49:107731
Tan J, Chen Y, Mo Z, Tan C, Wen R, Chen Z, Tian H (2022) Zinc oxide nanoparticles and polyethylene microplastics affect the growth, physiological and biochemical attributes, and Zn accumulation of rice seedlings. Environ Sci Pollut Res 29(40):61534–61546
Tang M, Huang Y, Zhang W, Fu T, Zeng T, Huang Y, Yang X (2022) Effects of microplastics on the mineral elements absorption and accumulation in hydroponic rice seedlings (*Oryza sativa* L.). Bull Environ Contam Toxicol 108(5):949–955
Tiwari N, Santhiya D, Sharma JG (2020) Microbial remediation of micro-nano plastics: current knowledge and future trends. Environ Pollut 265:115044
Tsiota P, Karkanorachaki K, Syranidou E, Franchini M, Kalogerakis N (2018) Microbial degradation of HDPE secondary microplastics: preliminary results. In: Proceedings of the international conference on microplastic pollution in the Mediterranean Sea. Springer International Publishing, pp 181–188
Ubalua AO (2011) Bioremediation strategies for oil polluted marine ecosystems. Aust J Agric Eng 2(6):160–168
Ugya AY, Ajibade FO, Hua X (2021) The efficiency of microalgae biofilm in the phycoremediation of water from river Kaduna. J Environ Manag 295:113109
Ullah R, Tsui MTK, Chen H, Chow A, Williams C, Ligaba-Osena A (2021) Microplastics interaction with terrestrial plants and their impacts on agriculture. J Environ Qual 50(5):1024–1041
Uwamungu JY, Wang Y, Shi G, Pan S, Wang Z, Wang L, Yang S (2022) Microplastic contamination in soil agro-ecosystems: a review. Environ Adv 9:100273
Vaid M, Sarma K, Gupta A (2021) Microplastic pollution in aquatic environments with special emphasis on riverine systems: current understanding and way forward. J Environ Manag 293:112860
van Kleunen M, Brumer A, Gutbrod L, Zhang Z (2020) A microplastic used as infill material in artificial sport turfs reduces plant growth. Plants, People, Planet 2(2):157–166
van Weert S, Redondo-Hasselerharm PE, Diepens NJ, Koelmans AA (2019) Effects of nanoplastics and microplastics on the growth of sediment-rooted macrophytes. Sci Total Environ 654:1040–1047
Varma CASL, Sahasrabudhe M, Ganguly S, Mallu MR (2023) Mycoremediation of micro-/nanoplastics-contaminated soils. In: Micro and nanoplastics in soil: threats to plant-based food. Springer International Publishing, Cham, pp 335–382
Venkatesan R, Santhamoorthy M, Alagumalai K, Haldhar R, Raorane CJ, Raj V, Kim SC (2022) Novel approach in biodegradation of synthetic thermoplastic polymers: an overview. Polymers 14(20):4271

Vimala PP, Mathew L (2016) Biodegradation of polyethylene using *Bacillus subtilis*. Proc Technol 24:232–239

Vivi VK, Martins-Franchetti SM, Attili-Angelis D (2019) Biodegradation of PCL and PVC: *Chaetomium globosum* (ATCC 16021) activity. Folia Microbiol 64:1–7

Volova TG et al (2011) Biodegradation of polyhydroxyalkanoates (PHAs) in the South China Sea and identification of PHA-degrading bacteria. Microbiology 80:252–260

Wang Z, Gao J, Li L, Jiang H (2012) Purification and characterization of an extracellular poly (3-hydroxybutyrate-co-3-hydroxyvalerate) depolymerase from *Acidovorax* sp. HB01. World J Microbiol Biotechnol 28:2395–2402

Wang L, Liu Y, Kaur M, Yao Z, Chen T, Xu M (2021) Phytotoxic effects of polyethylene microplastics on the growth of food crops soybean (*Glycine max*) and mung bean (Vigna radiata). Int J Environ Res Public Health 18(20):10629

Wang F, Feng X, Liu Y, Adams CA, Sun Y, Zhang S (2022) Micro (nano) plastics and terrestrial plants: up-to-date knowledge on uptake, translocation, and phytotoxicity. Resour Conserv Recycl 185:106503

Wang J, Li J, Liu W, Zeb A, Wang Q, Zheng Z, Shi R, Lian Y, Liu L (2023a) Three typical microplastics affect the germination and growth of amaranth (*Amaranthus mangostanus* L.) seedlings. Plant Physiol Biochem 194:589–599

Wang Y, Bai J, Wen L, Wang W, Zhang L, Liu Z, Liu H (2023b) Phytotoxicity of microplastics to the floating plant *Spirodela polyrhiza* (L.): plant functional traits and metabolomics. Environ Pollut 322:121199

Wang Y, Xiang L, Wang F, Redmile-Gordon M, Bian Y, Wang Z, Gu C, Jiang X, Schaffer A, Xing B (2023c) Transcriptomic and metabolomic changes in lettuce triggered by microplastics-stress. Environ Pollut:121081

Wei G, Yadan L, Chenguang L, Lin C, Yuhang Y, Xianghao Y, Fengmin L (2022) Research Progress on aging mechanism and environmental behavior of marine microplastics. Asian J Ecotoxicol 4:33–46

Weithmann N, Möller JN, Löder MG, Piehl S, Laforsch C, Freitag R (2018) Organic fertilizer as a vehicle for the entry of microplastic into the environment. Sci Adv 4(4):eaap8060

Wilkes RA, Aristilde L (2017) Degradation and metabolism of synthetic plastics and associated products by pseudomonas sp.: capabilities and challenges. J Appl Microbiol 123(3):582–593

Wu Y, Guo P, Zhang X, Zhang Y, Xie S, Deng J (2019) Effect of microplastics exposure on the photosynthesis system of freshwater algae. J Hazard Mater 374:219–227

Wu J, Liu W, Zeb A, Lian J, Sun Y, Sun H (2021) Polystyrene microplastic interaction with Oryza sativa: toxicity and metabolic mechanism. Environ Sci Nano 8(12):3699–3710

Wu X, Hou, H, Liu Y, Yin S, Bian S, Liang S, Wan C, Yuan S, Xiao K, Liu B, Yang J (2022) Microplastics affect rice (Oryza sativa L.) quality by interfering metabolite accumulation and energy expenditure pathways: a field study. J Hazard Mater 422:126834

Wu F, Guo Z, Cui K, Dong D, Yang X, Li J, Wu Z, Li L, Dai Y, Pan T (2023) Insights into characteristics of white rot fungus during environmental plastics adhesion and degradation mechanism of plastics. J Hazard Mater 448:130878

Xu Z, Zhang Y, Lin L, Wang L, Sun W, Liu C, Yu G, Yu J, Lv Y, Chen J, Wang Y (2022) Toxic effects of microplastics in plants depend more by their surface functional groups than just accumulation contents. Sci Total Environ 833:155097

Yang J, Yang Y, Wu WM, Zhao J, Jiang L (2014) Evidence of polyethylene biodegradation by bacterial strains from the guts of plastic-eating waxworms. Environ Sci Technol 48(23):13776–13784

Yang Y, Yang J, Wu WM, Zhao J, Song Y, Gao L, Yang R, Jiang L (2015a) Biodegradation and mineralization of polystyrene by plastic-eating mealworms: part 1. Chemical and physical characterization and isotopic tests. Environ Sci Technol 49(20):12080–12086

Yang Y, Yang J, Wu WM, Zhao J, Song Y, Gao L, Yang R, Jiang L (2015b) Biodegradation and mineralization of polystyrene by plastic-eating mealworms: part 2. Role of gut microorganisms. Environ Sci Technol 49(20):12087–12093

Yang H, He Y, Yan Y, Junaid M, Wang J (2021a) Characteristics, toxic effects, and analytical methods of microplastics in the atmosphere. Nano 11(10):2747
Yang T, Lupwayi N, Marc SA, Siddique KH, Bainard LD (2021b) Anthropogenic drivers of soil microbial communities and impacts on soil biological functions in agroecosystems. Glob Ecol Conserv 27:e01521
Yang W, Cheng P, Adams CA, Zhang S, Sun Y, Yu H, Wang F (2021c) Effects of microplastics on plant growth and arbuscular mycorrhizal fungal communities in a soil spiked with ZnO nanoparticles. Soil Biol Biochem 155:108179
Yang X, Wang H, Zhang L, Kong L, Chen Y, He Q, Li L, Grossart HP, Ju F (2021d) Marine algae facilitate transfer of microplastics and associated pollutants into food webs. Sci Total Environ 787:147535–147535
Yong JJY, Chew KW, Khoo KS, Show PL, Chang JS (2021) Prospects and development of algal-bacterial biotechnology in environmental management and protection. Biotechnol Adv 47:107684
Yoshida S, Hiraga K, Takehana T, Taniguchi I, Yamaji H, Maeda Y et al (2016) A bacterium that degrades and assimilates poly (ethylene terephthalate). Science 351(6278):1196–1199
Yu F, Yang C, Zhu Z, Bai X, Ma J (2019) Adsorption behavior of organic pollutants and metals on micro/nanoplastics in the aquatic environment. Sci Total Environ 694:133643
Yu H, Zhang X, Hu J, Peng J, Qu J (2020) Ecotoxicity of polystyrene microplastics to submerged carnivorous *Utricularia vulgaris* plants in freshwater ecosystems. Environ Pollut 265:114830
Yu ZF, Song S, Xu XL, Ma Q, Lu Y (2021) Sources, migration, accumulation and influence of microplastics in terrestrial plant communities. Environ Exp Bot 192:104635
Yu Q, Pei X, Wei Y, Naveed S, Wang S, Chang M et al (2022) The roles of bacteria in resource recovery, wastewater treatment and carbon fixation by microalgae-bacteria consortia: a critical review. Algal Res:102938
Yuan J, Ma J, Sun Y, Zhou T, Zhao Y, Yu F (2020) Microbial degradation and other environmental aspects of microplastics/plastics. Sci Total Environ 715:136968
Yuan F, Zhao H, Sun H, Sun Y, Zhao J, Xia T (2022) Investigation of microplastics in sludge from five wastewater treatment plants in Nanjing, China. J Environ Manag 301:113793
Zabala JA, Martínez-Paz JM, Alcon F (2021) A comprehensive approach for agroecosystem services and disservices valuation. Sci Total Environ 768:144859
Zantis LJ, Borchi C, Vijver MG, Peijnenburg W, Di Lonardo S, Bosker T (2023) Nano-and microplastics commonly cause adverse impacts on plants at environmentally relevant levels: a systematic review. Sci Total Environ:161211
Zhang Y, Gao T, Kang S, Sillanpää M (2019) Importance of atmospheric transport for microplastics deposited in remote areas. Environ Pollut 254:112953
Zhang J, Gao D, Li Q, Zhao Y, Li L, Lin H, Bi Q, Zhao Y (2020a) Biodegradation of polyethylene microplastic particles by the fungus *aspergillus flavus* from the guts of wax moth *galleria mellonella*. Sci Total Environ 704:135931. https://doi.org/10.1016/j.scitotenv.2019.135931
Zhang L, Xie Y, Liu J, Zhong S, Qian Y, Gao P (2020b) An overlooked entry pathway of microplastics into agricultural soils from application of sludge-based fertilizers. Environ Sci Technol 54(7):4248–4255
Zhang K, Hamidian AH, Tubić A, Zhang Y, Fang JK, Wu C, Lam PK (2021a) Understanding plastic degradation and microplastic formation in the environment: a review. Environ Pollut 274:116554
Zhang Q, Zhao M, Meng F, Xiao Y, Dai W, Luan Y (2021b) Effect of polystyrene microplastics on rice seed germination and antioxidant enzyme activity. Toxics 9(8):179
Zhang Y, Gao T, Kang S, Allen S, Luo X, Allen D (2021c) Microplastics in glaciers of the Tibetan plateau: evidence for the long-range transport of microplastics. Sci Total Environ 758:143634
Zhang S, Gao W, Cai K, Liu T, Wang X (2022) Effects of microplastics on growth and physiological characteristics of tobacco (Nicotiana tabacum L.). Agronomy 12(11):2692
Zhao K, Wei Y, Dong J, Zhao P, Wang Y, Pan X, Wang J (2022a) Separation and characterization of microplastic and nanoplastic particles in marine environment. Environ Pollut 297:118773

Zhao L, Ru S, He J, Zhang Z, Song X, Wang D, Li X, Wang J (2022b) Eelgrass (*Zostera marina*) and its epiphytic bacteria facilitate the sinking of microplastics in the seawater. Environ Pollut 292:118337

Zhao M, Xu L, Wang X, Li C, Zhao Y, Cao B, Zhang C, Zhang J, Wang J, Chen Y, Zou G (2023) Microplastics promoted cadmium accumulation in maize plants by improving active cadmium and amino acid synthesis. J Hazard Mater 447:130788

Zheng Y, Li J, Cao W, Jiang F, Zhao C, Ding H, Wang M, Gao F, Sun C (2020) Vertical distribution of microplastics in bay sediment reflecting effects of sedimentation dynamics and anthropogenic activities. Mar Pollut Bull 152:110885

Zhou J, Cao Y, Liu X, Jiang H, Li W (2020) Bladder entrance of microplastic likely induces toxic effects in carnivorous macrophyte Utricularia aurea Lour. Environ Sci Pollut Res 27:32124–32131

Zhu F, Zhu C, Wang C, Gu C (2019) Occurrence and ecological impacts of microplastics in soil systems: a review. Bull Environ Contam Toxicol 102:741–749

Zhu X, Zhao W, Chen X, Zhao T, Tan L, Wang J (2020) Growth inhibition of the microalgae *Skeletonemacostatum* under copper nanoparticles with microplastic exposure. Mar Environ Res 158:105005

Zhuang H, Liu X, Ma H, Li R, Liu B, Lin Z, Li Z (2023) Growth and physiological–biochemical characteristics of cucumber (*Cucumis sativus* L.) in the presence of different microplastics. Arab J Geosci 16(3):194

Ziajahromi S, Neale PA, Silveira IT, Chua A, Leusch FD (2021) An audit of microplastic abundance throughout three Australian wastewater treatment plants. Chemosphere 263:128294

Zolotova N, Kosyreva A, Dzhalilova D, Fokichev N, Makarova O (2022) Harmful effects of the microplastic pollution on animal health: a literature review. PeerJ 10:e13503

Zou J, Wang C, Li J, Wei J, Liu Y, Hu L, Bian H, Sun D (2022) Effect of polyethylene (LDPE) microplastic on remediation of cadmium contaminated soil by *Solanum nigrum* L. J Geosci Environ Protect 10(1):49–64

Chapter 6
Unravelling the Complex Relationship Amongst Microplastics and Microbes and Its Bioremediation

Nandigam Pardhipa and Sukhmeen Kaur Kohli

Abstract Low cost, mechanical strength, and resistance to heat and microorganisms are some factors that contributed to the popularity of plastics in industrial and everyday life. The stability of plastics is a consequence of various additives added to make them more durable. This durability, however, poses a problem as plastics are non-biodegradable and persist in ecosystems. These plastics, when exposed to various environmental factors, fragment to form microplastics. The small particle size and the hydrophobic nature of microplastics facilitate their interactions with microorganisms across multiple habitats. In this chapter, we focus on the ability of microorganisms to colonize microplastics. We discuss the various aspects of this interaction focusing on the alterations in microbial niche, environmental consequences, role of microplastics as vectors for pathogenic microbes, pollutants, and exploring various microbe-mediated bioremediation methods to tackle microplastic pollution.

Keywords Microplastics · Biofilms · Plastisphere · Bioremediation · Marine plastic · Debris

1 Introduction

'Microplastics' are defined as plastic particles having <5 mm in diameter, encompassing solid materials with numerous shapes, structures, density, colours, and origins. Different types of microplastics have varied influences on the environment based on above-mentioned properties and environmental factors like pH, temperature, etc. (Dey et al. 2021). Based on their origin, microplastics could be

N. Pardhipa · S. Kaur Kohli (✉)
Department of Earth and Climate Sciences, Indian Institute of Science Education and Research, Tirupati, Andhra Pradesh, India

K. Khanna et al. (eds.), *Microplastics and Soil Microbiome*,
https://doi.org/10.1007/978-981-96-4978-5_6

characterized into primary or secondary microplastics. The former are present in cosmetics, textiles, and other personalized products, while the latter is a product of the degradation of plastics (Padervand et al. 2020). Low cost, mechanical strength, and resistance to heat and microorganisms are some factors that contributed to the adverse use of plastics in industrial and everyday applications that have contributed to promotion in the production of plastics (Zhang et al. 2021; Laskar and Kumar 2019; Yuan et al. 2020). Zhang et al. (2021) revealed that by 2018, 359 million tonnes of plastic were produced worldwide. Plastic polymers, which are often derived from fossil fuels, are strengthened or have improved properties possibly due to addition of additives such as thermal stabilizers, dyes, and plasticizers, which are further more threatening to the environment (Law 2017; Millican and Agarwal 2021). On the other hand, the incomparable durability of plastics has made them a necessity for humans, where its long life causes plastic pollution and is a major threat to the environment. About 18% of plastics existing in the world is recycled, and 24% are incinerated, while the remaining are in landfills resulting in their entry in the ecosystem (Chamas et al. 2020). Figure 6.1 describes plastics types based on origin and classification of microplastics.

Plastics break into microplastics and nanoplastics due to their interaction with light, oxygen, microbes (Tuuri and Leterme 2023), temperature changes, and mechanical abrasions (Peeken et al. 2018). This increases microplastics and leaching of chemicals in the soil and eventually into the environment (Tuuri and Leterme 2023; Bhagwat et al. 2021; Jahnke et al. 2017). Due to their small size and low density, microplastics have become ubiquitous. Studies reporting ingestion of microplastics by aquatic animals and their impact on the food web, presence of microplastics in human blood and placenta, and the studies involving various animal models highlight the risk these particles carry to an organism's well-being.

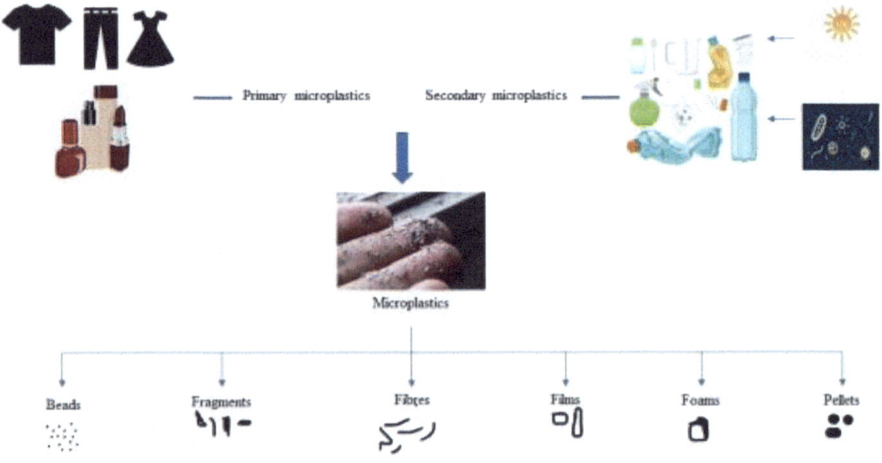

Fig. 6.1 Plastic types based on origin and classification of microplastics according to shape

Microplastics occur in different shapes and colours. These properties are studied using various tools like scanning electron microscopy, Fourier transform infrared (FTIR), and Raman spectroscopy. The research on microplastics gained momentum recently and is a relatively new field. However, the concentration of microplastics across various environments reveals their age-old presence. This chapter focuses on the interactions and relationship between microplastics and microorganisms. In situ and in vitro studies disclose the ability of microorganisms to colonize microplastics. Interestingly, there exists difference in the microbial composition of the microplastic community from their surrounding environment, revealing that origin of microplastic, material type, pre-treatment, and transport are some factors that influence microbial community biofouling the microplastics. These interactions bring changes to both the colonizing microbes and the microplastics. Here, we discuss the changes induced by microorganisms that interact with microplastics and explore the toxic effects of this interaction. At last, we focus on using microorganisms in bioremediation of microplastics to remove microplastics from the environment in a sustainable, non-toxic, and economical approach.

2 Composition and Distribution of Microplastics

Microplastics are studied by combining physical and chemical methodologies. Given their small particle size, a naked-eye examination is fruitless. Microscopy helps in studying the structural properties such as shape, size, and number of microplastics. On the other hand, scanning electron microscopy generates high-resolution images that help differentiate between microplastics and other organic materials. Von Moos et al. (2012) used polarized optical microscopy to identify polyethylene particles. They also suggest microplastics are hydrophobic and can be stained with Nile red and these particles can be visualized under a fluorescent microscope and the only drawback of employing Nile red is its ability to stain other hydrophobic organic materials and lipids. Chemical characterization is crucial for validating the nature and composition of the polymer. Fourier transform infrared (FTIR) and Raman spectroscopy are widely used for chemical characterization. Thermogravimetry (TGA), in addition to thermal desorption gas chromatography-mass spectroscopy (TDS-GC-MS), pyrolysis GC-MS, and atomic force microscopy are also applicable in microplastic studies (GESAMP 2019; Dehghani et al. 2017). Figure 6.2 is a pie graph representing percentage usage of various detection methods in microplastic research.

Based on their origin, microplastics could be either primary or secondary microplastics. Primary microplastics are micron-sized particles incorporated in glitters (Meral 2019) cosmetics and personal care products, paints, etc. The increasing plastic debris found in coastal areas, tourist places, and landfills is exposed to UV radiation, temperature fluctuation, microbial colonization, and physical abrasion that leads to the development of pits and cracks. Subsequently, the plastics disintegrate to form secondary microplastics (Kokalj et al. 2019). Studies worldwide report

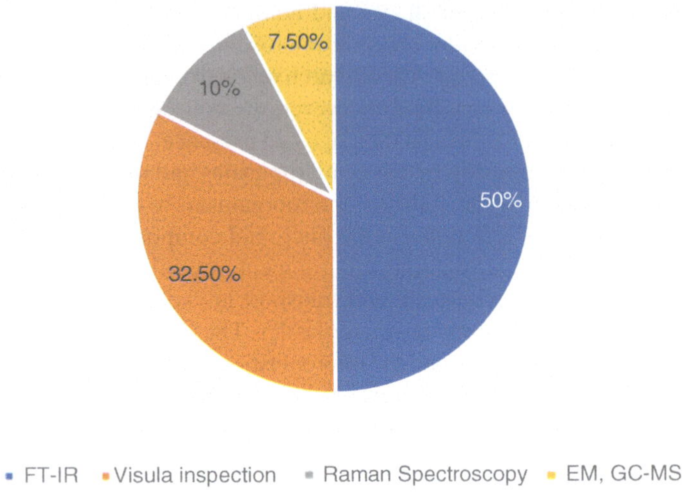

Fig. 6.2 Percentage prevalence of detection methods in microplastic research (*FTIR* fourier transform infrared, *EM* electron microscopy, *GC-MS* gas chromatography-mass spectroscopy)

microplastic shapes ranging from three-dimensional spheres, foams, and granules to two-dimensional films (Rosal 2021). Primary microplastics are typically beads or pellets. Fibres are the predominant type of secondary microplastics, followed by fragments (Masiá et al. 2020). Fragment, fibres, films, foam, and beads are the predominant microplastic types found in wastewater treatment plants (Bayo et al. 2020). Of these, low-density polyethylene films were predominant. Hartmann et al. (2019) classified planar microplastics as films, expanded porous materials to be foams, and identified thread like microplastics as fibres and filaments. Pellets, spheres, and spheroids are some other reported microplastic shapes (Rosal 2021).

Microplastics are added to the environment in response to certain anthropogenic activities. These are spread by airborne dust, precipitation, laundry waste, landfills, and wastewater treatment facilities (Dris et al. 2018; Tadsuwan and Babel 2021; Pikoli et al. 2022). Plastics have a relatively low density, which makes many of them buoyant in freshwater and marine environments, making them easily carried by currents (Sun et al. 2022). Wind currents carry the microplastics to the atmosphere, integrating them with tiny particles like dust (Dehghani et al. 2017). Currents also play a role in depositing these particles across terrestrial, marine, and aquatic environments (Xu et al. 2020). The spreading of microplastics to varied aquatic (marine or freshwater systems) and atmospheric regions are directly dependent on varied factors such as interaction between particles and fluids which further determines their buoyancy or settling down (Rosal 2021). Wastewater treatment plants (WWTPs) are hotspot for microplastics. Hot temperatures accelerate microplastic degradation. Bayo et al. (2020) showed seasonal variability as an important factor contributing to increased microplastics. Agricultural and industrial activities, instead of population, influence microplastic concentration in wastewater treatment plants. The microplastics removed from WWTPs pile up in sewage sludges. These sludges when applied to soil introduce microplastics to soil (Masiá et al. 2020).

Microfibres from textile sewage contribute to 35% of the microplastics identified in the marine environment (Xu et al. 2020). Moreover, plastic microbeads are incorporated as exfoliants in scrubbers, liquid soaps, and makeup items. The analysis of 37 personal care products by Habib et al. (2020) in UAE revealed the presence of polyethylene microbeads in 11 of the products. These microplastics pass through wastewater treatment plant and enter the sludge, where they are incinerated. Also, Law et al. (2010) demonstrated how plastic marine debris could reach lands far off human activity (North Atlantic Subtropical Gyre) within short periods (< 60 days) using drifter buoys and oceanographic models. Due to insufficient export chances, buoyant plastic particles accumulate in central ocean gyres and may exist for decades (Zettler et al. 2013).

Extensive use of plastic mulch and microplastics in irrigation water might also introduce microplastics to soil (Lu et al. 2019). Plastic mulch uses polyethylene films (<0.008 mm thick) to cover the soil. Plastic mulch decreases erosion and increases the soil's ability to hold moisture. The problem with plastic mulch is that it is easily damaged yet tricky to recycle (Chen et al. 2022). Sewage sludge particles, microplastics from landfills, and microplastics from tyre abrasion can be carried by wind and be deposited on the soil. Leaching allowed microplastics from top soil (0–30 cm) to travel vertically in deep soils via macrospores and mesopores (Kumar et al. 2022). Eventually microplastics reach underground water system by the activity of soil animals like earthworms, surface runoff, or infiltration of water (Sajjad et al. 2022). One of the imperative examples of microplastic pollution is fire outbreak at X-Press Pearl near the coast of Sri Lanka in the year 2021 that is a reminder that environment catastrophes also play a role in accelerating microplastic pollution. About 78 metric tonnes of plastic was deposited on the shores around the coastal areas of Sri Lanka. There is evidence of marine organisms like fishes ingesting this half burnt plastic pellets. The impact of this disaster is still under investigation, but, the massive amounts of plastic spell detrimental consequences on the environment (https://www.unep.org/news-and-stories/story/oil-acid-plastic-inside-shipping-disastergripping-sri-lanka).

3 Microbial Colonization and Biofilm Formation on Microplastics

Microorganisms are ubiquitous in nature. Their ability to adapt to diverse conditions makes them critical components of the ecosystem. Augmenting microplastics in the environment has promoted interactions between microplastics and microorganisms. These complex interactions vary depending on (i) the polymer morphology, material, additives, and spatial effects (De Tender et al. 2017), (ii) environmental conditions like pH and temperature, and (iii) characteristics of the microbes (Qiu et al.

2022). Microplastics are emerging as habitats for microorganisms (Lu et al. 2019). A biofilm is a consortium of microorganisms of one or more species anchored to a surface and encased in an extracellular polymeric matrix (Donlan 2002). This matrix is formed from the exopolysaccharides, proteins, lipids, and other compounds secreted by the microorganisms integral to the biofilm (Berlanga and Guerrero 2016). Furthermore, the matrix includes metabolites, particles, or inorganic substances in the environment (Donlan 2002). Mobile genetic elements (MGE) and horizontal gene transfer (HGT) in biofilms impact the diversity of microbes and various biochemical pathways between the microbes-microbes and microbes-environment interplay. HGT is also responsible for spreading antibiotic resistance genes (ARGs) in the surroundings (Arias-Andres et al. 2019). Microplastics provide a surface for forming biofilms in various environments affecting the ecosystem and posing a risk to public health. The hydrophobic environment of microplastics favours growth and colonization of early colonizers and accelerates the formation of biofilms and the subsequent acquisition of other microbes (Zettler et al. 2013). The formation of a distinct microbial niche on microplastics could also be influenced by their transportation. If a plastic colonizing microorganism is able to retain on the microplastics despite varying environmental conditions, it would lead to formation of a microbial community distinct from the surrounding environment from which the microplastic has been sampled from (De Tender et al. 2015), for example, a microplastic surface exposed to seawater in short period of time. Consequently, it results in formation of conditioning film which attracts more microbes and aid in usage of nutrients available on microplastic surfaces (Oberbeckmann et al. 2015).

Furthermore, formation of biofilm directly influences the chemical and physical characteristics of microplastics specifically surface morphology, its sturdiness, charge on surface area, as well as density. Alteration in these properties impacts the migration of pollutants, weathering of pollutants, sorption and desorption of pollutants and pathogens adhered to microplastics in marine ecosystem. Therefore, De Tender et al. (2015) investigated these wide aspects which affect bacterial colonization including (i) effect of environmental attributes such as physiological and chemical characteristics including available oxygen, temperature and salinity, (ii) effect of varied stages of biofilm formation process as it is an extremely dynamic phenomena, and (iii) effect of physio-chemical characteristics of plastic which directly influences the adherence of microbes to plastics. The interaction can either be a strongly hydrophobic surface providing anchorage to bacteria or provide essential nutrients for metabolically degrading a target species. For instance, high-density polyethylene (HDPE) has an irregular surface making it easier for microbes to adhere quickly compared to the flat surfaces of polystyrene and polypropylene (Chen et al. 2022). Corroborating with Chen and co-worker findings, Cai et al. (2019) affirmed that bacteria easily colonize polyethylene and polyvinylchloride compared to polypropylene and polyethylene terephthalate (PET).

Microplastics float in marine environments and are an attractive substrate for microorganisms and invertebrates and are shown to be a vector capable of transporting algal blooms and organic pollutants (Zettler et al. 2013). Zettler and co-worker further coined the term 'plastisphere' to address the diverse microbial habitat of heterotrophs, autotrophs, predators, and symbionts that colonize marine plastic debris. The microbial community available in seawater was compared with the ones found on polypropylene and polyethylene materials extracted from the seawater. Phototrophic cyanobacteria, *Phormidium* and *Rivularia*, absent in the seawater, were found in the plastisphere. DNA sequencing data revealed the existence of biofilm-forming diatoms like *Navicula* and *Nitzschia*. The presence of *Vibrio* operational taxonomical units (OTUs) in high concentrations on the plastic surface indicated the role of microplastics in spreading diseases. Also, Tu et al. (2020) observed increment in biofouling on polyethylene microplastics with increased time in seawater. This phenomenon, however, decreased with the increasing depth of seawater. The spatial arrangement of the biofilm showed a marked difference with exposure time, the number and type of living cells, extracellular polysaccharides (EPS), and dead cells increased by Day 135. Most of the microbes living on the polyethylene microplastics belong to the class *Alphaproteobacteria*, *Gammaproteobacteria*, and *Bacteroidia*, contributing to over 50% of the total microbial communities.

McCormick et al. (2014) collected microplastics from a river in Chicago, which received effluents from a nearby wastewater treatment plant. Scanning electron microscopy (SEM) imaging revealed the absence of fungal hyphae and algal cells in the biofilms formed on microplastics. Using 16S rRNA sequencing, the bacterial assemblages from upstream of the river (before entry of effluents into the river) and downstream of the river and concentration of organic matter and microplastics were compared. The bacterial communities varied significantly from one another, specifically *Pseudomonadaceae*, *Campylobacteraceae*, and *Proteobacteria*, which were mostly found to adhere to microplastics. *Arcobacter*, *Aeromonas*, and unclassified *Veillonellaceae* were higher on plastic substrate surfaces than non-extracellular polysaccharides plastic substrates. The diversity of microplastic biofilms was less than the bacterial diversity observed in water column and organic sediment. This could indicate selective colonization of bacteria on microplastics. Similarly, findings of Miao et al. (2019) and McCormick et al. (2016) affirmed that microplastic biofilms have lower alpha diversity than the ones formed on natural substrates. The higher abundance of *Pseudomonadaceae* and *Campylobacteraceae* in microplastics might be an indicative that microplastic transported through the domestic wastewater stream which creates an opportunity for colonization in water with wastewater-associated microorganisms and furthermore provides a vehicle for the transport of these organisms within aquatic ecosystems.

Oberbeckmann et al. (2021) conducted an in situ experiment, where polyethylene, polystyrene, and wooden beads were exposed to different coastal and estuarine locations in Baltic Sea, Germany. The locations varied in salinity, temperature, and nutrient regimes. The incubation period was restricted to 2 weeks to prevent formation of a secondary biofilm on the mature biofilm. This was to avoid the attachment of microbes to the biofilm rather than to the carrier. Metagenomic and metaproteomic profiles of the biofilm revealed that polyethylene and polystyrene served as colonization surface rather than as a nutrient source. The microbial profile of polystyrene and polyethylene shared greater similarity than with the wooden beads.

Also, the localities had significant influence on microbial community. Amaral-Zettler et al. (2013) showed species richness differs with latitude coordinates and their colonization. *Flavobacteriaceae*, *Rhodobacteraceae*, and *Erythrobacteraceae* were mostly found on the plastics from coastal region. Low-density polyethylene (LDPE) was pre-treated by UV irradiation before being introduced into lake water. The lake water contained surfactants including Triton X-100, Tween 20, or Tween 80. Analysis of microbial community associated with LDPE microplastics revealed that they were primarily colonized by *Proteobacteria*, *Bacteroidetes*, and *Firmicutes*. The surface roughness and hydrophobicity of microplastics were imperative factors shaping the LDPE microplastic-associated microbial community structure. Comparing untreated LDPE, UV pre-treated LDPE and LDPE exposed to surfactants, it was affirmed that *Moraxellaceae* were largely colonized in LDPE exposed to surfactants and Enterobacteriaceae was higher in UV irradiated sample compared to the untreated LDPE (Gong et al. 2019).

4 Effects of Microplastic-Microorganisms Interaction

Small particle size and easy transport make it easy for microplastics to settle in various habitats and integrate into the environment. The ubiquitous nature of microorganisms, in particular, bacteria, promote interaction between microorganisms and microplastics. This section focuses on the consequences of these interactions in various environments.

4.1 Dysbiosis in Gut Microbiota

Microplastics enter the human through inhalation and ingestion, and its small size facilitates to cross biological barriers. Microplastics in the gut interfere with nutrition uptake and gut epithelial villi. They promote dysbiosis that enhances the proliferation of certain microbes in hostile environments. For instance, the *Enterobacteriaceae* species are augmented in gut in response to inflammatory gut (Fackelmann and Sommer 2019). Studies on the interaction between the gut microbiota of zebrafish and polystyrene revealed that inflammation in the gut is due to alterations in the gut microbiota of zebrafish (Lu et al. 2019; Jin et al. 2018; Qiao et al. 2019). An increase in firmicutes and decrease in *Bacteroidetes* and *Proteobacteria* phyla was observed when polystyrene particles were exposed to the gut microbiota of adult male zebrafish (Jin et al. 2018). Similarly, intestinal barrier dysfunction and hepatic lipid metabolism disorder were observed in mice when the gut microbiota was exposed to polystyrene (Lu et al. 2018). Also, polystyrene microplastic exposure lowered the *Actinobacteria* content in the rat's microbiota, recording a significant change in 15 types of bacteria which was recorded at genus level (Jin et al. 2019). Ingestion of microplastics carrying microorganisms could induce gut dysbiosis by introducing a new species of microbe, in particular, bacteria to the gut microbiome. This creates competition for resources and could have a negative effect on the human health (Fackelmann and Sommer 2019).The resultant disturbances in gut microbial composition affect metabolite production and pathway. For instance, exposure of microplastics to mice gut microbiota ensued an increase in *Firmicutes* and decrease in *Bacteroidetes*. This dysbiosis is a feature associated with obesity. The ability of microplastics to act as a vector may introduce pathogens to the gut microbiota. Further, a disruption in production of bile acids and short chain fatty acids was observed in mice exposed to microplastics (Tu et al. 2023). Similarly, Nugrahapraja et al. (2022) studied the effects on microplastic contamination on human gut microbiota. Although some changes were observed in the abundance of certain bacterial species like *Clostridium*, the overall diversity remained the same. An interesting observation was the presence of plastic degrading genes like *feaB* that synthesizes phenylacetaldehyde dehydrogenase enzyme, capable of breaking down styrene in many bacteria.

4.2 Microplastics as Vectors of Pathogens

Harmful pathogenic species like *Vibrio* (Zettler et al. 2013; De Tender et al. 2015) and members of *Pseudoalteromonadaceae* are commonly detected on microplastics despite their limited presence in surrounding seawater or sediments (De Tender et al. 2015). *Vibrio parahaemolyticus* which, upon ingestion causes gastrointestinal illness in humans, was found to be a component of the plastisphere of North and Baltic Sea (Kirstein et al. 2016). Also, Masó et al. (2003) affirmed the occurrence of hazardous dinoflagellate species, i.e. *Alexandrium* specifically present on marine plastic waste and debris in the Mediterranean regions. Biofilms on polystyrene biofilms recovered from landfill leachates included opportunistic pathogens like *Ochrobactrum anthropi*, *Acinetobacter lwoffii*, *Afipia broomeae*, *Acinetobacter baumannii*, *Pseudomonas aeruginosa*, and pathogenic *Escherichia coli*. Transportation of microplastics across various trophic levels, evidence of human ingestion, and horizontal gene transfer amongst the members of biofilms may aggravate the disease-causing potential of the colonizing bacteria, and hence, microplastics are seen as potential vectors of pathogens (Bowley et al. 2021).

4.3 Influence of Microorganisms on Microplastic

Varied levels of sedimentation of microplastics occurs in response to fast and steady colonization of microorganisms and aggregation of microplastics with specifically microalgae and microbes which consequently leads to elevated buoyancy of aggregates and hence augments sedimentation (Lagarde et al. 2016). Similarly, when polyethylene was placed in seawater for a long time, the submerged piece had lower hydrophobicity and higher central buoyancy (Wang et al. 2021). Another report by Rummel et al. (2017) revealed that biofilms formed on microplastics influence their vertical transport and weathering rate in water columns. Furthermore wide array of reports suggests presence of polymer-degrading bacteria in biofilms (Tu et al. 2020). Biofilm formation on microplastics is known to enhance the hydrophilicity of microplastics (Tu et al. 2020). On the other hand, formation of hydrophilic functional groups is a marker for biodegradation.

4.4 Effects on Biogeochemical Cycling

Different microplastics have different effects on carbon and nitrogen cycling. The considerable rise in the relative abundance of *Burkholderia* indicates that microplastics may stimulate nitrogen-fixing bacteria (Huang et al. 2019). Microplastics affect the structure and diversity of microbial community in the sediment; this may directly influence the functions of microorganisms. The association between the microbial community and nitrogen-cycling generated by microplastics was investigated by Seeley et al. (2020) who suggested that different microplastic types had variable impacts. They further revealed that that polyvinylchloride (PVC) inhibits nitrification and denitrification. Contemporary studies also affirmed the addition of polyethylene microbeads raised the ammonium concentration and changed the nitrogen (N) cycling, which may result in eutrophication (Wang et al. 2021). The plentitude of biofilms also impacts phosphorous (P) in terrestrial river as well as marine ecosystems. Presently, a very scarce number of report are present on P cycling being affected by presence of microplastics. Chen et al. (2020) revealed occurrence of microplastics and microplastic biofilms on P cycle in microcosms specifically. They affirmed that an elevation in total P concentration of water after 25 days of cultivation was observed. Furthermore, they revealed augment in alkaline phosphatase activities in biofilms during the time of cultivation.

About 95% of organic carbon in ocean is present in the form of dissolved organic carbon (DOC) sequestered in the ocean. The 'microbial carbon pump' theory states that marine microbes play an imperative role in converting the dissolved organic carbon to recalcitrant dissolved organic carbon (Jiao and Zheng 2011). Biofouled 'microplastics' sink to the ocean floor due to the increase in density. Similarly, in the pelagic and benthic zones, surface-attached bacterial populations create flexible metabolic cooperation networks, and the biofilms are well-known metabolic hotspots capable of degrading dissolved organic carbon. Another study by Yu et al. (2022) showed that exposure to polyethylene microplastics alters the enzyme activity of N cycle-related enzymes including ammonia monooxygenase and nitrate reductase due to alterations in the composition and metabolic profile of microbes in freshwater ecosystems. Microplastic exposure results in a decrease in bacterial communities colonization and enzymatic activities (Yi et al. 2021; Zhang et al. 2021). Correlation analysis done by Zhang et al. (2021) indicated that microplastic richness decreased soil microbial biomass. For example, plastics and microplastics are susceptible to UV radiation, mechanical abrasion, and other physico-chemical factors. This results in production of nanoplastics that can be taken up by microbial cells, disturbing the metabolic interactions and inducing stress that is detrimental to the microbe. Further, the soil microorganisms will use more substrates for energy production and synthesis of redox active compounds under toxic stress as a result of the malfunction, protein denaturation, and cell membrane damage caused by augmented levels of microplastics. This subsequently speeds up C and P metabolism to lessen the toxicity of microplastics, hence aggravating nutrient limitations. To counteract these limitations, microbes take up nutrients from the external environment. This results in a deficiency of nutrients in the soil environment (Zhang et al. 2021).

The effects of low-density polyethylene (LDPE) on the microorganisms of the top soil layer by setting up a microcosm experiment were done by Rong et al. (2021). The results revealed an increase in *nifH* genes (codes for Nitrogenase), indicating an augment in *nifH* expressing bacteria. Also, ammonia oxidizing archeae were susceptible to LDPE, and this was recorded in the form of decrease in *amoA* genes. Furthermore enhancement in *amoA* genes corresponding to ammonia oxidizing bacteria indicated that these bacteria were predominant involved in oxidation of ammonia in the microplastic contaminated soil. Increase in denitrifying bacteria including *Pseudomonas*, *Streptomyces*, and *Methylobacterium* corresponded with an elevation in one of the denitrification marker gene *nirK*. There was a reduced expression of *nirS* (nitrite reductase gene). Hence, bacteria encoding *nirK* are more active in converting nitrite to nitrogen oxide in LDPE containing soil.

5 Microplastics as Carriers of Pollutants and Their Influence on Microorganisms

Microplastics act as carriers for two types of chemicals: (i) one is the additives that plastics acquire at the time of their manufacture. These include plasticizers, UV stabilizers, flame retardants, anti-oxidants, and dyes, and (ii) second is taken up from their surroundings (Campanale et al. 2020). Additives prolongs the lifespan of plastics and enhance the resin's properties, and they become contaminants once they enter the ecosystem due to inappropriate plastic disposal. For instance, most colorants include heavy metals, which are toxic to microorganisms and affect plant growth and influence the life of various organisms found in soil.

5.1 Effect of Chemical Pollutants

Flame retardants, bisphenol A (BPA) used in food packaging, are endocrine-disrupting chemicals humans can be ingested (Cingotti and Jensen 2019). Studies on rats have shown that exposure to BPA can cause long-term alterations in the

Table 6.1 Description of various additives which enhance durability and adversity of plastic and microplastics

S. no.	Additive	Function	Examples
1	Inert fillers	Enhance the strength and have a reinforcing effect	Graphite, silica, carbon black, kaolin
2	Plasticizer	Increased flexibility, plasticity, and impact resistance of the product	Phthalic acid esters
3	UV stabilizers	Protect from oxidation and thermal degradation	Organic and inorganic cadmium, barium, or lead salts
4	Dyes	Give colour to the resin	Azo pigments, anthraquinone chromophores
5	Flame retardants	Cool the material by forming a layer of ash or prevent oxidation of flammable gases	Triphenyl phosphate, hexabromocyclododecane (HBCD)
6	Lubricants	Prevents plastic from adhering to equipment to reduce friction	2% graphite, glyceryl monostearate

central nervous system, even at levels below what is considered safe for humans to consume daily (Fackelmann and Sommer 2019). Certain additives, specifically BPA, are endocrine disruptors that bind to microbial hormone receptors or interfere with hormone signalling between host and the gut microbiome, thus causing dysbiosis (Fackelmann and Sommer 2019). Table 6.1 demonstrates varied additives which are responsible for enhancement of adverse effect of plastics.

Various compounds present in the environment interact with microplastics and several studies have focused on the sorption of trace organic compounds (TrOCs) which are organic contaminants, typically released from pharmaceuticals, pesticides, and personal care products into the environment. These are detected in sub-ppm concentrations in soil, rivers, groundwater, and wastewater. However, the uptake of these chemicals by microplastics depends on multiple factors. Upon ingestion by different organisms, the pollutants cause toxic effects and affect the organism's health. These pollutants may reach higher trophic animals by bioaccumulation at multiple levels of the food chain (Rubin and Zucker 2022). Also, environmental factors (including salinity, pH, and temperature) and sorbent-sorbate properties (microplastic material, morphology, and size) influence the nature and extent of organic compounds carried by the microplastics. Physicochemical changes induced by weathering of microplastics are another phenomenon that dictates their interaction with various environmental pollutants. Microbial colonization, surface area to volume ratio, increased hydrophilicity, and chemical modifications like adding a carbonyl group influence the interaction between microplastics and pollutants (Rubin and Zucker 2022; Sarkar et al. 2021). An alternative cause of chemical induced dysbiosis is because chemicals like dichlorodiphenyltrichloroethane (DDT), polycyclic aromatic hydrocarbons, and other persistent organic pollutants (POPs) gain entry into gut by adsorption on microplastics. The bioavailability of these pollutants once they have reached the gut depends on a number of variables, including the fugacity gradient, the length of contact, and the desorption capacity, which is affected by the particle's size, shape, weathering stage, biofilms, polymer type, and surface polarity (Koelmans 2015; Fackelmann and Sommer 2019). Table 6.1 represents description of various additives which elevate the durability as well as adversity of plastics.

5.2 Microplastics as Carriers of Antibiotic Resistant Genes (ARGs)

Arias-Andres et al. (2018) demonstrated that bacteria associated with microplastics show greater horizontal gene transfer than planktonic bacteria or natural aggregates. Molecular analysis revealed the presence of antibiotic genes in microplastic biofilms (Pham et al. 2021). The sulfonamide-resistant genes *sul1*, *sul2*, and mobile genetic element *intl1* showed high abundance in polystyrene and polyethylene microplastics in activated sludge at wastewater treatment plant (Pham et al. 2021). Microplastics are carriers of pathogenic microbes; hence the acquisition of antibiotic resistance genes by these microbes would decrease the antibiotic susceptibility and affect treatment regimes. A study on adsorption of sulfamerazine, tetracycline, chloramphenicol, and tylosin on polyethylene microplastics in river, estuary, and marine waters revealed that microplastics promote antibiotic adsorption. It was also observed that greater salinity reduces the adsorption capacity. qPCR studies indicated the presence of ARGs-*sul1*, *tet*A, *tet*C, *tet*X, and *ermE* on plastic particles in both river and seawater (Wang et al. 2020). The genes *strB* and *blaTEM* were maximally enriched, and mefA, *ermB*, *tetM*, and *tetQ* were slightly enriched on microplastics recovered from landfill leachates (Shi et al. 2021).

6 Microbe-Mediated Bioremediation Strategies

Microplastic pollution needs to be tackled at various levels to put an end to the problem in an efficient manner. At the source level, adopting biodegradable and compostable plastics is the first step. The majority of plastic pollution occurs due to improper disposal of single-use plastics. An ideal method of removing microplastics from the environment should be cost- effective and sustainable. Long-term UV radiation on microplastics induce reactive oxygen species (ROS) that affect the morphology and induce chemical chain scission of microplastics (Wang et al. 2013; Guo and Wang 2019), eventually resulting in release of volatile organic compounds (VOC) or toxic intermediate compounds that may cause various environmental issues (La Nasa et al. 2021). The degradation of compounds and substrate, achieved by employing living organisms specifically microorganisms, is termed as biodegradation. Generally, the soluble components or by-products of biodegradation including low molecular weight constituents are adsorbed and sequestered by microorganisms. Degradation of pollutants by microbes can be categorized into complete and partial degradation. In case of complete degradation production of CO_2 occurs and is termed as biomineralization. And, as a result of degradation of organic matter in absence of terminal electron acceptor results in formation of methane and other short-chain hydrocarbons specifically under reduced environment (Zeghal et al. 2021). Naturally in environment, the process of biodegradation is catalysed by enzymes and biological compounds such as peroxides and acids secreted by microbes (Zeghal et al. 2021).

Microorganisms adapt to any environmental circumstance (Brooks et al. 2011; Aujoulat et al. 2012) and have the potential to utilize microplastic polymers. Therefore, a long-term polluted environment can be a source of microorganisms that break down microplastics (Pikoli et al. 2022). In laboratory experiments and natural environments, microbes form biofilm on the surface of the microplastics, treating the microplastic as a substrate. In many experiments, especially those using single culture of fungi, microplastic was often the primary source of carbon. Forming biofilm leads to pits and cavities that weaken the microplastics (Du et al. 2021). Secretion of extracellular enzymes by the microbial biofilm helps in weakening the microplastics and producing metabolites that can be taken up by the microbes embedded in the biofilm (Jeyavani et al. 2021). The biodegradation ability of microbes is studied based on gravimetric weight loss (%), and the structural and morphological changes to microplastic surfaces are analysed using FTIR or SEM (Yuan et al. 2020).

6.1 *Single Bacterium Cultures to Degrade Biofilms*

Bacteria are ubiquitous present, and several research studies testify the ability of plastics in interaction with bacteria to degrade pollutants. Pure bacterial cultures from plastic- rich zone like sludges, landfills, wastewater plants, or culture collections have been commonly employed in investigating bacterial-dependent biodegradation of microplastics. The advantage of using a single isolate is to investigate metabolic pathways and which are easily monitored and detect even minute alterations in microplastics (Yuan et al. 2020). However, a pure bacterial culture could produce toxic end products that limit bacterial growth (Dobretsov et al. 2013). *Bacillus paramycoides* isolated from leachate-filled landfills and identified by 16S rRNA sequencing degraded polystyrene microplastics with a degradation percentage of 11.12% in 42 days (Pikoli et al. 2022). Similarly, Auta et al. (2018) observed a significant increase in pH in the in Bushnell Haas (BH) broth when *Bacillus* sp. strain 27 and *Rhodococcus* sp. strain 36 were inoculated with polypropylene to assay microplastic biodegradation. The augment in pH, from 6.76 to 9.51 for *Bacillus* species and from 8.51 to 9.53 in *Rhodococcus*, could be due to the production of basic aromatic compounds.

Using C balance and 13-C tracer studies, Yang et al. (2015a, b) discovered that the larvae of *Tenebrio molitor* Linnaeus (commonly known as mealworms) could ingest Styrofoam. About 47.7% of the mineralized plastic produced CO_2, and a small fraction integrated into lipid biomass. The gut microbiota of these worms was suppressed using gentamicin antibiotic. It was observed that the gentamicin-treated mealworms could not mineralize polystyrene (Styrofoam). *Exiguobacterium* sp. strain YT2 isolated from the gut of mealworms could form biofilm on the surface of polystyrene particles in the liquid carbon-free basal medium. SEM imaging revealed the formation of pits and cavities in the polystyrene 28 days post biofilm formation. A degradation of $7.4 \pm 0.4\%$ of polystyrene was observed over 60 days of incubation (Yang et al. 2015a, b). In all the above studies, carbonyl bond formation was observed in the degraded polymers. Further, the hydrophobicity of microplastics decreased. Decreasing hydrophobicity indicates increasing susceptibility to microbial colonization and biodegradation (Yang et al. 2015a, b). Table 6.2 represents pure bacterial and fungal culture having role in biodegradation of microplastics in environment.

6.2 *Fungi in Biodegradation of Plastics*

Fungi have diversified lifestyles and can adapt to different redox conditions (Zeghal et al. 2021). They are morphologically classified as yeast, filamentous fungi, or dimorphic fungi. These organisms can be saprotrophs (decomposing dead material), obligatory (decomposers), or opportunistic organisms (obligatory, mutualists, or pathogens) (Sánchez 2020). Enzymes secreted by fungi degrade plastic polymers into monomers and produce carbon dioxide and water under aerobic conditions, whereas methane is produced under anaerobic conditions. Moreover, fungi secrete hydrophobins (surface proteins) that aid bioremediation by elevating substrate mobility and bioavailability (Solanki et al. 2022). The naturally occurring marine fungus *Zalerion maritimum* was observed to degrade polyethylene pellets (Paço et al. 2017). It was further suggested by Paco and co-workers that reduction in protein and carbohydrate augments in response to exposure time and it might be attributed to employment of microplastics as nutrient sources.

Aspergillus flavus isolated from the guts of the wax moth *Galleria mellonella* reduced the weight of high-density polyethylene (HDPE) by about 3.9% in 28 days. The FTIR data indicated the emergence of hydroxyl, carbonyl, and ether groups, proving that HDPE microplastics were bio-oxidized and aided in the biodegradation of polyethylene (Zhang et al. 2020). Extracellular polyurethenase enzyme was produced by *Pestalotiopsis microspora* to degrade polyurethane in anaerobic conditions (Russell et al. 2011).

Table 6.2 Pure culture bacteria and fungi reported in biodegradation of microplastics in the environment

S. no.	Microplastic type	Microbe	Microbe strain	Microbe type	Source	Duration (in days)	Weight-loss %	Mode of Action	References
1	Polystyrene (PS)	*Bacillus paramycoides*	CY1-1PE(I)	Bacteria	Leachate	40	11.12	Lipase	Pikoli et al. (2022)
2	Polyethylene (PE)	*Bacillus cereus*			Mangrove	40	1.6		Auta et al. (2017)
3		*Bacillus gottheilii*					6.2		
4	Polystyrene (PS)	*Bacillus cereus*					7.4		
5		*Bacillus gottheilii*					5.8		
6	Polyethylene terephthalate (PET)	*Bacillus cereus*					6.6		
7		*Bacillus gottheilii*					3		
8	Polypropylene (PP)	*Bacillus gottheilii*					3.6		
9		*Bacillus*	27			40	4		Auta et al. (2018)
10		*Rhodococcus*	36				6.4		
11	Polystyrene (PS)	*Exiguobacterium*	YT2		Gut of mealworm	60	7.4 ± 0.4	Biofilm formation	Yang et al. (2015a, b)
12	Polyethylene pellets (PE)	*Zalerion maritimum*		Fungi	Marine water	28		Multiple enzymes	Paço et al. (2017)
13	High-density polyethylene (HDPE)	*Aspergillus flavus*	PEDX3		Gut of wax worm	28	3.9025 ± 1.18%		Zhang et al. (2020)
14		*Aspergillus tubingensis*	VRKPT1		Marine coastal area	30	6.02 ± 0.2%		Devi et al. (2015)
15		*Aspergillus flavus*	VRKPT2		Marine coastal area	30	8.51 ± 0.1%		
16	Polyurethane (PUR)	*Pestalotiopsis microspora*			Anaerobic (25 °C)			Serine hydrolase	Russel et al. (2011), Dey et al. (2021)

6.3 *Other Bioremediation Technique*

Many studies demonstrate the capability of pure culture bacteria and fungi in microplastic degradation. Other than these, slowly different strategies are being developed to perform bioremediation efficiently. The biodegradation of low-density polyethylene (LDPE) microplastics and polystyrene microplastics by a mixed culture of *Bacillus* species and *Delftia acidovorans* yielded better biodegradation than pure cultures of *Bacillus licheniformis* and *Lysinibacillus massiliensis* (Kučić Grgić et al. 2021). Microplastics comprise different materials and most bioremediation approaches are specific to a particular polymer. This is a drawback as multiple polymer types are observed in an environment. A recent progress towards countering this drawback was experimented by Liu et al. (2021). In this experiment, *Pseudomonas aeruginosa* form bio aggregates with microplastics with the help of its sticky matrix, and following a 'trap and release' strategy, microplastics are discharged for recycling. The *P. aeruginosa* was engineered such that microplastics are trapped in the absence of arabinose, and presence of arabinose triggers biofilm dispersion and release of microplastics. This technique has several advantages such as it is (i) environment friendly and (ii) not limited to a microplastic polymer and (iii) helps free an environment from microplastic pollution. Bioengineering microbial strains to enhance their degradation efficiency in different environmental conditions is important to optimize remediation strategies (Jeyavani et al. 2021). More recently, co-metabolization of toxic chemicals and pollutants by microbial communities has been tested for petroleum hydrocarbons (Tran et al. 2021). Exploration of this method may help in developing better microplastic bioremediation strategies.

7 Conclusions and Future Prospects

Small size, hydrophobic nature, and widespread presence of microplastics make them easily accessible to microorganisms in landfills, soil, and aquatic and marine environments. Despite their ubiquitous presence, the research on microplastics is still in its infancy. Currently there are no reported threshold levels that can categorize different microplastic concentrations as safe or threat to organisms. The current research is indicative of the ability of microplastics-microorganisms interaction to play a role in gut microbiome dysbiosis, induce changes in the biogeochemical cycles, and play the role of vector carrying pathogens, heavy metals, and harmful chemicals. In addition, the interactions of microplastics with higher trophic organisms necessitate the removal of microplastics from various environments. Despite its economical and environment friendly approach, the bioremediation approach using single culture microbes is a slow process. More research focusing on microbial consortia and co-metabolism approach is necessary in this field. Bioengineering microorganisms need to be employed to increase the efficiency of bioremediation. Further, most research papers consider gravimetrical weight loss as a standard for checking bioremediation efficiency. Instead verifying complete mineralization of microplastics is a more valuable approach.

References

Arias-Andres M, Klümper U, Rojas-Jimenez K, Grossart HP (2018) Microplastic pollution increases gene exchange in aquatic ecosystems. Environ Pollut 237:253–261

Arias-Andres M, Rojas-Jimenez K, Grossart HP (2019) Collateral effects of microplastic pollution on aquatic microorganisms: an ecological perspective. TrAC Trends Anal Chem 112:234–240

Aujoulat F, Roger F, Bourdier A, Lotthé A, Lamy B, Marchandin H, Jumas-Bilak E (2012) From environment to man: genome evolution and adaptation of human opportunistic bacterial pathogens. Genes 3(2):191–232

Auta HS, Emenike CU, Fauziah SH (2017) Screening of bacillus strains isolated from mangrove ecosystems in peninsular Malaysia for microplastic degradation. Environ Pollut 231:1552–1559

Auta HS, Emenike CU, Jayanthi B, Fauziah SH (2018) Growth kinetics and biodeterioration of polypropylene microplastics by bacillus sp. and Rhodococcus sp. isolated from mangrove sediment. Mar Pollut Bull 127:15–21

Bayo J, Olmos S, López-Castellanos J (2020) Microplastics in an urban wastewater treatment plant: the influence of physicochemical parameters and environmental factors. Chemosphere 238:124593

Berlanga M, Guerrero R (2016) Living together in biofilms: the microbial cell factory and its biotechnological implications. Microb Cell Factories 15(1):1–11

Bhagwat G, O'Connor W, Grainge I, Palanisami T (2021) Understanding the fundamental basis for biofilm formation on plastic surfaces: role of conditioning films. Front Microbiol 12:687118

Bowley J, Baker-Austin C, Porter A, Hartnell R, Lewis C (2021) Oceanic hitchhikers–assessing pathogen risks from marine microplastic. Trends Microbiol 29(2):107–116

Brooks AN, Turkarslan S, Beer KD, Yin Lo F, Baliga NS (2011) Adaptation of cells to new environments. Wiley Interdiscip Rev Syst Biol Med 3(5):544–561

Cai L, Wu D, Xia J, Shi H, Kim H (2019) Influence of physicochemical surface properties on the adhesion of bacteria onto four types of plastics. Sci Total Environ 671:1101–1107

Campanale C, Massarelli C, Savino I, Locaputo V, Uricchio VF (2020) A detailed review study on potential effects of microplastics and additives of concern on human health. Int J Environ Res Public Health 17(4):1212. https://doi.org/10.3390/ijerph17041212

Chamas A, Moon H, Zheng J, Qiu Y, Tabassum T, Jang JH et al (2020) Degradation rates of plastics in the environment. ACS Sustain Chem Eng 8(9):3494–3511

Chen X, Chen X, Zhao Y, Zhou H, Xiong X, Wu C (2020) Effects of microplastic biofilms on nutrient cycling in simulated freshwater systems. Sci Total Environ 719:137276

Chen Y, Wang X, Wang X, Cheng T, Fu K, Qin Z, Feng K (2022) Biofilm structural and functional features on microplastic surfaces in greenhouse agricultural soil. Sustain For 14(12):7024

Cingotti N, Jensen GK (2019) Health and Environment Alliance (HEAL). Food contact materials and chemical contamination. Health and Environment Alliance: Brussels, Belgium

De Tender CA, Devriese LI, Haegeman A, Maes S, Ruttink T, Dawyndt P (2015) Bacterial community profiling of plastic litter in the Belgian part of the North Sea. Environ Sci Technol 49(16):9629–9638

De Tender C, Devriese LI, Haegeman A, Maes S, Vangeyte J, Cattrijsse A et al (2017) Temporal dynamics of bacterial and fungal colonization on plastic debris in the North Sea. Environ Sci Technol 51(13):7350–7360

Dehghani S, Moore F, Akhbarizadeh R (2017) Microplastic pollution in deposited urban dust, Tehran metropolis, Iran. Environ Sci Pollut Res 24:20360–20371

Devi RS, Kannan VR, Nivas D, Kannan K, Chandru S, Antony AR (2015) Biodegradation of HDPE by aspergillus spp. from marine ecosystem of gulf of Mannar, India. Mar Pollut Bull 96(1–2):32–40

Dey TK, Uddin ME, Jamal M (2021) Detection and removal of microplastics in wastewater: evolution and impact. Environ Sci Pollut Res 28:16925–16947

Dobretsov S, Abed RM, Teplitski M (2013) Mini-review: inhibition of biofouling by marine microorganisms. Biofouling 29(4):423–441

Donlan RM (2002) Biofilms: microbial life on surfaces. Emerg Infect Dis 8(9):881
Dris R, Gasperi J, Tassin B (2018) Sources and fate of microplastics in urban areas: a focus on Paris megacity. In: Freshwater microplastics: emerging environmental contaminants? pp 69–83
Du H, Xie Y, Wang J (2021) Microplastic degradation methods and corresponding degradation mechanism: research status and future perspectives. J Hazard Mater 418:126377
Fackelmann G, Sommer S (2019) Microplastics and the gut microbiome: how chronically exposed species may suffer from gut dysbiosis. Mar Pollut Bull 143:193–203
GESAMP (2019) Report and studies No. 99. Guidelines for the monitoring and assessment of plastic litter in the ocean
Gong M, Yang G, Zhuang L, Zeng EY (2019) Microbial biofilm formation and community structure on low-density polyethylene microparticles in lake water microcosms. Environ Pollut 252:94–102
Guo X, Wang J (2019) Sorption of antibiotics onto aged microplastics in freshwater and seawater. Mar Pollut Bull 149:110511
Habib RZ, Abdoon MMS, Al Meqbaali RM, Ghebremedhin F, Elkashlan M, Kittaneh WF et al (2020) Analysis of microbeads in cosmetic products in The United Arab Emirates. Environ Pollut 258:113831
Hartmann NB, Huffer T, Thompson RC, Hassellöv M, Verschoor A, Daugaard AE et al (2019) Are we speaking the same language? Recommendations for a definition and categorization framework for plastic debris. Environ Sci Technol 53(3):1039–1047
Huang Y, Zhao Y, Wang J, Zhang M, Jia W, Qin X (2019) LDPE microplastic films alter microbial community composition and enzymatic activities in soil. Environ Pollut 254:112983
Jahnke A, Arp HPH, Escher BI, Gewert B, Gorokhova E, Kühnel D et al (2017) Reducing uncertainty and confronting ignorance about the possible impacts of weathering plastic in the marine environment. Environ Sci Technol Lett 4(3):85–90
Jeyavani J, Sibiya A, Shanthini S, Ravi C, Vijayakumar S, Rajan DK, Vaseeharan B (2021) A review on aquatic impacts of microplastics and its bioremediation aspects. Curr Pollut Rep 7:286–299
Jiao N, Zheng Q (2011) The microbial carbon pump: from genes to ecosystems. Appl Environ Microbiol 77(21):7439–7444. https://doi.org/10.1128/AEM.05640-11
Jin Y, Xia J, Pan Z, Yang J, Wang W, Fu Z (2018) Polystyrene microplastics induce microbiota dysbiosis and inflammation in the gut of adult zebrafish. Environ Pollut 235:322–329
Jin Y, Lu L, Tu W, Luo T, Fu Z (2019) Impacts of polystyrene microplastic on the gut barrier, microbiota and metabolism of mice. Sci Total Environ 649:308–317
Kirstein IV, Kirmizi S, Wichels A, Garin-Fernandez A, Erler R, Löder M, Gerdts G (2016) Dangerous hitchhikers? Evidence for potentially pathogenic vibrio spp. on microplastic particles. Mar Environ Res 120:1–8
Koelmans AA (2015) Modeling the role of microplastics in bioaccumulation of organic chemicals to marine aquatic organisms. A critical review. In: Marine anthropogenic litter, pp 309–324
Kokalj AJ, Kuehnel D, Puntar B, Gotvajn AŽ, Kalčikova G (2019) An exploratory ecotoxicity study of primary microplastics versus aged in natural waters and wastewaters. Environ Pollut 254:112980
Kučić Grgić D, Miloloža M, Lovrinčić E, Kovačević A, Cvetnić M, Ocelić Bulatović V et al (2021) Bioremediation of MP-polluted waters using bacteria bacillus licheniformis, Lysinibacillus massiliensis, and mixed culture of bacillus sp. and Delftia acidovorans. Chem Biochem Eng Q 35(2):205–224
Kumar A, Mishra S et al (2022) Microplastics in terrestrial ecosystems: un-ignorable impacts on soil characterises, nutrient storage and its cycling. TrAC Trends Anal Chem:116869
La Nasa J, Lomonaco T, Manco E, Ceccarini A, Fuoco R, Corti A et al (2021) Plastic breeze: volatile organic compounds (VOCs) emitted by degrading macro- and microplastics analyzed by selected ion flow-tube mass spectrometry. Chemosphere 270:128612
Lagarde F, Olivier O, Zanella M, Daniel P, Hiard S, Caruso A (2016) Microplastic interactions with freshwater microalgae: hetero-aggregation and changes in plastic density appear strongly dependent on polymer type. Environ Pollut 215:331–339

Laskar N, Kumar U (2019) Plastics and microplastics: a threat to environment. Environ Technol Innov 14:100352
Law KL (2017) Plastics in the marine environment. Annu Rev Mar Sci 9:205–229
Law KL, Morét-Ferguson S, Maximenko NA, Proskurowski G, Peacock EE, Hafner J, Reddy CM (2010) Plastic accumulation in the North Atlantic subtropical gyre. Science 329(5996):1185–1188
Liu SY, Leung MML, Fang JKH, Chua SL (2021) Engineering a microbial 'trap and release' mechanism for microplastics removal. Chem Eng J 404:127079
Lu L, Wan Z, Luo T, Fu Z, Jin Y (2018) Polystyrene microplastics induce gut microbiota dysbiosis and hepatic lipid metabolism disorder in mice. Sci Total Environ 631:449–458
Lu L, Luo T, Zhao Y, Cai C, Fu Z, Jin Y (2019) Interaction between microplastics and microorganism as well as gut microbiota: a consideration on environmental animal and human health. Sci Total Environ 667:94–100
Masiá P, Sol D, Ardura A, Laca A, Borrell YJ, Dopico E et al (2020) Bioremediation as a promising strategy for microplastics removal in wastewater treatment plants. Mar Pollut Bull 156:111252
Masó M, Garcés E, Camp J (2003) Drifting plastic debris as a potential vector for dispersing harmful algal bloom (HAB) species. Sci Mar 67(1)
McCormick A, Hoellein TJ, Mason SA, Schluep J, Kelly JJ (2014) Microplastic is an abundant and distinct microbial habitat in an urban river. Environ Sci Technol 48(20):11863–11871
McCormick AR, Hoellein TJ, London MG, Hittie J, Scott JW, Kelly JJ (2016) Microplastic in surface waters of urban rivers: concentration, sources, and associated bacterial assemblages. Ecosphere 7(11):e01556
Meral Y (2019) Glitters as a source of primary microplastics: an approach. J Agric Environ Ethics 32:459–478
Miao L, Wang P, Hou J, Yao Y, Liu Z, Liu S, Li T (2019) Distinct community structure and microbial functions of biofilms colonizing microplastics. Sci Total Environ 650:2395–2402
Millican JM, Agarwal S (2021) Plastic pollution: a material problem? Macromolecules 54(10):4455–4469
Nugrahapraja H, Sugiyo PWW, Putri BQ, Huang L, Hafza N, Götz F et al (2022) Effects of microplastic on human gut microbiome: detection of plastic- degrading genes in human gut exposed to microplastics – preliminary study. Environments 9(11):140
Oberbeckmann S, Löder MG, Labrenz M (2015) Marine microplastic-associated biofilms-a review. Environ Chem 12(5):551–562
Oberbeckmann S, Bartosik D, Huang S, Werner J, Hirschfeld C, Wibberg D et al (2021) Genomic and proteomic profiles of biofilms on microplastics are decoupled from artificial surface properties. Environ Microbiol 23(6):3099–3115
Paço A, Duarte K, da Costa JP, Santos PS, Pereira R, Pereira ME et al (2017) Biodegradation of polyethylene microplastics by the marine fungus Zalerion maritimum. Sci Total Environ 586:10–15
Padervand M, Lichtfouse E, Robert D, Wang C (2020) Removal of microplastics from the environment. A review. Environ Chem Lett 18:807–828
Peeken I, Primpke S, Beyer B, Gütermann J, Katlein C, Krumpen T et al (2018) Arctic Sea ice is an important temporal sink and means of transport for microplastic. Nat Commun 9(1):1505
Pham DN, Clark L, Li M (2021) Microplastics as hubs enriching antibiotic-resistant bacteria and pathogens in municipal activated sludge. J Hazard Mater Lett 2:100014
Pikoli MR, Astuti P, Rahmah FA, Sari AF, Solihat NA (2022) Biodegradation of microplastics by microorganisms isolated from two mature landfill leachates. CMU J Nat Sci 21(1):e2022005
Qiao R, Deng Y, Zhang S, Wolosker MB, Zhu Q, Ren H, Zhang Y (2019) Accumulation of different shapes of microplastics initiates intestinal injury and gut microbiota dysbiosis in the gut of zebrafish. Chemosphere 236:124334
Qiu X, Qi Z, Ouyang Z, Liu P, Guo X (2022) Interactions between microplastics and microorganisms in the environment: modes of action and influencing factors. Gondwana Res 108:102–119

Rong L, Zhao L, Zhao L, Cheng Z, Yao Y, Yuan C et al (2021) LDPE microplastics affect soil microbial communities and nitrogen cycling. Sci Total Environ 773:145640
Rosal R (2021) Morphological description of microplastic particles for environmental fate studies. Mar Pollut Bull 171:112716
Rubin AE, Zucker I (2022) Interactions of microplastics and organic compounds in aquatic environments: a case study of augmented joint toxicity. Chemosphere 289:133212
Rummel CD, Jahnke A, Gorokhova E, Kühnel D, Schmitt-Jansen M (2017) Impacts of biofilm formation on the fate and potential effects of microplastic in the aquatic environment. Environ Sci Technol Lett 4(7):258–267
Russell JR, Huang J, Anand P, Kucera K, Sandoval AG, Dantzler KW et al (2011) Biodegradation of polyester polyurethane by endophytic fungi. Appl Environ Microbiol 77(17):6076–6084
Sajjad M, Huang Q, Khan S, Khan MA, Yin L, Wang J et al (2022) Microplastics in the soil environment: a critical review. Environ Technol Innov 27:102408
Sánchez C (2020) Fungal potential for the degradation of petroleum-based polymers: an overview of macro-and microplastics biodegradation. Biotechnol Adv 40:107501
Sarkar AK, Rubin AE, Zucker I (2021) Engineered polystyrene-based microplastics of high environmental relevance. Environ Sci Technol 55(15):10491–10501. https://doi.org/10.1021/acs.est.1c02196
Seeley ME, Song B, Passie R, Hale RC (2020) Microplastics affect sedimentary microbial communities and nitrogen cycling. Nat Commun 11(1):2372
Shi J, Wu D, Su Y, Xie B (2021) Selective enrichment of antibiotic resistance genes and pathogens on polystyrene microplastics in landfill leachate. Sci Total Environ 765:142775
Solanki S, Sinha S, Singh R (2022) Myco-degradation of microplastics: an account of identified pathways and analytical methods for their determination. Biodegradation 33(6):529–556
Sun XL, Xiang H, Xiong HQ, Fang YC, Wang Y (2022) Bioremediation of microplastics in freshwater environments: a systematic review of biofilm culture, degradation mechanisms, and analytical methods. Sci Total Environ:160953
Tadsuwan K, Babel S (2021) Microplastic contamination in a conventional wastewater treatment plant in Thailand. Waste Manag Res 39(5):754–761
Tran KM, Lee HM, Thai TD, Shen J, Eyun SI, Na D (2021) Synthetically engineered microbial scavengers for enhanced bioremediation. J Hazard Mater 419:126516
Tu C, Chen T, Zhou Q, Liu Y, Wei J, Waniek JJ, Luo Y (2020) Biofilm formation and its influences on the properties of microplastics as affected by exposure time and depth in the seawater. Sci Total Environ 734:139237
Tu P, Xue J, Niu H, Tang Q, Mo Z, Zheng X, Wu L, Chen Z, Cai Y, Wang X (2023) Deciphering gut microbiome responses upon microplastic exposure via integrating metagenomics and activity-based metabolomics. Meta 13(4):530. https://doi.org/10.3390/metabo13040530
Tuuri EM, Leterme SC (2023) How plastic debris and associated chemicals impact the marine food web: a review. Environ Pollut 321:121156
Von Moos N, Burkhardt-Holm P, Köhler (2012) A uptake and effects of microplastics on cells and tissue of the blue mussel Mytilus edulis L. after an experimental exposure. Environ Sci Technol 46(20):11327–11335
Wang Y, Wang H, Li X, Liu D, Jiang Y, Sun Z (2013) O3/UV synergistic aging of polyester polyurethane film modified by composite UV absorber. J Nanomater 2013:4–4
Wang S, Xue N, Li W, Zhang D, Pan X, Luo Y (2020) Selectively enrichment of antibiotics and ARGs by microplastics in river, estuary and marine waters. Sci Total Environ 708:134594
Wang J, Peng C, Li H, Zhang P, Liu X (2021) The impact of microplastic-microbe interactions on animal health and biogeochemical cycles: a mini-review. Sci Total Environ 773:145697
Xu C, Zhang B, Gu C, Shen C, Yin S, Aamir M, Li F (2020) Are we underestimating the sources of microplastic pollution in terrestrial environment? J Hazard Mater 400:123228
Yang Y, Yang J, Wu WM, Zhao J, Song Y, Gao L et al (2015a) Biodegradation and mineralization of polystyrene by plastic-eating mealworms: part 1. Chemical and physical characterization and isotopic tests. Environ Sci Technol 49(20):12080–12086

Yang Y, Yang J, Wu WM, Zhao J, Song Y, Gao L et al (2015b) Biodegradation and mineralization of polystyrene by plastic-eating mealworms: part 2. Role of gut microorganisms. Environ Sci Technol 49(20):12087–12093

Yi M, Zhou S, Zhang L, Ding S (2021) The effects of three different microplastics on enzyme activities and microbial communities in soil. Water Environ Res 93(1):24–32

Yu H, Liu M, Gang D, Peng J, Hu C, Qu J (2022) Polyethylene microplastics interfere with the nutrient cycle in water-plant-sediment systems. Water Res 214:118191

Yuan J, Ma J, Sun Y, Zhou T, Zhao Y, Yu F (2020) Microbial degradation and other environmental aspects of microplastics/plastics. Sci Total Environ 715:136968

Zeghal E, Vaksmaa A, Vielfaure H, Boekhout T, Niemann H (2021) The potential role of marine fungi in plastic degradation–a review. Front Mar Sci 8:738877

Zettler ER, Mincer TJ, Amaral-Zettler LA (2013) Life in the "plastisphere": microbial communities on plastic marine debris. Environ Sci Technol 47(13):7137–7146

Zhang J, Gao D, Li Q, Zhao Y, Li L, Lin H et al (2020) Biodegradation of polyethylene microplastic particles by the fungus aspergillus flavus from the guts of wax moth galleria mellonella. Sci Total Environ 704:135931

Zhang K, Hamidian AH, Tubić A, Zhang Y, Fang JK, Wu C, Lam PK (2021) Understanding plastic degradation and microplastic formation in the environment: a review. Environ Pollut 274:116554

Chapter 7
Microplastic Alters Rhizosphere Microbiome: A Review

Anil Kumar Singh

Abstract Rhizosphere is the hotspot of plant-microbe interaction. The healthy rhizosphere microbiome is essential for plant as it assists in nutrient uptake and offers protection against pathogen attack. Accumulation of microplastics in soil impacts rhizosphere microbiome by altering the soil physicochemical characteristics such as bulk density, pH, porosity, electric conductivity, enzymatic activities, and nutrient bioavailability. The size, shape, and chemical nature of plastics directly influence interaction between microorganisms and microplastics. The present chapter discusses the primary sources, fate, and impact of microplastics on microbial communities of rhizosphere and soil nutrient cycling. Further, the chapter discusses the applicability of the generated knowledge in formulating strategies for removing small-size plastic from soil. Challenges and future perspectives are also discussed in the present chapter to pave the path for future courses of research.

Keywords Sustainable · Enzyme · Biodegradable Plastic · Ecotoxicity · Nanoplastics

1 Introduction

The rhizosphere is a unique, integrated, and ever-changing region of soil. This narrow region of soil around the root has physical, chemical, and biological properties significantly different from the bulk soil. The rhizosphere soil has a vibrant habitat teeming with microorganisms, organic matter, and numerous biochemical interactions. Soil-dwelling organisms such as bacteria, archaea, fungi, protozoa, arthropods, nematodes, mollusks, and earthworms contribute significantly toward physical, chemical, and biological properties of rhizosphere. Understanding the

A. K. Singh (✉)
Department of Agriculture, Sant Baba Bhag Singh University, Khiala, Punjab, India

K. Khanna et al. (eds.), *Microplastics and Soil Microbiome*,
https://doi.org/10.1007/978-981-96-4978-5_7

dynamics of rhizosphere is essential for understanding nutrient uptake, soil health, plant growth, and resilience against diseases (Zhuang et al. 2024).

Rhizosphere plays a vital role in plant health by supporting plant growth-promoting microorganisms that enhance nutrient absorption, nitrogen fixation, and disease resistance. Roots release nutrient rich exudates in its close vicinity. Root exudates have both low and high molecular weight compounds. Nutrient-rich exudates not only attract beneficial microorganisms in rhizosphere but also help solubilize nutrients in the soil, making them more accessible to plants (Zhuang et al. 2024). The interactions within the rhizosphere improve soil structure, water retention, and nutrient availability and promote plant resilience against environmental stress. In one of the recent study, Wei et al. (2023) compared microbial communities in the bulk soils and rhizosphere of a Chinese cabbage. They observed that rhizosphere soil have higher abundance of beneficial bacteria that supported cabbage growth as compared with the bulk soil. The rhizosphere soil of Chinese cabbage had higher amount of organic matter, available phosphorus, total phosphorus, available potassium, total potassium, total nitrogen, and alkaline hydrolyzable nitrogen in comparison to bulk soil. The rhizosphere soil also demonstrated better catalase, urease, sucrase, and phosphatase activity in comparison to bulk soil. They observed that rhizosphere soil have *Proteobacteria*, *Acidobacteriota*, *Actinobacteriota*, and *Bacteroidota* as the prominent bacterial phyla. Dominant fungal phyla of rhizosphere soil were *Olpidiomycota*, *Ascomycota*, *Mortierellomycota*, and *Basidiomycota*. The bulk soil was dominated by bacterial phyla, namely, *Proteobacteria*, *Acidobacteriota*, *Chloroflexi*, and *Actinobacteriota*. *Ascomycota*, *Olpidiomycota*, *Mortierellomycota*, and *Basidiomycota* were prominent fungal phyla in the bulk soil. The research group also reported that the microbial networks were less complicated in rhizosphere soil as compared to the bulk soil (Wei et al. 2023). The symbiotic relationship between plants and soil microbes is vital for maximizing nutrient uptake and promoting optimal plant growth (Bowya and Balachandar 2020).

The rhizosphere also acts as a buffer zone, protecting plants from various stressors, including pathogens and environmental extremes. Beneficial microorganisms, such as mycorrhizal fungi and rhizobacteria, colonize the root surface and form a protective barrier against harmful pathogens. They can outdo the pathogenic organisms for resources or even produce antimicrobial compounds that inhibit their growth. This natural defense mechanism is crucial in enhancing resilience, reducing the need for chemical pesticides, and promoting sustainable agricultural practices. Several researchers have highlighted the importance of rhizospheric microbial biota in supporting sustainable agriculture practices (Sharma et al. 2024; Wei et al. 2023; Bowya and Balachandar 2020). Sharma et al. (2024) believed that altering the rhizosphere and its microbial community have great potential for mitigating the negative effects of high-input agricultural practices and supporting both agricultural and environmental sustainability. In another study, Bowya and Balachandar (2020) engineered rhizosphere of rice plant to improve the colonization by plant growth-promoting rhizobacterium *Pseudomonas chlororaphis* ZSB15-M2. Foliar application of salicylic acid, *Corynebacterium glutamicum* cell extract (CGCE), and

Saccharomyces cerevisiae cell extract led to nearly tenfold increase in the population of strain ZSB15-M2 in rhizosphere as compared to the control. CGCE exposure also enhanced soil organic carbon, microbial biomass, and soil protein content in the rhizosphere. Foliar application of CGCE improved soil enzymes activities such as dehydrogenase, acid phosphatase, and alkaline phosphatase in the rice rhizosphere. The engineered rhizosphere improved the availability of nutrients like nitrogen, phosphorus, potassium, and zinc in the rice rhizosphere as compared to the control (Bowya and Balachandar 2020). A healthy rhizosphere fosters symbiotic relationships, enabling plants to thrive in diverse environmental conditions, ultimately contributing to sustainable agriculture and ecosystem stability. Most researchers are of view that vigor of the rhizosphere can be maintained or promoted by healthy agricultural practices such as crop rotation, cover cropping, and tillage minimization (Sharma et al. 2024; Bowya and Balachandar 2020). Healthy rhizosphere contributes to a more balanced soil ecosystem around root (Zhuang et al. 2024).

The rhizosphere serves as a critical interface for various ecological interactions. It is here that plant roots and soil microorganisms communicate through signaling molecules, facilitating processes such as nutrient exchange, stress responses, and even plant growth regulation. This intricate web of interactions highlights the importance of maintaining a balanced rhizosphere to support not only individual plant health but also broader ecosystem functions (Sharma et al. 2024). Xenobiotic contaminates of soil is known to alter various ecological interactions in rhizosphere. Environmental contaminants such as heavy metals, pesticides, and industrial chemicals reduce the diversity of rhizosphere microbial community and change ecological functioning of the rhizosphere (Aryal 2024). These environmental contaminants has potential to disrupt beneficial microbial communities, hinder root growth, and reduce nutrient uptake, leading to poor plant health and decreased agricultural productivity. Furthermore, toxic substances in rhizosphere soils may inhibit essential biological processes such as nitrogen fixation and carbon cycle. Ultimately, soil pollution vitiates the natural balance in rhizosphere, threatening plant ecosystems, soil fertility, and long-term sustainability of agricultural systems (Lü et al. 2024).

In the recent past, small-size plastics, namely, microplastics (MPs) and nanoplastics (NPs), have emerged as prominent contaminate of soil, water, and air. There is ample evidence showing the harmful impact of MPs on living organisms, particularly aquatic species that serve as important nutrient sources for humans. MP contamination is becoming a rapidly growing global concern and is now recognized as the second-most pressing environmental and ecological issue after global warming. In the past decade, the majority of research on MPs has concentrated on environmental science and ecology, particularly in aquatic environments. However, more recent studies are starting to shift their focus toward investigating MPs in terrestrial ecosystems. These concerns are reinforced by estimates indicating that the buildup of MPs in land ecosystems is significantly higher than in the ocean (Zeb et al. 2024). MP contaminants of terrestrial environment have also reached the narrow region around the plant root. Researchers are of view that small-size plastic contaminates such as MPs/NPs most likely to alter with microbial diversity of rhizosphere, restructure microbial community, and change functioning of the rhizosphere (Bouaicha et al. 2022). In one of the study, Riaz et al. (2024) compared distribution of MPs in mangrove sediments of rhizosphere and non-rhizosphere regions. Rhizosphere region have less MP count (7112 particles) as compared to non-rhizosphere regions (7848 particles) (Riaz et al. 2024).

This chapter aims to emphasize the severity of the risk posed by MP/NP contamination in rhizosphere soil, as well as to guide future research on the interaction between MPs/NPs and the rhizosphere microbiome. A comprehensive overview of the current state of research on the interaction between MPs/NPs and the rhizosphere microbiome is provided in the chapter. The chapter also discusses the impact of MPs on the physicochemical characteristics of soil, the microbial communities within the rhizosphere, and soil nutrient cycling. The effects of soil MPs on crop plants are also examined, as they strongly influence the environment of rhizosphere. Additionally, the chapter explores the applicability of available knowledge in formulating strategies to remove MPs from soil. Finally, challenges and future perspectives are discussed to help pave the path for future research.

2 Source of Small-Size Plastics

Plastic plays a critical role in modern life due to its versatility, durability, and cost-effectiveness. It is used in a wide range of industries, including healthcare, transportation, electronics, and packaging. In packaging, plastic preserves food and goods, reducing waste and spoilage. Its lightweight nature improves fuel efficiency in vehicles and airplanes. Improper plastic waste disposal and insufficient waste management result in the release of plastics into the environment, where they persist for extended periods and can readily accumulate in various environmental systems. Non-biodegradable nature of plastic further aggravates the challenges associated with plastic pollution. According to the United Nations Environment Program (UNEP), approximately 300 million tons of plastic are produced each year, with a significant portion ending up in oceans, rivers, and soil. It is believed that plastic waste burden on earth surface will reach to over 13.23 billion tons by 2050 (Seo et al. 2024). Over time, exposure to sunlight, wind, high temperature, UV rays, and waves causes larger plastic wastes to fragment into smaller-size plastics such as microplastics (MPs) and nanoplastics (NPs) (Ziani et al. 2023; Zeb et al. 2024).

The term "microplastic" was first reported in 2004 to designate plastic fragment of nearly 20μm in size. The systematic definition of MP was formulated in the first international workshop on microplastics organized by US National Oceanographic and Atmospheric Agency (NOAA) in 2008. Small-size plastics with size less than 5 mm are called microplastics (MPs). The small plastic particles in range of 1–1000 nm are referred as nanoplastics (NPs). Some authors view environmental MPs as a broad term as it encompasses various particle shapes, sizes, and polymer nature, along with the diverse physical and chemical characteristics (Ziani et al. 2023; Li et al. 2024). The primary sources of MPs/NPs are plastic particles originating from the textile industry, plastic manufacturing industry, cosmetics, and scrubbing process. Secondary MPs/NPs are produced by weathering of plastic particles (Ziani et al. 2023). The chemical nature of most common MPs reported from environment are polyethylene (PE), ethylene vinyl acetate (EVA), polyvinyl chloride (PVC), polypropylene (PP), polyethylene terephthalate (PET), polymethyl methacrylate (PMMA), and polystyrene (PS). Variations in the chemical composition of MPs directly influence their ability to interact with other pollutants and the environmental risks (Jia et al. 2023).

The MPs/NPs are capable of short- and long-range atmospheric transportation, thus reaching far off places from point of origin. In recent past, MPs/NPs have emerged as a significant environmental concern owing to its presence in ecosystems, food chains, and human health. Presence of MPs has been reported in marine environment, freshwater bodies, atmosphere, and terrestrial system (Ziani et al. 2023). The MPs/NPs have also been detected in living organisms (Chae and An 2020). The prevalence of MPs in our environment has prompted widespread attention due to their potential harm and the complexity of their impacts. One of the most concerning aspects of MP pollution is their impact on marine life. Studies have shown that marine organisms, from plankton to fish and even larger mammals, ingest microplastics, mistaking them for food. This ingestion can lead to physical harm, such as blockages or injury to internal organs, and can disrupt feeding behaviors and growth. Furthermore, microplastics can act as carriers for harmful chemicals, including persistent organic pollutants (POPs), which accumulate on their surfaces (Ziani et al. 2023). When ingested, these toxins can bioaccumulate in the food chain, posing risks to higher trophic levels, including humans (Li et al. 2024; Ziani et al. 2023; Chae and An 2020). Recent reports have shown presence of MPs in various food products, including seafood, honey, and even table salt (Li et al. 2024). This raises alarming questions about the safety of our food supply and the potential health risks associated with MP consumption. While the full extent of health impacts is still being studied, early evidence suggests that MPs may trigger inflammation, disrupt endocrine functions, and potentially lead to long-term health issues (Ziani et al. 2023).

3 MP/NP Pollution in Vicinity of Crop Plants

The MP/NP contamination of agricultural soils is global problem thus receiving increasing levels of attention from scientists and stakeholders. Small plastic contaminants have reached agriculture field through different input pathways, such as irrigation with sewage water, air deposition, landfilling with plastic wastes degradation of plastic residues, application of organic manure, compost, polymer-based fertilizers, and pesticides. Modern agriculture practices heavily rely on plastic products (such as mulching sheets, greenhouse sheets, polyhouse sheets, foam, plastic pipes, and packing materials) right from nursery stage to post-harvest stage (Verma et al. 2023).

Seo et al. (2024) considered application of sewage sludge as a major reason for MP pollution of agricultural land. Generally, the sewage sludge obtained from wastewater treatment plants contains high concentration of MPs. The sewage sludge may contain over 7000 MP particles per kilogram of sludge. Greenhouse/polyhouse plastic sheets are another important source of plastic pollution. Plastic sheet used in the greenhouse/polyhouse has very short lifespan of around 2–3 years. Climatic factors like elevated temperatures, solar radiation, precipitation, and wind are important contributors toward the physical weathering, aging, and quality degradation of plastic films. Chemically most common MP/NP contaminants present in agriculture fields are low-density PE, high-density PE, PVC, EVA, PET, PP, and PMMA (Seo et al. 2024).

Biodegradable plastics such as polylactide (PLA) and starch-based films were introduced to overcome concerns associated with plastic pollution. These plastics were created to provide the same advantages as traditional polyethylene film, with the added ability to biodegrade, enabling them to be incorporated into the soil and decompose into carbon dioxide, water, and microbial biomass. However, despite being intended to decompose in agricultural fields, research has shown that biodegradable plastics like PLA and starch-based films may not fully degrade in natural conditions. Instead, biodegradable plastics usually fragment into smaller plastic particles, leading to the buildup of bio-MPs in the soils (Seo et al. 2024).The ecological and environmental risks of these biodegradable plastics are still not well understood.

Generally, MP/NP contamination of agriculture soil adversely affects both quality and quantity of agriculture output. Yi et al. (2023) studied effect of PE MPs on grain yield of common rice cultivar Nangeng 5055 (NG) and hybrid rice Jiafengyou 6 (JFY). MPs in soil did not alter rice grain yield of common rice cultivar. Contrary, MPs reduced quantity as well as quality of hybrid rice. Presence of MPs reduced number of grains per panicle (13%), total grain amino acid content (1.7%), and rice grain yield (23%) of hybrid rice (Yi et al. 2023). Presence of PS-MP and lead in soil reduced 100 seed weight as well as seed yield per plant in mung bean (Chen et al. 2023). In another study, Chen et al. (2022) did not observe any significant negative effects on rice yield in presence of PET-MP, PAN-MP, and PE-MP. Contrary, MPs increased total nitrogen content of grains and rice yield by improving nitrogen uptake ability of rice plants (Chen et al. 2022). In an interesting work, Chae and An (2020) demonstrated transfer of NPs from the soil to a mung bean and then to a consumer. Mung bean growing in plastic contaminated soil accumulated NPs in leaves. Snails feeding on these leaves exhibited histologically damaged tissues for digestive organ. This also decreased viability of gut microbiota in the snail. These findings encouraged researchers to explore how MPs/NPs move from soil to animals through plants and the eventual impact this transfer has on humans (Chae and An 2020). Studies on plastic pollution in agricultural soils are being conducted with increasing frequency owing to the known effects of MPs/NPs on the soil ecosystem and agricultural output.

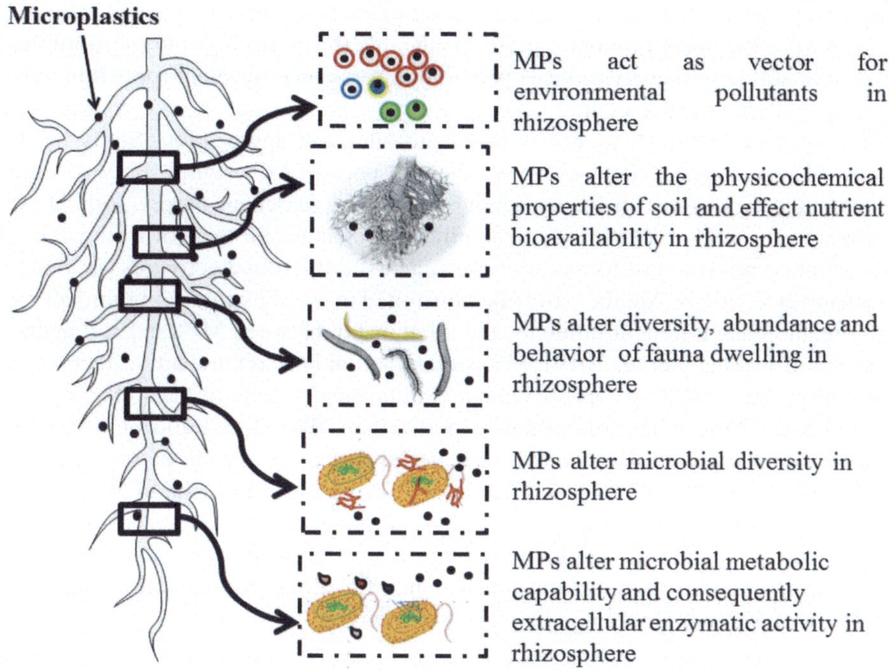

Fig. 7.1 Changes induced by presence of MPs in the rhizosphere

4 MPs/NPs Alter Properties of Rhizosphere Soil

Small-size plastic such as MPs and NPs are capable to alter physical, chemical, and biological properties of rhizosphere soil. Usually presence of MPs and NPs in soil adversely affects functioning and structure of rhizosphere. Further, these changes in rhizosphere adversely affect growth, development, and yield of crop plant. Figure 7.1 summarizes changes induced by MPs in the rhizosphere.

4.1 Soil Physical Properties

Physical properties of soil such as texture, arrangement of soil particles, permeability, bulk density, and compaction have significant influence on structure of rhizosphere. Presence of MPs in soil alter physical features such as texture, structure, bulk density, water aggregate stability, water holding capacity, hydraulic conductivity, and rainwater infiltration (Hanif et al. 2024). Plastics have lower density than most of the minerals components of the soil and thus presence of MPs/NPs in soil reduces bulk density of soil. The reduction in soil bulk density improves soil aeration by forming additional pathways for air movement, which in turn increases water evaporation from the soil. Enhanced water evaporation from the soil is bound to have negative impact on plant performance due to soil drying. Furthermore this change in soil bulk density reduces penetration resistance for plant roots thus supporting root growth (Seo et al. 2024).

Presence of MPs/NPs in soil is known to alter soil aggregation. Study by de Souza Machado et al. (2019) exhibited that presence of MPs significantly reduced water-stable aggregates in loamy sandy soil and increased in clayey soil. These changes in physical properties of soil significantly changed biomass of spring onion, elemental composition of tissue, root traits, and soil microbial activities (de Souza Machado et al. 2019). Another study demonstrated that soil texture, soil mineralogy, and organic matter content influence the relationship between MPs and soil aggregate stability (Zhang et al. 2019). MP contamination is also known to affect water infiltration, water holding capacity, and hydraulic conductivity of the soil by blocking soil pores. Most of the studies have reported that MPs reduce infiltration and soil water-holding capacity. However, it is believed that the response of soils against MPs may vary with soil types and nature of MPs (de Souza Machado et al. 2019; Seo et al. 2024). Shift in a major soil physical parameter are expected to induce changes in soil microbial community composition. Predicting shifts in rhizosphere microbial community composition and also their functional consequences under the influence of MP soil physical parameter is challenging.

4.2 Soil Chemical Properties

The presence MPs in soil can alter chemical properties such as pH, nutrient availability, and sorption capacity of soil. The soil pH is important indicator of soil health. The pH value has significant influence on soil chemical properties of soil. It regulates elemental speciation, bioavailability of nutrient, soil fertility, and agricultural productivity. Microorganisms in the rhizosphere are also directly affected by change in soil pH. Soil contamination by MPs alters pH of soil. The pH of soil may increase or decrease depending upon the type of MPs, concentration of MPs, shape, size, and aging period of MPs. The ways in which MPs affect soil pH remain unclear, and the interactions between MPs and soil amendments have not yet been studied (Bouaicha et al. 2022; Seo et al. 2024).

The sorption capacity of soil enables it to retain nutrients and prevent groundwater contamination from organic contaminates through the leaching process. Leaching leads to groundwater pollution and decreases soil fertility by depleting essential nutrients in the soil. Presence of MPs in the soil has been reported to decrease sorption capacity of the soil. However, influence of various MPs and soil type on sorption capacity of soil is still not fully known (Seo et al. 2024).

Soil microbes particularly microorganisms residing in rhizosphere play a crucial role in nutrient utilization and cycling. MPs have carbon as main element and thus MP contamination leads to increased soil organic carbon concentration in soil. High MP contamination of soil increases organic carbon, inorganic nitrogen, and total phosphorus concentration in the soil. MP contamination of rhizosphere changes C/N ratio in favor of carbon (Seo et al. 2024). When an organic material with a high C/N ratio is added to soil, soil microbes are likely to search soil to obtain enough nitrogen. Most of the research has demonstrated that MPs in the soil adversely affect the biologically available N concentration of the soil and leads to N depletion in plants (Zang et al. 2020; Seo et al. 2024). Zang et al. (2020) demonstrated that MPPVC and MP PE significantly changed the carbon pool in the soil suggesting the possibility of negative affect on N and P nutrients.

MPs/NPs also act as sources and carriers of pollutants. Several additives are added to plastics for introducing desirable characteristic in plastics. MPs/NPs generated from these plastics also contain additives that leach into the soil, raising further environmental concerns. MPs also act as carrier for other soil pollutants such as heavy metal, persistent organic pollutants, and petroleum aromatic hydrocarbon. Pollutants adsorbed on the surface of MPs/NPs have more toxic effect on living organism than those originating directly from smaller plastic particles (Wu et al. 2024). It is believed that adsorption capacity of MPs increases over time. Thus, older MPs have more capability to adsorb more contaminants than newly added MPs. The interaction between heavy metals and MPs depends on the morphology of MPs, electrostatic attraction, and organic-bound forms (Wu et al. 2024). Li et al. (2023) reported synergistic toxic effect of PE MPs and Cd on maize photosynthesis. PE-MPs did not have significant impact on the leaf Cd content of both maize cultivars Xianyu 335 (XY) and Zhengdan 958 (ZD). However, PE-MPs and Cd adversely altered photosynthesis machinery by increasing the malondialdehyde content and damaging photosynthetic components such as chlorophyll and photosystem I and II. Synergistic toxic effect of PE MPs and Cd suppressed expression of the several genes (such as PSII, PSI, F-type ATPase, cytochrome b6/f complex) linked to photosynthesis. Contrary, Zong et al. (2021) reported that the combination of PS-MP and heavy metals (copper and cadmium) increased the chlorophyll content, enhanced photosynthesis, and reduced the accumulation of reactive oxygen species (ROS) in wheat seedling. The presence of PS-MP also reduced the accumulation of copper and cadmium in wheat seedlings. PS-MP adsorbed heavy metals on the surface and reduced bioavailability as well as toxic effect of copper and cadmium on wheat (Zong et al. 2021).

4.3 Soil Biological Properties

Biological properties of soil are directly influenced by the soil dwelling flora and fauna. There physiological activity drives nutrient cycling and organic matter decomposition within the ecosystem. Presence of MPs in the soil has adverse effect on the diversity and count of soil-dwelling organisms particularly the rhizosphere microbiome. Physiological activity of soil-dwelling organisms is also altered under the influence of small-size plastic, thus changing the soil enzyme activity.

Soil fauna includes variety of soil animals, which play a vital role in soil processes and influence ecology of rhizosphere. Soil fauna are diverse in size, shape, and function and include organisms such as earthworms, ants, termites, beetles, millipedes, spiders, and slugs. Soil fauna is known to the transport and distribution of MPs in the soil. Influence of MPs on soil fauna is dependent on species of fauna, type, and concentration of MPs. The studies have demonstrated that earthworms can directly ingest MPs/NPs and contribute toward degradation small-size plastics. Most researchers view that 1% of MP concentration in the soil does not have any significant toxic effect on earthworm (Khaldoon et al. 2022). However, high concentration of MPs in soil has adverse effect on soil fauna.

Microorganisms residing in the rhizosphere have a significant effect on crop plant performance. As previously discussed, MPs alter the physical properties of soil, and these changes are likely to influence microbial communities in the rhizosphere. In addition to these changes, the presence of MPs/NPs in rhizosphere soil exerts stress on microorganisms. Most soil-dwelling microorganisms are unable to cope with this stress, as MPs/NPs are relatively difficult for them to metabolize. Only a very few soil-dwelling microorganisms have the ability to metabolize MPs and NPs. Microorganisms that utilize MPs and NPs as a source of nutrients increase their biomass. There are several reports showing effect of NPs/MPs on the microbial community structures and function of rhizosphere (Zhu et al. 2022; Meng et al. 2023).

Zhu et al. (2022) reported that MPs, namely, polyethylene (PE), polyvinylchloride (PVC), and polystyrene (PS), can influence soil bacterial community structures and function of wheat rhizosphere. MPs reduced alpha diversities of bacteria in rhizosphere soil. Interestingly, the relative abundance of phyla and genera directly involved in degradation of hydrophobic pollutant increased under the influence of MPs. Phylogenetic Investigation of Communities by Reconstruction of Unobserved States indicated that PS reduced functional profile of a rhizosphere soil microbial community by decreasing number of bacterial genus. Contrary, PE and PVC improved functional profile of a rhizosphere soil and consequently biodegradation of xenobiotic contaminants (Zhu et al. 2022). Meng et al. (2023) studied the effect of low density polyethylene (LDPE) MPs and biodegradable MPs (Bio-MPs) on rhizosphere soil bacterial communities of *Phaseolus vulgaris*. Presence of MPs in rhizosphere increased alpha diversity of bacterial community as compared to control (without contamination of MPs). Increase in concentration of MPs in rhizosphere decreased alpha diversity of bacterial communities. The results also showed that the amount of microplastics had a significant impact on the changes in microbial community composition, more so than the nature of microplastics or their interactions. In comparison to the control, the family *Comamonadaceae* was found to be more abundant in all treatments involving LDPE microplastics and biodegradable microplastics. Rhizosphere soil with 2.5% Bio-MPs and 2.5% LDPE MPs has family *Micrococcaceae* most abundant family. The family Sphingomonadaceae and Xanthobacteraceae was least abundant in 2.5% LDPE-MPs and 1.0% Bio-MPs, respectively (Meng et al. 2023).

Changes in rhizosphere microbial communities are likely to alter mineralization rates and affect communities of root-colonizing symbionts, such as arbuscular mycorrhizal fungi. MPs may also impact soil-borne pathogens. Alterations in the physical properties of rhizosphere soil can influence the physiological activity of root symbionts. Root symbionts are beneficial microorganisms, including fungi and bacteria that form mutualistic associations with plant roots, enhancing plant growth and health. Key among these are mycorrhizal fungi and nitrogen-fixing bacteria. Mycorrhizal fungi connect to plant roots, extending the root's reach and boosting water and nutrient uptake, especially phosphorus. Nitrogen-fixing bacteria, such as *Rhizobium* in legumes, convert atmospheric nitrogen into a form plants can use, providing essential nutrients. In return, plants supply these symbionts with carbohydrates produced during photosynthesis. Root symbionts are critical to sustainable agriculture, soil health, and plant resilience, supporting both natural ecosystems and agricultural productivity.

Rhizosphere microbial community plays crucial role in biogeochemical cycle. MP contamination increases relatively slow-degrading carbon content in rhizosphere soil (Fig. 7.2). When MPs enter the soil environment, they can alter the physicochemical properties, modify microbial habitats, and influence the composition and activities of existing microbial communities, ultimately affecting soil

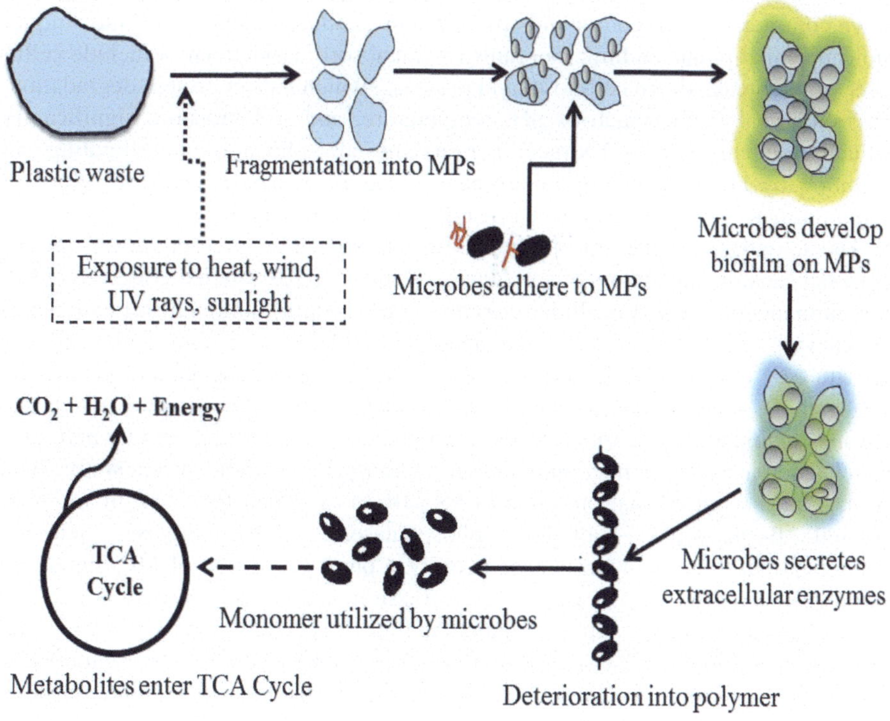

Fig. 7.2 Biodegradation of MPs in the soil

function and nutrient cycling. MPs have ability to influence carbon cycle in soils by directly influencing the composition of soil carbon pools. Soil MPs add both stable and easily decomposable carbon components, such as soluble additives. Soil microorganisms, including bacteria and fungi, have a dual function in the soil carbon cycle. They promote carbon release through their catabolic activities while also inhibiting carbon release through stabilization mechanisms. Both fungi (such as *Penicillium* sp., *Aspergillus* sp., *Zalerion maritimum*, and *Engyodontium album*) and bacteria (such as *Rhodococcus* sp., *Pseudomonas* sp., *Ideonella sakaiensis*, and *Bacillus* sp.) have been proven to have the potential to promote the degradation of microplastics (Miri et al. 2022). Microbes secrete extracellular enzymes that can depolymerize microplastics through hydrolysis reactions. Enzymes like hydrolases are responsible for breaking down MPs into smaller molecules. Zhang et al. (2023) observed that bioplastics like PHA, PBS, and PLA increased microbial biomass carbon and dissolved organic carbon levels. This suggests that biodegradable microplastics can simultaneously promote microbial uptake and facilitate the transformation of soil organic matter into dissolved organic compounds. These extracellular enzymes play a crucial role in making plastic polymers more hydrophilic by introducing functional groups, such as carbonyl or alcohol groups, which enhance microbial attachment and subsequent biodegradation (Miri et al. 2022). Soil extracellular enzymes are critical for nutrient cycling and organic matter decomposition within ecosystem. Produced by soil microbes and plant roots, these enzymes break down complex organic compounds like proteins, cellulose, and lignin into simpler molecules that plants and microorganisms can absorb. Key enzymes include cellulase, which breaks down cellulose, and protease, which aids in protein degradation. Environmental factors such as pH, temperature, and soil moisture significantly influence enzyme activity. Extracellular enzymes also play a role in mitigating soil pollution by breaking down contaminants. Overall, these enzymes are vital for maintaining soil health, plant growth, and ecosystem productivity.

The activities of soil extracellular enzymes produced by microorganisms are the potential sensitive indicators of soil quality. Yu et al. (2020) studied effect of PE MP contaminants on soil extracellular enzymes. MP (100μm) inhibited soil extracellular enzyme (catalase, laccase, manganese peroxidaseurease, and β-glucosidase) activities by changing soil nutritional substrates and physicochemical properties. The authors suggest that microplastics competed with soil microorganisms for physicochemical niches, which leads to a reduction in microbial activity and, ultimately decreasing the extracellular enzyme activity. In one of the recent study, Shah et al. (2023) reported that presence of PS-MP, PE-MP, and PVC-MP in soil suppressed β-glucosidase, alkaline phosphatase, cellobiohydrolase, leucine-aminopeptidase, and dehydrogenase activity. Contrary addition of MPs enhanced soil urease activity (Shah et al. 2023). Decrease in soil enzyme activity is liable to delay degradation of organic pollutants leading to accumulation of pollutant and increase in soil toxicity. Another important change observed under the influence of MPs is presence of antibiotic resistance bacteria (ARB) that can become profound source of antibiotic resistance genes (ARGs). In general rhizosphere have environment conducive to harbor ARB. However, presence of MPs in rhizosphere further aggravates the problem associated with movement of ARGs. Bacteria develop specific microbial biofilm (referred as plastisphere) on MPs. The plastisphere possibly create a supporting environment for ARB proliferation and ARG exchange via both vertical and horizontal transfer of ARGs (Luo et al. 2023).

5 Effect of MPs/NPs on Crop Plants

MPs and NPs are known to negatively impact the growth and development of crop plants, ultimately reducing overall yield. MPs can effect plant growth and development by either direct or indirect effect. Direct effect is associated with the physical obstruction, while indirect effects linked with a reduction in soil productivity. Seeds and roots are primary parts of plant that experience direct and mechanical damage in presence of MPs in the soil. The negative effects of MPs on plant growth and development vary based on the size, type of MPs, plant species, and dosage and duration of exposure. The presence of MPs in soil stresses crop plants by disrupting essential physiological processes, including ionic balance, redox regulation, nutrient absorption, and photosynthesis. MPs can also have adverse effect on crop plant by altering soil fertility.

MP stress hinders seed germination by blocking the pores in the seed capsule, which limits water absorption and slows down the imbibition process. In one of the studies, Guo et al. (2022) reported that PS-MP stress reduced seed germination rates and germination potential of herbaceous ornamental plants. In another study, Dong et al. (2022) observed that exposure to polymethyl methacrylate (PMMA) MPs and As (V) reduced the germination index and root length of rapeseed.

MPs/ NPs are available in rhizosphere soil in different forms. Bioavailability of MPs/NPs depends mainly on soil properties, particle size, chemical nature of MPs/ NPs, and nature of crop plants. MPs cannot penetrate plant tissues directly as expected, since their large particle size prevents them from passing through plant cell walls. Negatively charged mucilage and exudates form a primary protective layer and have been found to prevent the uptake of positively charged MPs on the outer surface of the cell wall. The root rhizodermis layer prevents entry of MPs into the soil. However, NPs can easily enter root and accumulate within plant parts. NPs can easily be adsorbed on root hairs. Movement of small-size plastic has been linked to apoplastic and symplastic transportation. Apoplastic transport takes place outside the plasma membrane through the extracellular spaces, cell walls of adjacent cells, and xylem vessels, whereas symplastic transport involves movement between the cytoplasm of adjacent cells through specialized structures called plasmodesmata and sieve plates. The apoplastic pathway is important for radial movement within plant tissues and allows nanosize particles to reach the root central cylinder and the vascular tissues, for further movement upward the aerial part. Once inside the central vascular tissues, nanosize particles can move toward the aerial part though the xylem, following the transpiration stream. Symplastic transportation follows the casparian strip and endodermal cells of root. The sieve tube elements in the phloem may also contribute in symplastic transportation and allow distribution of nanosize particles in non-photosynthetic tissues and organs of plants such as fruits and grains. The uptake and transport of small-size plastic particles in plants heavily rely on transpiration pull. Transpiration stream allows movement of plastic particles from the roots to the shoots via the vascular system. Lie et al. (2021) demonstrated that PS NPs can accumulate in tissue of aerial parts of cucumber plant. MP accumulated in in leaves, flowers, and fruits. The 300 nm PS NPs significantly enhanced physiological activity of root, malondialdehyde (MDA) production, and proline content of the roots (Lie et al. 2021).

Accumulation of NPs in plant can damage plant cell membrane, alter intracellular molecules, and become source of oxidative stress. PS NPs (100 nm) inhibited growth of *Vicia faba* by blocking cell wall pores and inhibiting nutrient transport (Jiang et al. 2019). Maity et al. (2020) observed that PS NPs exerted morpho- and cytogenotoxic effects in *Allium cepa* by prompting production reactive oxygen species. ROS generation-induced chromosomal abnormalities and lowers the expression of the cdc2 gene (Maity et al. 2020).

Liao et al. (2019) observed that larger particle-sized microplastics have a greater impact on reducing seed germination than smaller particles. Large-size PS-MPs (100 nm) significantly reduced tomato seed germination as compared to small-size PS microplastics (5 mm). In one of the study, Lian et al. (2020) found that high concentration of MPs in the soil promoted wheat seed germination, while low and medium concentration of MPs inhibited germination. This was attributed to variations in particle size, agglomeration, and charge of microplastic particles, which can affect the wheat seed's ability to germinate and grow under microplastic stress (Lian et al. 2020). Mondal et al. (2022) exposure to PET-MPs increased seed germination of chickpea. PET microplastics may have a priming effect on seed germination, possibly by breaking the seed coat and improving water uptake through microscopic pores, thereby enhancing the imbibition process (Mondal et al. 2022).

Interaction of root tissues with MPs adversely affect root physiological activity, root elongation, and fresh and dry biomass of root. Root system and depth might be critical factors in the internalization of MPs from soils. Crop plants with shallow root system are more prone to experience a high risk of internalization of MPs. In the soil culture, increased temperature and higher transpiration rate can accelerate the internalization of MPs in shoots, and leaves tend to increase with increase in temperature rate. MPs have been known to reduce root growth in several crops wheat, broad bean, and onion. Gao et al. (2021) are of view that MPs adhere to the root surface and block to root pores. Blockage of root pores disturbs water and nutrient uptake from soil leading to reduce root growth. Apart from physical damage, MPs are also capable of inducing oxidative stress in root (Dong et al. 2020). Small-size MPs having large specific surface area gets easily adsorbed on the root surface and thus have more potential to damage root. Wu et al. (2024) believed that small-size MPs accumulate in root via process of endocytosis. In addition to particle size, surface charge, functional group types, and shape of MPs, the zeta potential of MPs also affects the adsorption of MPs and their movement from soil to roots. In an interesting study Lian et al. (2022) exhibited that sharp-edged PLA MPs induce more severe adverse effects on plant growth and bacterial communities in the soil as compared to elliptical-shaped PE MPs.

Literatures suggest MPs may alter several physiological activity of plant such as photosynthesis, ionic homeostasis, and hormonal and redox regulation. Photosynthesis is a biological process that converts light energy, water, carbon dioxide, and minerals into energy-rich organic compounds. Organic compounds synthesized during photosynthesis are utilized by organism to carry out its day to day metabolic activities. Lian et al. (2021) exhibited that the presence of MPs in soil reduced photosynthetic pigments in economically important plants such as lettuce, cucumber, tomato, cabbage, lettuce, and pakchoi. Nonetheless, MP stress also amends leaf gas exchange ability, Rubisco activity, and efficiency of electron transport chain (ETC) ultimately altering photosynthetic process. It is thought that MPs interfere with the molecular structure of thylakoid membranes, lowering the efficiency of the ETC by hindering electron transfer activity in chloroplasts, which in turn reduces photosynthesis. They found that superoxide dismutase (SOD), catalase (CAT), and/or peroxidases (POD) activities increased regardless crop species (i.e., wheat, fava bean, rice, and maize) and then decreasing response at root tissue. Soil contaminants such MPs/NPs have ability to induce oxidative stress by production of reactive oxidation species (ROS). Balanced levels of ROS are essential because they contribute significantly to the regulation of the transition between cellular proliferation and differentiation. Elevated levels of ROS can lead to DNA damage and disrupt the timing of programmed cell death, posing serious risks to plant health. MP stress disrupts the balance between ROS production and ROS scavenging by increasing the ROS production beyond the capacity of plants to activate the antioxidant defense system, which subsequently disrupts the integrity of membranes, decreases mitotic cell division, and even causes cytogenetic anomalies and micronuclei. MPs in agricultural soils are estimated to have further enhanced toxicity due to the application of pesticides. Studies have shown that antioxidant enzymes and antioxidants are generated by plants to scavenge ROS and protect them from oxidative damage by PMs. Exposure to MPs increases activity of antioxidant enzymes such as superoxide dismutase (SOD), catalase (CAT), and/or peroxidases (POD). The size and type of MPs have significant impact on the oxidative stress response of plants. Larger-size MPs do not enter the roots but can increase ROS response by blocking root pores in root tips. The penetration and buildup of MPs in plants might not trigger a ROS response at the root tip but can enhance the ROS response within the root tissues. MPs that accumulate within plant tissues are gradually broken down, leading to an initial rise in the ROS response in root tissues, followed by a decrease in the ROS response in the root tips. Biba et al. (2023) observe that PS-MP treatments increased ROS production along with lipid and protein oxidation as compared to the same concentrations of PMMA-MP exposure. A marked increase in the activity of H_2O_2-scavenging enzymes, including catalase and peroxidases, was observed following exposure to PMMA-MP and PS-MP in onion roots (Biba et al. 2023). When the concentration of MPs is high enough to attach to the root surface, it can cause stress by disrupting the absorption of water and nutrients. Few studies have reported exposure to MPs increase in the root length and biomass. Several plants tend to lengthen their root system under stressful condition (Biba et al. 2023).

6 Removal of MPs from Soils

Removing MPs/NPs from soils is challenging due to complex composition of soil, random accumulation in soil, and limited understanding of MP/NP accumulation in the soil. The MP/NP removal strategies are required to protect soil health in all possible ways. Single strategies are insufficient and thus require to combining multiple approaches to effectively remove MPs/NPs from soils. Technologies such as filtration (such as membrane bioreactor and rapid sand filtration) and electrocoagulation have been demonstrated to remove small-size particle from the wastewater facilities but cannot be applied to MPs in soils. Aforementioned technologies can help reduce MPs in biosolids and sewage sludge, thereby lowering the risk of MP contamination on agricultural land.

Photocatalytic degradation can decompose plastic particles into CO_2 and H_2O. However, this approach encounters obstacles since soil restricts sunlight penetration and shields MPs with its particles. Ferromagnetic biochar can hinder MPs from seeping through the soil and can even be partially recovered. Nonetheless, challenges arise with the installation of ferromagnetic biochar and its interactions with clay and soil organisms. Therefore, integrating soil microorganisms with soil amendments holds potential for effective MP removal from soils.

Applying soil fauna for biodegradation of MPs/NPs is another strategy that can be applied for eliminate from small-size plastic from soil. Terrestrial invertebrates such as earthworms, waxworms, and mealworms are strong candidate to remove plastic contaminates from soil. Waxworms and mealworms have been reported to utilize and degrade plastic. However the primary challenge in using waxworms and mealworms is that they are not well adapted to the soil environment as since their sensitivity to moisture and temperature. Bioremediation of MPs in the soil by earthworm can be another promising strategy to remove plastic contaminant from soil. Earthworms have ability to improve soil properties and promote microbial biodegradation of MPs. Seo et al. (2024) believe that sustainable bioremediation relies on maintaining a healthy earthworm life cycle and thorough examination under real-world conditions as well as field-scale studies.

7 Conclusions

In the face of global challenges such as climate change, soil degradation, and food security, understanding the effect of emerging pollutants on the rhizosphere has become increasingly essential. MPs/NPs are among the emerging soil pollutants of concern. A critical analysis of the literature shows that MP/NP contamination in the rhizosphere can alter its ecology by influencing the diversity and function of microbial communities, plant physiology, and soil quality parameters. Interestingly, most research on the effects of MPs on the rhizosphere has evaluated only single aspects. In nature, however, multiple factors combine to influence the ecology of rhizosphere. Future research must take a holistic approach to understand how MP contamination alters microbe diversity and functioning of the rhizosphere. Additionally, using omics approaches to study the interactions between MPs/NPs and the rhizosphere microbiome can help bridge existing knowledge gaps. Understanding the interaction between MPs and microbiome can help in establishing microbial biodegradation as effective and environment friendly way to remove MPs from soil. Overall chapter exhibits that there are still lots of gaps that needs to be explored, and future work must direct the research to fill the gaps.

References

Aryal M (2024) Rhizomicrobiome dynamics: a promising path towards environmental contaminant mitigation through bioremediation. J Environ Chem Eng 12(2):112221

Biba R, Cvjetko P, Jakopčić M, Komazec B, Tkalec M, Dimitrov N, Begović T, Balen B (2023) Phytotoxic effects of polystyrene and polymethyl methacrylate microplastics on allium cepa roots. Plants (Basel) 12(4):747

Bouaicha O, Mimmo T, Tiziani R, Praeg N, Polidori C, Lucini L, Vigani G, Terzano R, Sanchez-Hernandez JC, Illmer P, Cesco S, Borruso L (2022) Microplastics make their way into the soil and rhizosphere: a review of the ecological consequences. Rhizosphere 22:100542

Bowya T, Balachandar D (2020) Rhizosphere engineering through exogenous growth-regulating small molecules improves the colonizing efficiency of a plant growth-promoting rhizobacterium in rice. 3 Biotech 10(6):277

Chae Y, An Y-J (2020) Nanoplastic ingestion induces behavioral disorders in terrestrial snails: trophic transfer effects via vascular plants. Environ Sci Nano 7:975–983

Chen S, Feng Y, Han L, Li D, Feng Y, Jeyakumar P, et al (2022) Responses of rice (Oryza sativa L.) plant growth, grain yield and quality, and soil properties to the microplastic occurrence in paddy soil. J Soils Sediments 22:2174–2183

Chen F, Aqeel M, Khalid N, Nazir A, Irshad MK, Akbar MU, Alzuaibr FM, Ma J, Noman A (2023) Interactive effects of polystyrene microplastics and Pb on growth and phytochemicals in mung bean (Vigna radiata L.). J Hazard Mater 449:130966

de Souza Machado, Lau AA, Kloas CW, Bergmann J, Bachelier JB, Faltin E et al (2019) Microplastics can change soil properties and affect plant performance. Environ Sci Technol 53:6044–6052

Dong Y, Gao M, Song Z, Qiu W (2020) Microplastic particles increase arsenic toxicity to rice seedlings. Environ Pollut 259:113892

Dong R, Liu R, Xu Y, Liu W, Wang L, Liang X et al. (2022) Single and joint toxicity of polymethyl methacrylate microplastics and As (V) on rapeseed (Brassia campestris L.). Chemosphere 291:133066

Hanif MN, Aijaz N, Azam K et al (2024) Impact of microplastics on soil (physical and chemical) properties, soil biological properties/soil biota, and response of plants to it: a review. Int J Environ Sci Technol. https://doi.org/10.1007/s13762-024-05656-y

Gao M, Liu Y, Dong Y, Song Z (2021) Effect of polyethylene particles on dibutyl phthalate toxicity in lettuce (Lactuca sativa L.). J Hazard Mater 401:123422

Guo Z, Li P, Yang X, Wang Z, Lu B, Chen W, Wu Y, Li G, Zhao Z, Liu G, Ritsema C, Geissen V, Xue S (2022) Soil texture is an important factor determining how microplastics affect soil hydraulic characteristics. Environ Int 165:107293

Jia L, Liu L, Zhang Y, Fu W, Liu X, Wang Q, Tanveer M, Huang L (2023) Microplastic stress in plants: effects on plant growth and their remediations. Front Plant Sci 14:1226484

Jiang X, Chen H, Liao Y, Ye Z, Li M, Klobu car G (2019) Ecotoxicity and genotoxicity of polystyrene microplastics on higher plant Vicia faba. Environ Pollut 250:831–838

Khaldoon S, Lalung J, Maheer U, Kamaruddin MA, Yhaya MF, Alsolami ES, Alorfi HS, Hussein MA, Rafatullah M (2022) A review on the role of earthworms in plastics degradation: issues and challenges. Polymers 14(21):4770

Li Z, Li Q, Li R et al (2021) The distribution and impact of polystyrene nanoplastics on cucumber plants. Environ Sci Pollut Res 28:16042–16053

Li Y, Feng HY, Xian ST, Wang JW, Zheng XB, Song XL (2023) Phytotoxic effects of polyethylene microplastics combined with cadmium on the photosynthetic performance of maize (Zea mays L.). Plant Physiol Bioch 203:108065

Li X, Shi F, Zhou M, Wu F, Su H, Liu X, Wei Y, Wang F (2024) Migration and accumulation of microplastics in soil-plant systems mediated by symbiotic microorganisms and their ecological effects. Environ Int 191:108965

Li Z, Li Q, Li R et al (2021) The distribution and impact of polystyrene nanoplastics on cucumber plants. Environ Sci Pollut Res 28:16042–16053

Liao Y C, Nazygul J, Li M, Wang XL, Jiang LJ (2019) Effects of microplastics on the growth, physiology, and biochemical characteristics of wheat (Triticum aestivum). Huan jing ke xue 40:4661–4667

Lian J, Wu J, Xiong H, Zeb A, Yang T, Su X, Su L, Liu W (2020) Impact of polystyrene nanoplastics (PSNPs) on seed germination and seedling growth of wheat (Triticum aestivum L.). J Hazard Mater 385:121620

Lian J, Liu W, Meng L, Wu J, Chao L, Zeb A et al. (2021) Foliar-applied polystyrene nanoplastics (PSNPs) reduce the growth and nutritional quality of lettuce (Lactuca sativa L.). Environ Pollut 280:116978

Lian Y, Weitao Liu, Ruiying Shi, Aurang Zeb, Qi Wang, Jiantao Li, Zeqi Zheng, Jingchun Tang (2022) Effects of polyethylene and polylactic acid microplastics on plant growth and bacterial community in the soil. J Hazard Mater 435:129057

Lü H, Tang GX, Huang YH, Mo CH, Zhao HM, Xiang L, Li YW, Li H, Cai QY, Li QX (2024) Response and adaptation of rhizosphere microbiome to organic pollutants with enriching pollutant-degraders and genes for bioremediation: a critical review. Sci Total Environ 912:169425

Luo T, Dai X, Wei W, Xu Q, Ni B-J (2023) Microplastics enhance the prevalence of antibiotic resistance genes in anaerobic sludge digestion by enriching antibiotic-resistant bacteria in surface biofilm and facilitating the vertical and horizontal gene transfer. Environ Sci Technol 57(39):14611–14621

Maity S, Chatterjee A, Guchhait R, De S, Pramanick K (2020) Cytogenotoxic potential of a hazardous material, polystyrene microparticles on Allium cepa L. J Hazard Mater 385:121560

Meng F, Harkes P, van Steenbrugge JJM, Geissen V (2023) Effects of microplastics on common bean rhizosphere bacterial communities. Appl Soil Ecol 181:104649

Miri S, Saini R, Davoodi SM, Pulicharla R, Brar SK, Magdouli S (2022) Biodegradation of microplastics: better late than never. Chemosphere 286 (Pt 1):131670

Mondal N K, Kundu S, Debnath P, Mondal A, Sen K (2022) Effects of polyethylene terephthalate microplastic on germination, biochemistry and phytotoxicity of Cicer arietinum L. and cytotoxicity study on Allium cepa L. Environ Toxicol Pharmacol 94:103908

Riaz S, Sahar R, Qader I, Burhan ZU, Alvi SK, Rasool SG, Siddiqui PJA, Shafique S (2024) Preliminary assessment of microplastic in rhizosphere and non-rhizosphere region of mangrove at four locations along Karachi coast, Pakistan. Mar Pollut Bull 205:116608. https://doi.org/10.1016/j.marpolbul.2024.116608

Seo Y, Zhou Z, Lai Y, Chen G, Pembleton K, Wang S, He JZ, Song P (2024) Micro- and nanoplastics in agricultural soils: assessing impacts and navigating mitigation. Sci Total Environ 931:172951

Shah T, Ali A, Haider G, Asad M, Munsif F (2023) Microplastics alter soil enzyme activities and microbial community structure without negatively affecting plant growth in an agroecosystem. Chemosphere 322:138188

Sharma IP, Giri K, Kumar N (2024) Editorial: rhizosphere manipulation for sustainable agriculture. Front Plant Sci 15

Verma KK, Song XP, Xu L, Huang HR, Liang Q, Seth CS, Li YR (2023) Nano-microplastic and agro-ecosystems: a mini-review. Front Plant Sci 14:1283852

Wei X, Fu T, He G, Zhong Z, Yang M, Lou F, He T (2023) Characteristics of rhizosphere and bulk soil microbial community of Chinese cabbage (*Brassica campestris*) grown in *Karst* area. Front Microbiol 14:1241436

Wu X, Lin L, Lin Z, Deng X, Li W, He T, Zhang J, Wang Y, Chen L, Lei Z, Liu C, Zhimin X (2024) Influencing mechanisms of microplastics existence on soil heavy metals accumulated by plants. Sci Total Environ 926:171878

Yi Z, Zhang Z, Chen G, Rengel Z, Sun H (2023) Microplastics have rice cultivar-dependent impacts on grain yield and quality, and nitrogenous gas losses from paddy, but not on soil properties. J Hazard Mater 446:130672

Yu H, Fan P, Hou J et al (2020) Inhibitory effect of microplastics on soil extracellular enzymatic activities by changing soil properties and direct adsorption: an investigation at the aggregate-fraction level. Environ Pollut 267:115544

Zang H, Zhou J, Marshall MR, Chadwick DR, Wen Y, Jones DL (2020) Microplastics in the agroecosystem: are they an emerging threat to the plant-soil system? Soil Biol Biochem 148:107926

Zeb A, Liu W, Ali N, Shi R, Wang Q, Wang J, Li J, Yin C, Liu J, Yu M, Liu J (2024) Microplastic pollution in terrestrial ecosystems: global implications and sustainable solutions. J Hazard Mater 461:132636

Zhang G, Zhang F, Li X (2019) Effects of polyester microfibers on soil physical properties: perception from a field and a pot experiment. Sci Total Environ 670:1–7

Zhang G, Liu D, Lin J, Kumar A, Jia K, Tian X, Yu Z, Zhu B (2023) Priming effects induced by degradable microplastics in agricultural soils. Soil Biol Biochem 180:109006

Zhu J, Liu S, Wang H, Wang D, Zhu Y, Wang J, He Y, Zheng Q, Zhan X (2022) Microplastic particles alter wheat rhizosphere soil microbial community composition and function. J Hazard Mater 436:129176

Zhuang Y, Wang H, Tan F, Wu B, Liu L, Qin H, Yang Z, He M (2024) Rhizosphere metabolic cross-talk from plant-soil-microbe tapping into agricultural sustainability: current advance and perspectives. Plant Physiol Biochem 210:108619

Ziani K, Ioniță-Mîndrican CB, Mititelu M, Neacşu SM, Negrei C, Moroşan E, Drăgănescu D, Preda OT (2023) Microplastics: a real global threat for environment and food safety: a state of the art review. Nutrients 15(3):617

Zong X, Zhang J, Zhu J, Zhang L, Jiang L, Yin Y, Guo H (2021) Effects of polystyrene microplastic on uptake and toxicity of copper and cadmium in hydroponic wheat seedlings (*Triticum aestivum* L.). Ecotoxicol Environ Saf 217:112217

Chapter 8
Bioremediation of Microplastic Pollution Using Microorganisms: Opportunities and Challenges

Ronika Thakur, Vedna Kumari, Neelam Bhardwaj, Tarun Sharma, and V. Nareshkumar

Abstract Plastic has emerged as an essential material in the modern world due to its cost-effectiveness, ease of processing, low density, water-resistance, and durability. These plastic wastes are managed by land filing, incineration, and recycling, while the rest accumulate in the aquatic and terrestrial environments impacting biotic and abiotic systems extensively as they further deteriorate to microplastics and nanoplastics contaminating food, human, and environment. Microorganisms have ability to break down plastics either by forming biofilms on the surface which decrease the hydrophobicity or they accelerate the biodegradation process by exchanging metabolic intermediates. Genetic engineering and gene-editing tools such as CRISPR/Cas9 are novel biotechnological approaches used for identification of the genes encoding enzymes involved in degradation pathways and modify and clone the concerned genes in suitable host microorganism which might accelerate the degradation of plastics. Considering the significance, this chapter emphasizes the role of modern biotechnological techniques for managing microplastic pollution.

Keywords Biodegradation · Microplastic contamination · Microorganisms · Genetic engineering · CRISPR/Cas9 · Biotechnology

---

R. Thakur · V. Kumari · N. Bhardwaj · V. Nareshkumar (✉)
Department of Genetics and Plant Breeding, CSK HPKV, Palampur, HP, India

T. Sharma
Department of Agronomy, CSK HPKV, Palampur, HP, India

K. Khanna et al. (eds.), *Microplastics and Soil Microbiome*,
https://doi.org/10.1007/978-981-96-4978-5_8

1 Introduction

Plastics have long existed and have benefited humanity in all aspects of existence. Due to their tough, brittle character, plastics are ideal as alternatives for any substance. Although plastics have countless benefits, they are considered to be humankind's worst inventions. Conventional, chemical, or thermal processes require a significant amount of energy to dissolve plastics, and they also leave behind a sizable number of issues once the degradation process is complete. The by-products of plastics' conventional degradation are likewise exceedingly bad for the environment. Because it requires a lot of money to decompose, plastic is typically not destroyed properly. They are left unaltered for the following 100 years after being dumped in open holes or piled high on mountains. However, the degradation of plastics over time have resulted into microplastics, which are particles of size smaller than 5 mm (Sharma et al. 2022).

Contamination from microplastics is a growing problem globally. Due to plastic pollution, more than 1 million seabirds and 100,000 aquatic species have perished (Othman et al. 2021). Microplastics have a huge surface area and are better able to permeate into biological cells because of their microscopic size. The hydrophobic surface of microplastics draws co-contaminants such heavy metals, drug-related toxins, flame retardants, and other plasticizers, which can subsequently penetrate living things. In the environment, microplastics are frequently resistant, which causes them to move up the food chain and eventually reach humans as the last consumers and results in biomagnification.

Both biological and non-biological methods are being investigated in the study of microplastic disintegration. As a method for treating microplastic, scientists have so far focused on different types and combinations of microorganisms, such as bacteria, fungi, bacterial consortia, and biofilms, that can degrade different microplastics (Yuan et al. 2020). Additionally, the presence of naturally occurring microbes, such as diatoms, dinoflagellates, cyanobacteria, barnacles, bryozoans, and isopods, has aided in the degradation of microplastics as they use plastics as a carbon source in their respiratory metabolic pathways. However, these microbes are not constant and have been observed to vary from one ecological niche to another depending on the local environmental conditions (Sharma et al. 2022). Moreover, recently the microbial enzymes are being exploited for their potential of degrading polymers.

Despite having many striking features, the manufacturing of plastic products leads to serious environmental consequences that necessitate immediate solutions. Microorganisms may break down and metabolize plastics, indicating the possible use of biotechnologies based on the plastic biodegradation mechanisms (Qin et al. 2021). Gene-editing tools such as zinc finger proteins, transcription activator-like effector nucleases, and, more recently, the clustered regularly interspaced palindromic repeats (CRISPR)/Cas9 have been applied for genome engineering of plants, animals, and microorganisms to alter the expression of specific genes (Bhattacharyya et al. 2023; Biswas et al. 2022; Mandal et al. 2022; Jiang et al. 2023). These strategies can be efficiently used to incorporate genes encoding enzymes like polyethylene terephthalate hydrolase, esterase, dehalogenase, laccase, and depolymerase associated with microplastic degradation. Three different CRISPR sequences were identified for oxygenase enzyme associated with polyethylene degradation in *Streptomyces albogriseolus* LBX-2 (Shao et al. 2019).

Furthermore, these promising technologies have facilitated biodegradation by creating genetically engineered PETase, a polyethylene terephthalate digesting enzyme through the mutation of two active site residues and achieved an enhanced PET degradation capacity enzyme (Austin et al. 2018). In addition, genomics/proteomics and genetic tools have escalated the understanding degradation mechanisms and modification of microorganisms/enzymes for efficient biodegradation use (Arkatkar et al. 2009). Recently, Yan et al. (2020) have engineered a thermophilic anaerobic bacterium *Clostridium thermocellum* which is capable of secreting a thermophilic cutinase enzyme, helpful in degrading polyethylene terephthalate polymer into soluble monomers under high temperatures up to 70 °C indicating its promising ability to process microplastic waste.

Additionally, nanotechnology has begun to be used in plastic deterioration due to its extraordinary ability and range of capabilities (Chandran et al. 2023). Nanotechnology offers not only a cost-effective way to address the issue of plastic degradation but also a sustainable and environmentally beneficial one. The production of brittle, thermally stable biopolymers to replace conventional plastic products, which was made possible by nanotechnology, had a positive effect in this area.

Therefore, this review underlines recent advancements in the biotechnology-based biodegradation of polymers and focuses on the mechanisms of biodegradation that are workable solutions to the plastic waste problem, as this is one of the enlightened solutions for enduring issues that plastics provide globally.

2 Microplastics in the Environment

Microplastics are small plastic particles that measure less than 5 mm in diameter (Grünzner et al. 2023). They can come from various sources, including cosmetics, clothing, and industrial processes. Microplastics can enter the environment through a variety of routes, such as wastewater treatment plants, runoff from urban areas, and agricultural runoff (Yu et al. 2020). Once in the environment, microplastics can have a significant impact on ecosystems and wildlife (Susanti et al. 2020). They can be ingested by aquatic and terrestrial animals, leading to physical harm and even death (Law and Thompson 2014). Microplastics can also absorb and concentrate toxins, which can then be transferred up the food chain (Cverenkárová et al. 2021). In addition to the direct impact on wildlife, microplastics can also have indirect effects on the environment. For example, they can alter the composition of sediments and soil, affecting nutrient cycling and the health of plants (Wu et al. 2019; Shen et al. 2022). Reducing the amount of microplastics in the environment requires a multi-faceted approach. This includes reducing the use of plastic products, improving waste management practices, and developing new technologies to capture and remove microplastics from the environment. It is also important to raise public awareness about the issue and promote individual actions that can reduce microplastic pollution.

3 Microplastics Come in a Variety of Shapes and Sizes and Can Be Classified Into Two Main Categories (An et al. 2020)

1. Primary microplastics: These are microplastics that are intentionally produced and used for various applications, such as microbeads in personal care products or as microfibers in clothing. Primary microplastics are typically smaller than 5 mm in diameter (Fu and Wang 2019).
2. Secondary microplastics: These are microplastics that are formed from the degradation and breakdown of larger plastic items such as bags, bottles, and other plastic debris that have been discarded into the environment. Secondary microplastics can range in size from a few millimeters to microscopic particles (Galafassi et al. 2019; Peng et al. 2020).

Microplastics can also be classified based on their shape and composition. Some common types of microplastics include:

1. Microbeads: These are spherical plastic particles commonly found in personal care products, such as exfoliating scrubs and toothpaste.
2. Microfibers: These are thin, elongated plastic particles that are shed from synthetic clothing, such as polyester and nylon.
3. Microfragments: These are small fragments of plastic that have broken off from larger plastic items, such as plastic bottles or bags.
4. Microfilms: These are thin plastic films that can be found in various industrial applications, such as packaging and agriculture.

Regardless of their type or origin, microplastics can have negative impacts on the environment and human health, and efforts are underway to reduce their production and release into the environment.

4 Primary Sources of Microplastics

There are many primary sources of microplastics in the environment. Some of the most common include:

1. Synthetic textiles: Clothing made from synthetic materials such as polyester, nylon, and acrylic shed microfibers when washed (Siegfried et al. 2017). These microfibers can make their way into wastewater and ultimately end up in rivers, lakes, and oceans.

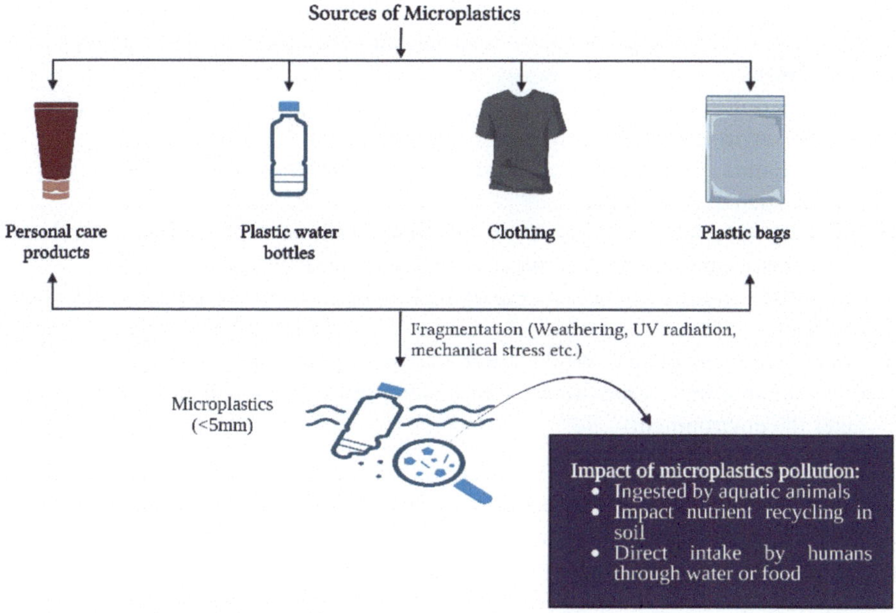

Fig. 8.1 Sources of microplastics and its impact on environment

2. Plastics in the environment: Plastics that have been discarded or littered can break down over time into smaller pieces, eventually becoming microplastics.
3. Personal care products: Some personal care products, such as toothpaste and exfoliating scrubs, contain microbeads made of plastic. These microbeads can be washed down the drain and enter the environment (Barnes et al. 2009).
4. Industrial processes: Industrial processes that involve the use of plastic pellets or powders can result in the release of microplastics into the environment.
5. Agriculture: The use of plastic mulch and other agricultural plastics can contribute to the release of microplastics into the environment (Fig. 8.1).

Reducing the amount of microplastics in the environment requires addressing these primary sources through a combination of regulatory action, industry initiatives, and individual behavior change.

5 Secondary Sources of Microplastics

Secondary sources of microplastics in the environment are those that result from the breakdown of larger plastic items or the weathering of other plastic-containing materials (Barnes et al. 2009; Browne et al. 2011). Some examples of secondary sources of microplastics include:

1. Tires: As tires wear down, small particles of rubber are released into the environment. Some of these particles are microplastics (Knight et al. 2020; Parker-Jurd et al. 2021).
2. Road markings: Some road markings contain microplastics that can be released into the environment as the markings wear down.
3. Paints: Paints can contain microplastics, particularly those used for marine applications such as boat hulls.
4. Building materials: Some building materials, such as vinyl siding, can release microplastics into the environment as they weather over time.
5. Landfills: Landfills can be a source of microplastics as plastic waste breaks down over time.
6. Water treatment plants: While water treatment plants are designed to remove contaminants from wastewater, some microplastics may still pass through and enter the environment.

It is important to address both primary and secondary sources of microplastics in order to reduce the amount of these particles in the environment. This can involve a range of approaches, from reducing plastic use to improving waste management and promoting the development of more sustainable materials.

6 Microplastic Degradation: Fungi

Fungi primarily comprise a diverse group of organisms, most of which are saprotrophs, or opportunistic or obligate parasites. They can thrive in a variety of terrestrial and aquatic environments and climatic conditions due to their remarkable adaptability (Anand et al. 2023). Various fungal enzymes are involved in microplastic degradation. They generate a wide array of intra- and extracellular enzymes capable of catalyzing different reactions and breaking down petroleum-based polymers. The intracellular enzymes play a crucial role in fungal adaptation and detoxification processes (Schwartz et al. 2018). Furthermore, extracellular enzymes like hydrolases play a role in breaking down complex polymers (Sánchez 2009) and enhancing the solubility of pollutants, thereby reducing bioaccumulation (Olicón-Hernández et al. 2017). The fungal species such as *Aspergillus niger*, *Cladosporium*, *Fusarium solani*, *Penicillium simplicissimum*, and *Zalerion maritimum* are linked to the breakdown of various types of polymers, viz., polyethylene terephthalate polypropylene, polyethylene, etc. (de Oliveira et al. 2020; Magnin et al. 2020). Laccase, an enzyme produced by lignin-degrading fungi, catalyzes the oxidation of both aromatic and non-aromatic substrates, such as chlorophenolic and nonphenolic compounds, as well as lignin itself (Straub et al. 2017). Since PP can be degraded under elevated temperatures and rapid rates in large-scale reactors, these enzymes' thermostability may encourage their use in these systems. These fungal species use microplastics as a carbon source after degradation by extracellular enzymes, promoting the formation of various chemical groups, such as carbonyl, carboxyl, and ester, while reducing their hydrophobicity. Fungal biodegradation of microplastics plays a crucial role in addressing microplastic pollution. Regrettably, there are still a lot of obstacles to overcome before microbes can effectively remove microplastics from the environment. The biodegradation process is influenced by factors such as microbial species, initial biomass, the hydrophobicity of the plastic, and the size of the plastic particles. A study conducted by Paço et al. (2019) concluded that the basidiomycete *Zalerion maritimum* has shown high removal efficiencies of polyethylene microplastics; in contrary, *Nia vibrissa*, another basidiomycete fungus, demonstrates reduced biodegradation efficiency under identical conditions. Besides, Hadad et al. (2005) have explained the importance of *Brevibacillus borstelensis* in microplastics degradation.

Devi et al. (2015) found that the fungal strains *Aspergillus tubingensis* VRKPT1 and *Aspergillus flavus* VRKPT2 degraded high-density polyethylene from marine coastal regions. Additionally, promising degradation of low-density polyethylene was observed by *Mucor circinelloides*, and *Aspergillus flavus* isolated from a municipal landfill was reported by Kunlere et al. (2019). Along with this, the pretreatment of microplastics with concentrated chemicals such as sodium hydroxide and nitric acid enhances the rate of polyethylene biodecomposition by *Aspergillus niger* (Nwachukwu et al. 2010). Another physical pretreatment process, thermo-oxidation at 80 °C for 15 days, was necessary to induce degradation of low-density polyethylene mediated by *Aspergillus niger* and *Penicillium pinophilum* (Volke-Sepúlveda et al. 2002; Anand et al. 2023).

7 Microplastic Degradation: Bacteria

Multiple researches on the breakdown of MPs utilizing bacteria have been conducted all over the world. MP biodegradation can be accomplished using both pure cultures and a consortium (Habib et al. 2020). Most of MP-degrading isolates are separated from contaminated sediments, mangroves, raw sewage, crop residues, cow manure, contaminated water, and land polluted with plastic debris. Microplastic biodegradation starts with a decrease in polymer length and structure, followed by alterations in functional groups that make them more prone to microbial degradation. The breakdown and assimilation of both hydrolysable and non-hydrolysable polymers involve a wide array of metabolic pathways and enzymes. In bacteria, *Pseudomonas* represented 21% of the genera associated with microplastic degradation, followed by *Bacillus* at approximately 15% and a combination of these two genera at 17% (Matjasic et al. 2021).

A newly isolated *Lysinibacillus* sp. *JJY0216* broke down polypropylene and polyethylene by 4% and 9%, respectively. The continuous activity of microbial depolymerizing enzymes is likely essential for this degradation (Jeon et al. 2021). Over time, the population (cfu/g soil) of five *Bacillus* strains, viz., *B. cereus* (P3 and P13), *B. thuringiensis* (P8 and P10), and *B. licheniformis* (P6) were found to increase with the polymer blend consortium showing the greatest growth led directly to bacterial biofilm formation on the surface of polymers attributed by carbonyl bond breakage in polymeric blends, as well as a decrease in thermal stability. *Bacillus* species promote polymeric blend degradation, implying that bacterially treated samples have lower mechanical strength and thermal stability than controls (Jain et al. 2022).

It was later confirmed *Rhodococcus rhodochrous* is capable of biodegrading metal stearates, including polypropylene (PP) films which act as pro-oxidants (Bonhomme et al. 2013; Fontanella et al. 2013). In a subsequent study, Kowalczyk et al. (2016) discovered *Achromobacter xylosoxidans*, which not only formed a thin biofilm on high-density polymers but also degraded them. Auta et al. (2017) identified eight potential bacterial strain from mangrove sediment capable of degrading UV-treated microplastics such as PP, PS, and PE in their study. *Pseudomonas*, *Chelatococcus*, and *Lysinibacillus fusiformis* are among additional bacterial strains contributing to PP degradation. Diverse types of plastics have been reported to be broken down in the guts of arthropods, especially in insect larvae of mealworms (*Tenebrio molitor*) (Yang et al. 2015), Indian meal moths (*Plodia interpunctella*) (Yang et al. 2014), and wax moths (*Galleria mellonella*) (Brandon et al. 2018; Kong et al. 2019).

Several studies, in contrast, have found that utilizing a bacterial consortium is a more effective choice for the full mineralization of toxic by-products (Skariyachan et al. 2018; Muenmee et al. 2016). The bacterial isolates in the consortium exhibit symbiotic, synergistic, and mutualistic relationships, enhancing tolerance and activity throughout the treatment process (Singh and Wahid 2015). A mesophilic bacterial consortium consisting of *Paenibacillus* sp. and *Bacillus* sp. degraded polyethylene, leading to a 14.7% reduction in the dry mass of microplastic particles after 60 days (Park and Kim 2019).

8 Microplastic Degradation: Algae

With the benefits of a quick growth cycle, efficient photosynthetic activity, eco-friendliness, simplicity of use, and biomass production, algae have created opportunities for environmental purification. In photoautotrophic conditions, microalgae don't need organic carbon sources, and they typically don't have endotoxins, so they are far better suited for environmental applications and have the ability to degrade the complex polymers (Yan et al. 2016).

Two novel enzymes, namely, PET hydrolase and mono(2-hydroxyethyl) terephthalic acid hydrolase, both of which exhibit PET-degrading activity from *I. sakaiensis* 201-F6, were discovered. These enzymes break down PET through hydrolysis to produce nontoxic molecular components (monomers) such as ethylene and terephthalic acid. A wide range of studies have been conducted since PETase's discovery in an effort to confirm its structure and improve its activity (Austin et al. 2018; Yoshida et al. 2016).

Recently, *Phaeodactylum tricornutum* was used in the generation of PETase. Liquid chromatography and electron microscopy were used to show that PETase was released, while *P. tricornutum* served as the host. But for *P. tricornutum* to grow and thrive, it needs low temperatures, silica as a nutrient, and high salinity. Because of these restrictions on application, a different microalgal host is required to be used in order to produce PETase (Moog et al. 2019). In a freshwater system, Chen et al. (2019) carried out an experiment that demonstrated how the development of microalgae biofilms affected the flotation of polypropylene sheets (squares measuring 5 and 10 mm on each side). It has been reported that hetero-aggregates, composed of algae and tiny plastic materials, could potentially incorporate microplastics, but this process is still poorly understood and therefore largely unpredictable.

The effects of the relationships between microplastic and microalgae on ecosystems represent the largest knowledge gap at the moment. When experimental findings indicate a reciprocal relationship between microalgae and plastics, it is challenging to determine how this relationship will play out in an ecosystem. Particularly, it is imperative to investigate the possible impacts of microplastic-microalgae interactions on the primary productivity of aquatic ecosystems, which could result in either positive or negative outcomes. Modeling techniques can be helpful to test the wide-ranging effects of various plastic pollution scenarios. To improve the performance of these models, accurate on-site observations are required to "validate" them.

9 Biotechnological Tools for Microplastic Degradation

Biotechnology has recently demonstrated significant potential, particularly in addressing plastic pollution through two key bio-based strategies: the production of biodegradable plastics and the utilization of organisms capable of degrading plastics. The development of biodegradable plastics aims to mitigate the accumulation of plastics in the environment, thereby reducing the formation of microplastics, as these materials are believed to be more readily decomposed by microorganisms (Vroman and Tighzert 2009). These materials may be created by pro-oxidants or by combining regular plastics with biopolymers, such as starch. Unlike macroplastics, which possess a lower surface-to-volume ratio, microplastics may be more effectively addressed and remediated through biotechnological approaches.

To access environmental "food resources" and space, biotechnological tools must be integrated with microbiota specifically adapted to microplastics. In order to find yet another species (or assemblages) that may be able to promote or facilitate the degradation of plastics, it is equally worthwhile to investigate this interaction. The cloning and transformation of suitable host microorganisms under promoter control, employing techniques such as PCR, site-directed mutagenesis, and antisense RNA technology (Kumar et al. 2018), enable microbes to efficiently metabolize xenobiotic compounds (Hussain et al. 2018).

The activity of Cutinase (Cut190), a genetically modified enzyme derived from *Saccharomonospora viridis* AHK190, depends on calcium binding, with inactivity observed in calcium's absence. Engineering a disulfide bond at the second of its three calcium-binding sites increased the melting temperature, indicating the disulfide bond mimics calcium stabilization. Furthermore, replacing surface Asn and Gln residues with Asp, Glu, and His residues resulted in an increase in the melting temperature of the enzymes (Oda et al. 2018). Compared to the native enzyme, mutant variants displayed enhanced capability in breaking down polyethylene terephthalate (PET), with the thermophilic cutinase from the plant compost metagenome functioning at temperatures as high as 70 °C. The signal peptide of exoglucanase Cel48S was utilized to clone Cutinase into *Clostridium thermocellum* for robust gene expression and extracellular secretion. Although thermostable PET hydrolases have been expressed in *E. coli*, *B. subtilis*, *B. megaterium*, and *Pichia pastoris*, only recombinant *C. thermocellum* is capable of cutinase expression and PET degradation at 60 °C (Yan et al. 2021). It is evident that the microbe-produced enzyme cutinase acts best at a temperature of 75 °C and breaks down polyester linkages, to degrade polyethylene terephthalate. According to Shirke et al. (2018), genetically modified yeast generates bacterial cutinase that aids in the breakdown of polyethylene terephthalate by preventing formation in key locations with sugars. Islam et al. (2019) reported that the genetically modified version of the cutinase enzyme shortens the degradation time significantly, reducing it from 41.8 to 6.2 h.

Using plasmids to incorporate a novel biosynthetic pathway, *E. coli* was genetically engineered to convert plastic-derived terephthalic acid into vanillin, a valuable ingredient in the food and cosmetics industries. Key benefits of this genetic engineering method include its eco-friendly nature, with no hazardous waste generated, its operation in an aqueous medium at room temperature with a pH of 5.5–7, and the lack of any requirement for added cofactors or reagents (Sadler and Wallace 2021). Genetic engineering of *Pseudomonas aeruginosa*, involving the removal of the *wspF* gene, increased its biofilm's ability to accumulate microplastics by enhancing exopolymeric substance production, while the *yhjH* gene was inserted under an arabinose-induced promoter. By lowering cyclic dimeric guanosine monophosphate levels, the induced expression of the *yhjH* gene reduced biofilm formation enough to release trapped microplastics, allowing the development of bioremediation tools for aquatic ecosystems (Liu et al. 2021).

The primary digestive enzymes, such as manganese-dependent peroxidase, responsible for the breakdown of microplastics were produced by genetically modified strains of *E. coli* and *S. cerevisiae* BY4741. Laccase enzymes synthesized by genetically engineered strains of *E. coli* BL21 and *P. chrysosporium* have shown improved efficiency in the degradation of polyethylene terephthalate (Sharma et al. 2018; Paco et al. 2019).

The manipulation of genomes in plants, animals, and microorganisms to achieve targeted gene expression has been made possible by tools like TALENs, meganucleases, zinc finger proteins, CRISPR/Cas9, and base editing technologies. By altering a gene of interest, genome editing helps facilitate loss- and gain-of-function studies that modulate the expression of multiple genes.

Meganucleases/homing endonucleases is a double-stranded DNase that target large deoxyribonucleic acid (DNA) recognition sites present in the genome and termed as molecular DNA scissors. Gene targeting is significantly enhanced by homing endonucleases, which induce double-strand breaks repaired through homologous recombination. As robust platforms for engineering, meganucleases allow the creation of enzymes that cut DNA accurately and facilitate recombination for genome editing purposes. Compared to other sequence-specific nucleases, engineered meganucleases are less versatile due to the complexity of redesigning them for new DNA specificities (Daboussi et al. 2015; Zess and Begemann 2021).

A ZFN is a synthetic endonuclease made of a zinc finger protein (ZFP) linked to the *FokI* restriction enzyme's cleavage domain. The components generally consist of amino acid modules, no longer than 30 residues, that attach to DNA triplets. All 64 trinucleotide combinations have been recognized by ZNFs, and any particular sequence of DNA triplets can be recognized by ZNFs by stringing together various zinc finger moieties. A single ZNF typically recognizes 3–6 nucleotide triplets, and 2 ZNFs are necessary for targeting because the nucleases they work with require dimerization. One ZNF targets the upstream sequence, while the other targets the downstream region. By creating ZFPs with new sequence specificities, a ZFN can be reconfigured to cleave new targets (Guo et al. 2010).

TALENs, first described in 2009 (Boch et al. 2009), are derived from *Xanthomonas* bacteria and feature TALE proteins that bind DNA, making them valuable tools in genetic engineering due to their adaptability to diverse functional domains. By associating with transcriptional activators, repressors, or endonucleases, TALE proteins can shift their role from transcriptional regulation to genome editing (Thakore and Gersbach 2016). It is a potent technique with excellent precision and specificity while exhibiting minimal cytotoxicity, making it perfect for microbial DNA manipulation to produce enzymes that can be used efficiently for microplastics degradation.

CRISPR/Cas9 (clustered regularly interspaced short palindromic repeats) has emerged as a breakthrough site-specific genome-editing technique that has been widely used in a variety of organisms. CRISPR/Cas9 allows scientists to create transgene-free genome-edited organisms in a short span of time, allowing them to target and modify specific DNA sequences (Osakabe and Osakabe 2015). Genome-editing technologies introduce mutations in three main stages. Initially, an engineered nuclease comprising a recognition module and nuclease domain recognizes the specific target DNA sequence. The nuclease then binds to this sequence and creates double-strand breaks (DSBs) either at or near the target site. Once the DSB is created, the cell's endogenous repair mechanisms, NHEJ or HDR, step in to fix it. NHEJ frequently introduces small insertions or deletions (Indels) due to its error-prone nature, while HDR achieves accurate repair of the DSB (Wang et al. 2016). *Streptomyces albogriseolus* LBX-2, with its three distinct CRISPR sequences, is a promising candidate for genetic engineering and precise genome editing. This bacterium also plays a role in polyethylene degradation, facilitated by the oxygenase enzyme, which catalyzes the oxidation process. Its ability to break down a major environmental pollutant, combined with its genetic engineering potential, makes it valuable for both environmental and industrial applications (Shao et al. 2019).

As a more recent method of genome editing, base editing employs elements of CRISPR systems along with additional enzymes to introduce point mutations directly into cellular DNA or RNA without creating double-stranded DNA breaks (DSBs). Combining a disabled nuclease with a nucleobase deaminase enzyme, and sometimes a DNA glycosylase inhibitor, DNA base editors enable precise base pair conversions. The nuclease ensures the editor is directed to the target DNA, while the deaminase enables targeted editing, avoiding double-strand breaks and reducing off-target edits. RNA base editors use components that target RNA to make analogous changes. Base editors enable the efficient installation of point mutations in non-diverging cells without producing an excessive amount of undesirable editing by-products. They do this by directly converting one base or base pair into another (Rees and Liu 2018). The described approaches can incorporate genes encoding microplastic-degrading enzymes, such as polyethylene terephthalate hydrolase, dehalogenase, esterase, depolymerase, and laccase, into suitable host organisms. These enzymes break down plastics and pollutants, facilitating the degradation of microplastics in environmental settings. Introducing these genes into bacteria, fungi, or algae enhances their plastic-degrading capabilities. This strategy holds promise for bioremediation and sustainable waste management, helping address microplastic pollution (Anand et al. 2023).

10 Role of Bioinformatics

Bioinformatics is an emerging branch of biological science which deals with analysis, collection, classification, storage, and visualization of all biological information using information technology. The term was given by two Dutch biologists named Paulien Hogweed and Ben Hesper in 1960. Bioinformatics generates the database, which are organized collections of structured information managed by a database management system. Bioinformatics has developed into a powerful instrument for accelerating the breakdown of plastic waste, including microscopic plastic particles (Purohit et al. 2020).

A number of databases (Table 8.1) of biodegradation pathways are publicly accessible online to assess the degradation process by providing insights into the metabolic pathways, microbial enzymes, and genes involved (Karp et al. 2019; Caspi et al. 2020). These databases and computational tools help identify enzymes involved in specific metabolic pathways and predict the biodegradation pathways of toxic chemicals, laying the groundwork for developing innovative strategies to degrade plastic (Ali et al. 2021; Anand et al. 2023).

Table 8.1 List of various databases used for microplastic degradation by microorganisms

Sr. no.	Database	Function	Source/reference
1.	Biocatalysis/biodegradation database	Information on microbial enzyme-catalyzed reactions that are valuable for synthetic applications or play a role in biodegradation pathways to eliminate chemicals from the environment	http://eawag-bbd.ethz.ch
2.	Plastic microbial biodegradation database (PMBD)	Database for identifying predicted plastic-active enzymes and microbial breakdown of plastics (Gan and Zhang 2019) 949 microorganisms and their relationships with plastics, 79 genes associated with plastic biodegradation, and over 8000 enzyme sequences automatically annotated as potentially involved in plastic degradation	http://pmbd.genome-mining.cn/home
3.	Pathway prediction system	To forecast potential metabolic pathways through which microorganisms may biodegrade chemical compounds	http://eawag-bbd.ethz.ch/predict/
4.	RAPID	Information on enzyme catalysis with main focus on identification of substrate-enzyme compatibilities	http://rapid.umn.edu/rapid/
5.	BRENDA	To gain a deeper understanding of the sources, substrates, inhibitors, kinetics, and other defining characteristics of the enzymes	https://www.brenda-enzymes.org
6.	BioCyc	Specific biocatalysis reactions, metabolism of specific microorganisms	https://biocyc.org
7.	Accelrys	Main focus on the use of enzymes and microorganisms for chemical synthesis for selectivity, efficiency, and greater environmental friendliness	http://accelrys.com/products/datasheets/biocatalysis.pdf
8.	Poymerdatabase	General information on plastics	http://polymerdatabase.com/main.html
9.	Plastics-active enzyme database (PAZy)	Only microbial enzymes that have been exhaustively characterized	https://doi.org/10.1002/prot.26325
10.	PlasticDB	Information on 579 microorganism species with plastic-degrading abilities and 182 proteins identified for plastic breakdown Information of microorganism, proteins, annotate gene, annotate genome, facility to compare genomes and availability of Fasta file with all protein sequence (Gambarini et al. 2022)	https://plasticdb.org/
11.	MetaCyc database	A database covering fatty acid biosynthesis, lipid metabolism, carotenoid metabolism, antibiotic and cytotoxin biosynthesis, bioluminescence, and more	http://metacyc.org/

11 Future Perspective

Microplastics may be harmful to the ecosystem and poses a risk to human health. Through the use of residues to create environmentally friendly polymers and the use of micro(nano)plastic waste to create products with added value, biotechnology is paving the way for circular economy practices. Intense research into the relationship between biodegradability and microbial activity is required. Furthermore, the transcriptomic and proteomic evaluations of different strain should be evaluated in order to identify critical enzymes and metabolic pathways for biodegradation.

12 Biosafety Issues

The use of genetically modified microorganisms for on-site experiment is hampered by a number of regulatory barriers. Additionally, these genetically altered microbes have demonstrated their effectiveness in laboratory conditions, but field tests are necessary to confirm this. An effort has already been made to create microorganisms that can effectively serve as self-eliminating microbial scavengers for the bioremediation of various toxic environmental pollutants using synthetic biology and metabolic engineering (French et al. 2020).

13 Conclusion

About 400 metric tons of plastic waste are produced annually, of which 180 metric tons are released into the atmosphere. Although microplastics are found everywhere, their concentrations are noticeably higher in agricultural soil, shorelines, beaches, estuaries, and industrial settings. Recent studies have documented the scope and severity of plastic pollution, which affects both living things and the environment, including humans. In the end, biotechnology can aid in the creation of circular economy policies because (bio-based) biodegradable plastics could serve as a viable sustainable substitute. However, since the conditions for their efficient biodegradation may not be met, it is important to take into account how they perform in environmental settings as well as the fate and effects of degradation products.

Acknowledgments The authors of this paper are grateful for providing the office space and other resources that made it possible to draught this manuscript.

References

Ali SS, Elsamahy T, Koutra E, Kornaros M, El-Sheekh M, Abdelkarim EA et al (2021) Degradation of conventional plastic wastes in the environment: a review on current status of knowledge and future perspectives of disposal. Sci Total Environ 771:144719

An L, Liu Q, Deng Y, Wu W, Gao Y, Ling W (2020) Sources of microplastic in the environment. In: Microplastics in terrestrial environments: emerging contaminants and major challenges, pp 143–159

Anand U, Dey S, Bontempi E, Ducoli S, Vethaak AD, Dey A, Federici S (2023) Biotechnological methods to remove microplastics: a review. Environ Chem Lett:1–24

Arkatkar A, Arutchelvi J, Sudhakar M, Bhaduri S, Uppara PV, Doble M (2009) Approaches to enhance the biodegradation of polyolefins. Open Environ Eng J 2:68–80

Austin HP, Allen MD, Donohoe BS, Rorrer NA, Kearns FL, Silveira RL et al (2018) Characterization and engineering of a plastic-degrading aromatic polyesterase. Proc Natl Acad Sci 115(19):E4350–E4357

Auta HS, Emenike CU, Fauziah SH (2017) Screening of Bacillus strains isolated from mangrove ecosystems in Peninsular Malaysia for microplastic degradation. Environ Pollut 231:1552–1559

Barnes DK, Galgani F, Thompson RC, Barlaz M (2009) Accumulation and fragmentation of plastic debris in global environments. Philos Trans R Soc B Biol Sci 364(1526):1985–1998

Bhattacharyya N, Anand U, Kumar R, Ghorai M, Aftab T, Jha NK et al (2023) Phytoremediation and sequestration of soil metals using the CRISPR/Cas9 technology to modify plants: a review. Environ Chem Lett 21(1):429–445

Biswas P, Anand U, Ghorai M, Pandey DK, Jha NK, Behl T et al (2022) Unraveling the promise and limitations of CRISPR/Cas system in natural product research: approaches and challenges. Biotechnol J 17(7):2100507

Boch J, Scholze H, Schornack S, Landgraf A, Hahn S, Kay S, Bonas U (2009) Breaking the code of DNA binding specificity of TAL-type III effectors. Science 326(5959):1509–1512

Bonhomme S, Brusson JM, Pitteri S, Samuel G, Pichon G et al (2013) Comparison of biodegradability of various polypropylene films containing pro-oxidant additives based on Mn, Mn/Fe or Co. Polym Degrad Stab 98(4):875–884

Brandon AM, Gao SH, Tian R, Ning D, Yang SS, Zhou J et al (2018) Biodegradation of polyethylene and plastic mixtures in mealworms (larvae of Tenebrio molitor) and effects on the gut microbiome. Environ Sci Technol 52(11):6526–6533

Browne MA, Crump P, Niven SJ, Teuten E, Tonkin A, Galloway T, Thompson R (2011) Accumulation of microplastic on shorelines worldwide: sources and sinks. Environ Sci Technol 45(21):9175–9179

Caspi R, Billington R, Keseler IM, Kothari A, Krummenacker M, Midford PE, Ong WK, Paley S, Subhraveti P, Karp PD (2020) The MetaCyc database of metabolic pathways and enzymes-a 2019 update. Nucleic Acids Res 48(D1):D445–D453. Fontanella, S

Chandran RR, Thomson BI, Natishah AJ, Mary J, Nachiyar V (2023) Nanotechnology in plastic degradation. Biosci Biotechnol Res Asia 20(1):53–68

Chen X, Xiong X, Jiang X, Shi H, Wu C (2019) Sinking of floating plastic debris caused by biofilm development in a freshwater lake. Chemosphere 222:856–864

Cverenkárová K, Valachovičová M, Mackuľak T, Žemlička L, Bírošová L (2021) Microplastics in the food chain. Life 11(12):1349

Daboussi F, Stoddard TJ, Zhang F (2015) Engineering meganuclease for precise plant genome modification. In: Advances in new technology for targeted modification of plant genomes, pp 21–38

de Oliveira TA, Barbosa R, Mesquita AB, Ferreira JH, de Carvalho LH, Alves TS (2020) Fungal degradation of reprocessed PP/PBAT/thermoplastic starch blends. J Mater Res Technol 9(2):2338–2349

Devi RS, Kannan VR, Nivas D, Kannan K, Chandru S, Antony AR (2015) Biodegradation of HDPE by Aspergillus spp. from marine ecosystem of Gulf of Mannar, India. Mar Pollut Bull 96(1–2):32–40

Fontanella S, Bonhomme S, Brusson JM, Pitteri S, Samuel G, Pichon G et al (2013) Comparison of biodegradability of various polypropylene films containing pro-oxidant additives based on Mn, Mn/Fe or Co. Polym Degrad Stab 98(4):875–884

French KE, Zhou Z, Terry N (2020) Horizontal 'gene drives' harness indigenous bacteria for bioremediation. Sci Rep 10(1):1–11

Fu Z, Wang J (2019) Current practices and future perspectives of microplastic pollution in freshwater ecosystems in China. Sci Total Environ 691:697–712

Galafassi S, Nizzetto L, Volta P (2019) Plastic sources: a survey across scientific and grey literature for their inventory and relative contribution to microplastics pollution in natural environments, with an emphasis on surface water. Sci Total Environ 693:133499

Gambarini V, Pantos O, Kingsbury JM, Weaver L, Handley KM, Lear G (2022) PlasticDB: a database of microorganisms and proteins linked to plastic biodegradation. Database 2022

Gan Z, Zhang H (2019) PMBD: a comprehensive plastics microbial biodegradation database. Database 2019

Grünzner M, Pahl S, White MP, Thompson RC (2023) Exploring expert perceptions about microplastics: from sources to potential solutions. Microplast Nanoplast 3(1):1–17

Guo J, Gaj T, Barbas CF III (2010) Directed evolution of an enhanced and highly efficient FokI cleavage domain for zinc finger nucleases. J Mol Biol 400(1):96–107

Habib S, Iruthayam A, AbdShukor MY, Alias SA, Smykla J, Yasid NA (2020) Biodeterioration of untreated polypropylene microplastic particles by Antarctic bacteria. Polymers 12(11):2616

Hadad D, Geresh S, Sivan A (2005) Biodegradation of polyethylene by the thermophilic bacterium *B revibacillus borstelensis*. J Appl Microbiol 98(5):1093–1100

Hussain I, Aleti G, Naidu R, Puschenreiter M, Mahmood Q, Rahman MM et al (2018) Microbe and plant assisted-remediation of organic xenobiotics and its enhancement by genetically modified organisms and recombinant technology: a review. Sci Total Environ 628:1582–1599

Islam S, Apitius L, Jakob F, Schwaneberg U (2019) Targeting microplastic particles in the void of diluted suspensions. Environ Int 123:428–435

Jain K, Bhunia H, Reddy MS (2022) Degradation of polypropylene-poly-L-lactide blends by Bacillus isolates: a microcosm and field evaluation. Biorem J 26(1):64–75

Jeon JM, Park SJ, Choi TR, Park JH, Yang YH, Yoon JJ (2021) Biodegradation of polyethylene and polypropylene by Lysinibacillus species JJY0216 isolated from soil grove. Polym Degrad Stab 191:109662

Jiang W, Yan W, Tan Q, Xiao Y, Shi Y, Lei J et al (2023) The toxic differentiation of micro-and nanoplastics verified by gene-edited fluorescent Caenorhabditis elegans. Sci Total Environ 856:159058

Karp PD, Billington R, Caspi R, Fulcher CA, Latendresse M, Kothari A et al (2019) The BioCyc collection of microbial genomes and metabolic pathways. Brief Bioinform 20(4):1085–1093

Knight LJ, Parker-Jurd FN, Al-Sid-Cheikh M, Thompson RC (2020) Tyre wear particles: an abundant yet widely unreported microplastic? Environ Sci Pollut Res 27:18345–18354

Kong HG, Kim HH, Chung JH, Jun J, Lee S, Kim HM et al (2019) The Galleria mellonella hologenome supports microbiota-independent metabolism of long-chain hydrocarbon beeswax. Cell Rep 26(9):2451–2464

Kowalczyk A, Chyc M, Ryszka P, Latowski D (2016) Achromobacter xylosoxidans as a new microorganism strain colonizing high-density polyethylene as a key step to its biodegradation. Environ Sci Pollut Res 23:11349–11356

Kumar NM, Muthukumaran C, Sharmila G, Gurunathan B (2018) Genetically modified organisms and its impact on the enhancement of bioremediation. In: Bioremediation: applications for environmental protection and management. Springer, Singapore, pp 53–76

Kunlere IO, Fagade OE, Nwadike BI (2019) Biodegradation of low density polyethylene (LDPE) by certain indigenous bacteria and fungi. Int J Environ Stud 76(3):428–440
Law KL, Thompson RC (2014) Microplastics in the seas. Science 345(6193):144–145
Liu SY, Leung MML, Fang JKH, Chua SL (2021) Engineering a microbial 'trap and release' mechanism for microplastics removal. Chem Eng J 404:127079
Magnin A, Pollet E, Phalip V, Avérous L (2020) Evaluation of biological degradation of polyurethanes. Biotechnol Adv 39:107457
Mandal S, Ghorai M, Anand U, Roy D, Kant N, Mishra T et al (2022) Cytokinins: a genetic target for increasing yield potential in the CRISPR era. Front Genet 13
Matjašič T, Simčič T, Medvešček N, Bajt O, Dreo T, Mori N (2021) Critical evaluation of biodegradation studies on synthetic plastics through a systematic literature review. Sci Total Environ 752:141959
Moog D, Schmitt J, Senger J, Zarzycki J, Rexer KH, Linne U et al (2019) Using a marine microalga as a chassis for polyethylene terephthalate (PET) degradation. Microb Cell Factories 18(1):1–15
Muenmee S, Chiemchaisri W, Chiemchaisri C (2016) Enhancement of biodegradation of plastic wastes via methane oxidation in semi-aerobic landfill. Int Biodeterior Biodegradation 113:244–255
Nwachukwu S, Obidi O, Odocha C (2010) Occurrence and recalcitrance of polyethylene bag waste in Nigerian soils. Afr J Biotechnol 9(37):6096–6104
Oda M, Yamagami Y, Inaba S, Oida T, Yamamoto M, Kitajima S, Kawai F (2018) Enzymatic hydrolysis of PET: functional roles of three Ca 2+ ions bound to a cutinase-like enzyme, Cut190*, and its engineering for improved activity. Appl Microbiol Biotechnol 102:10067–10077
Olicón-Hernández DR, González-López J, Aranda E (2017) Overview on the biochemical potential of filamentous fungi to degrade pharmaceutical compounds. Front Microbiol 8:1792
Osakabe Y, Osakabe K (2015) Genome editing with engineered nucleases in plants. Plant Cell Physiol 56(3):389–400
Othman AR, Hasan HA, Muhamad MH, Ismail NI, Abdullah SRS (2021) Microbial degradation of microplastics by enzymatic processes: a review. Environ Chem Lett 19:3057–3073
Paço A, Jacinto J, da Costa JP, Santos PS, Vitorino R, Duarte AC, Rocha-Santos T (2019) Biotechnological tools for the effective management of plastics in the environment. Crit Rev Environ Sci Technol 49(5):410–441
Park SY, Kim CG (2019) Biodegradation of micro-polyethylene particles by bacterial colonization of a mixed microbial consortium isolated from a landfill site. Chemosphere 222:527–533
Parker-Jurd FN, Napper IE, Abbott GD, Hann S, Thompson RC (2021) Quantifying the release of Tyre wear particles to the marine environment via multiple pathways. Mar Pollut Bull 172:112897
Peng L, Fu D, Qi H, Lan CQ, Yu H, Ge C (2020) Micro-and nano-plastics in marine environment: source, distribution and threats – a review. Sci Total Environ 698:134254
Purohit J, Chattopadhyay A, Teli B (2020) Metagenomic exploration of plastic degrading microbes for biotechnological application. Curr Genomics 21(4):253–270
Qin ZH, Mou JH, Chao CYH, Chopra SS, Daoud W, Leu SY et al (2021) Biotechnology of plastic waste degradation, recycling, and valorization: current advances and future perspectives. ChemSusChem 14(19):4103–4114
Rees HA, Liu DR (2018) Base editing: precision chemistry on the genome and transcriptome of living cells. Nat Rev Genet 19(12):770–788
Sadler JC, Wallace S (2021) Microbial synthesis of vanillin from waste poly (ethylene terephthalate). Green Chem 23(13):4665–4672
Sánchez C (2009) Lignocellulosic residues: biodegradation and bioconversion by fungi. Biotechnol Adv 27(2):185–194

Schwartz M, Perrot T, Aubert E, Dumarçay S, Favier F, Gérardin P et al (2018) Molecular recognition of wood polyphenols by phase II detoxification enzymes of the white rot Trametes versicolor. Sci Rep 8(1):8472
Shao H, Chen M, Fei X, Zhang R, Zhong Y, Ni W et al (2019) Complete genome sequence and characterization of a polyethylene biodegradation strain, Streptomyces albogriseolus LBX-2. Microorganisms 7(10):379
Sharma B, Dangi AK, Shukla P (2018) Contemporary enzyme based technologies for bioremediation: a review. J Environ Manag 210:10–22
Sharma P, Mahongnao S, Nanda S (2022) Microbial degradation of micro-plastics. In: Plastic and microplastic in the environment: management and health risks, pp 222–244
Shen M, Song B, Zhou C, Almatrafi E, Hu T, Zeng G, Zhang Y (2022) Recent advances in impacts of microplastics on nitrogen cycling in the environment: a review. Sci Total Environ 815:152740
Shirke AN, White C, Englaender JA, Zwarycz A, Butterfoss GL, Linhardt RJ, Gross RA (2018) Stabilizing leaf and branch compost cutinase (LCC) with glycosylation: mechanism and effect on PET hydrolysis. Biochemistry 57(7):1190–1200
Siegfried M, Koelmans AA, Besseling E, Kroeze C (2017) Export of microplastics from land to sea. A modelling approach. Water Res 127:249–257
Singh L, Wahid ZA (2015) Methods for enhancing bio-hydrogen production from biological process: a review. J Ind Eng Chem 21:70–80
Skariyachan S, Patil AA, Shankar A, Manjunath M, Bachappanavar N, Kiran S (2018) Enhanced polymer degradation of polyethylene and polypropylene by novel thermophilic consortia of Brevibacillus sps. and Aneurinibacillus sp. screened from waste management landfills and sewage treatment plants. Polym Degrad Stab 149:52–68
Straub S, Hirsch PE, Burkhardt-Holm P (2017) Biodegradable and petroleum-based microplastics do not differ in their ingestion and excretion but in their biological effects in a freshwater invertebrate Gammarus fossarum. Int J Environ Res Public Health 14(7):774
Susanti NKY, Mardiastuti A, Wardiatno Y (2020) Microplastics and the impact of plastic on wildlife: a literature review. IOP Conf Ser Earth Environ Sci 528(1):012013
Thakore PI, Gersbach CA (2016) Design, assembly, and characterization of TALE-based transcriptional activators and repressors. TALENs Methods Protoc:71–88
Volke-Sepúlveda T, Saucedo-Castañeda G, Gutiérrez-Rojas M, Manzur A, Favela-Torres E (2002) Thermally treated low density polyethylene biodegradation by Penicillium pinophilum and Aspergillus Niger. J Appl Polym Sci 83(2):305–314
Vroman I, Tighzert L (2009) Biodegradable polymers. Materials 2(2):307–344
Wang H, La Russa M, Qi LS (2016) CRISPR/Cas9 in genome editing and beyond. Annu Rev Biochem 85(1):227–264
Wu P, Huang J, Zheng Y, Yang Y, Zhang Y, He F et al (2019) Environmental occurrences, fate, and impacts of microplastics. Ecotoxicol Environ Saf 184:109612
Yan N, Fan C, Chen Y, Hu Z (2016) The potential for microalgae as bioreactors to produce pharmaceuticals. Int J Mol Sci 17(6):962
Yan F, Wei R, Cui Q, Bornscheuer UT, Liu YJ (2020) Microb Biotechnol 14(2):374–385
Yan F, Wei R, Cui Q, Bornscheuer UT, Liu YJ (2021) Thermophilic whole-cell degradation of polyethylene terephthalate using engineered Clostridium thermocellum. Microb Biotechnol 14(2):374–385
Yang J, Yang Y, Wu WM, Zhao J, Jiang L (2014) Evidence of polyethylene biodegradation by bacterial strains from the guts of plastic-eating waxworms. Environ Sci Technol 48(23):13776–13784
Yang Y, Yang J, Wu WM, Zhao J, Song Y, Gao L et al (2015) Biodegradation and mineralization of polystyrene by plastic-eating mealworms: part 2. Role of gut microorganisms. Environ Sci Technol 49(20):12087–12093
Yoshida S, Hiraga K, Takehana T, Taniguchi I, Yamaji H, Maeda Y et al (2016) A bacterium that degrades and assimilates poly (ethylene terephthalate). Science 351(6278):1196–1199

Yu Q, Hu X, Yang B, Zhang G, Wang J, Ling W (2020) Distribution, abundance and risks of microplastics in the environment. Chemosphere 249:126059

Yuan J, Ma J, Sun Y, Zhou T, Zhao Y, Yu F (2020) Microbial degradation and other environmental aspects of microplastics/plastics. Sci Total Environ 715:136968

Zess E, Begemann M (2021) CRISPR-Cas9 and beyond: what's next in plant genome engineering. In Vitro Cell Dev Biol Plant 57(4):584–594

Chapter 9
Microplastic in Soil Ecosystem: Transition and Ecological Impacts

Raman Tikoria, Mohd Ali, Deepak Kumar, Roohi Sharma, Parkirti, Priyanka Saini, Anureet Kaur, and Puja Ohri

Abstract Microplastics are a sort of developing contaminant that are present in large quantities in soil and are still moving and changing. The possible dangers that microplastics (MPs) entail to the ecological system and soil welfare have recently drawn a lot of attention. The main ways that MPs got into soil were through organic fertilizer and mulch breakdown and landfilling of plastics and their distribution in soil systems displayed notable geographical heterogeneity. Leaching and bioturbation speed up the movement of MPs both vertically and horizontally via soil systems, extending their area of contamination. The primary factors that encourage behind the physical, chemical, and biological alterations of movement of MPs in topsoil include mechanical wear, the oxidation process, and biological processes. In addition, movement of MPs has an impact on soil structure, nutrient cycling, and the microbial population. The shape, kind of polymer, size, and degree of MP movement all affect these effects. MPs have an impact on the soil biota by obstructing oxygen flux, enzyme activity, and associated coding genes in addition to microorganism diversity and structure. In addition, chemicals produced by MPs may have an impact on microbial activity. This chapter will aid in a better understanding of how MPs behave in soil considering the possible risks they present to biotic communities and soil health.

Keywords Microplastic · Soil ecosystem · Transformation · Ecological impacts Dynamics

R. Tikoria (✉)
Department of Zoology, Guru Nanak Dev University, Amritsar, Punjab, India

Department of Zoology, Centra University Punjab, Ghudda Bathinda, Punjab, India

M. Ali · D. Kumar · R. Sharma · Parkirti · P. Ohri
Department of Zoology, Guru Nanak Dev University, Amritsar, India

P. Saini
Department of Zoology, Deshbandhu College, University of Delhi, Delhi, India

A. Kaur
Department of Zoology, School of Basic Sciences, Central University of Punjab, Bathinda, India

K. Khanna et al. (eds.), *Microplastics and Soil Microbiome*,
https://doi.org/10.1007/978-981-96-4978-5_9

1 Introduction

Microplastics (MPs) are frequently characterized as fragments smaller than five millimeters in dimensions. The majority of these come from tiny plastic particles that are either directly or indirectly discharged into the environment through the decomposition of big pieces of plastic (Yano et al. 2021). MPs are increasingly recognized as a serious environmental problem, as they are nearly pervasive in the surroundings, water, soil, and other media of the environment (Qi et al. 2020). Because of their tiny size of the small particles and moderate rate of decomposition, they may be readily taken up by living things and spread via food webs (Lehel and Murphy 2021). Furthermore, they might serve as a vector for other pollutants. Because of these, the United Nations Environment Program (UNEP) recently named MP contamination one of the top ten environmental issues (Wang 2023). Landfills hold approximately 79% of the world's plastic garbage (Jung et al. 2022). The dumpsite soil may act as an extended plastic sink and collect significant quantities of MPs (Huang et al. 2022). One of the most vital resources for human society and terrestrial ecosystems which is more likely to be contaminated by MPs. Researchers all over the world reported that soil-buried MPs is exceedingly difficult to decompose (Shahnawaz et al. 2019).

More and more people believe that plastic pollution is a key factor in the global decline in biodiversity and poses a serious threat to human health and the functioning of Earth's systems (Persson et al. 2022). Considering the original built size particle, MPs are classified as primary or secondary. Plastic microbeads and nanoparticles are primary microplastics that are directly employed in a number of manufacturing procedures (Smith et al. 2018). Secondary microplastics are formed at the soil surface as a result of solar UV irradiation, or it can happen inside the soil profile as a result of physical degradation (abiotic) and biological threat (Tsiota et al. 2018; Barreto et al. 2020). The plastic shape will determine its classification as fibers, pieces, thin films, or particles. Fibers are frequently the dominating shape if they reach topsoil through biosolids or irrigation fluids produced from wastewater from cities, depending on the source (Tang et al. 2023). The present chapter mainly concentrates on the sources of microplastic along with the transformation into the soil and how microplastic affects soil characteristics and soil microbiota.

2 Sources of Microplastic in Soil

Mulching films, municipal garbage, biosolids, plastic-coated chemical fertilizers, and atmospheric precipitation are the main ways that macro- and microplastics enter in the soils used for agriculture (Qi et al. 2020; Petersen and Hubbart 2021; Urli et al. 2023). The various sources of microplastic in soil are summarized in Fig. 9.1.

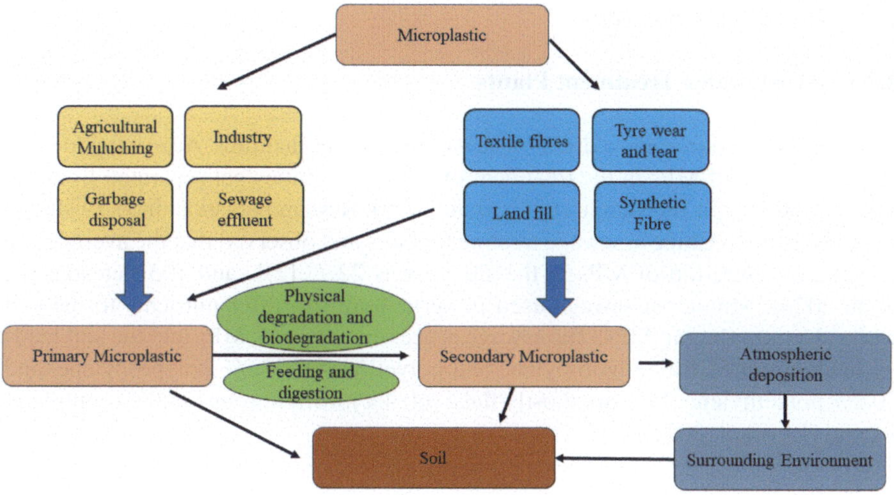

Fig. 9.1 Various sources of microplastic to the soil

2.1 *Organic Fertilizer and Compost Application*

Nutrients include organic acids, peptides, and inorganic components abundant in organic fertilizer, and it is essential in agricultural development (Abdullahi et al. 2021). The majority of these tiny substances (MPs) identified in organic fertilizer-contaminated flake and fiber-containing soil pieces are evident as the predominant morphology, granule size more than 0.5 mm, and concentrations ranging from 1 to 182 mg/kg (Henríquez-Hernández et al. 2020). Because of their tiny diameter, organic fertilizers may contain more MPs (Smith et al. 2018). Swiss researchers' estimation implies the sum quantity of plastic transported to Swiss ground through compost was 1390 tons in 2017 using the material flow analysis (MFA) model (Kawecki et al. 2021).

2.2 *Plastic-Contained Irrigation Water*

Currently, the world's irrigated land area is 270 million hectares, making up 18% of all agricultural land (Willer et al. 2018). MPs are going to get into the dirt via irrigation using sewage. According to reports, typical soil samples contain fewer than 320 n/kg concentration of MPs. However, sewage-irrigated soil has more than 7000 n/kg of MPs (Huang et al. 2021). MP concentrations in sewage are 4200–15,800 n/m^3, 1000–24,000 n/m^3, 627,000 n/m^3, and 15,000 n/m^3 in Ireland, Germany, Russia, and Sweden, respectively (Huang et al. 2022).

2.3 *Industrial Sources*

2.3.1 Wastewater Treatment Plants

MPs are both a source and a drain for sewage treatment facilities. According to studies, about 90% of MPs in the treatment of sewage facilities get separated from the sewage and stay in the sludge (Long et al. 2019). Researchers examined 79 sludge samples from 28 Chinese sewage treatment plants and observed that the average dry weight concentration of MPs in the sludge was 22.7, 12.1, and 103 items/kg (Li et al. 2023). Municipal sewage used in agriculture can offer nutrients to the soil while also introducing MPs. The sewage processing releases MPs into the soil facility made up of 52.6% polypropylene and polyethylene, 27.2% polypropylene, and 20.2% polyethylene. MPs are mostly flake (60.4%), with fibrous (35.0%) coming in second (Liu et al. 2021).

2.3.2 Improper Factory Management and Accidental Leakage

Due to improper plastic product management in the manufacturing and inadequate post-use recycling, when bigger plastics are released into the soil, they can decompose into MPs. These MPs access the shoreline and the beach by horizontal motion brought on by biological activity, wind, waves, or tides (Gorman et al. 2020). Because of the aging and deterioration of the devices, plenty of polystyrene drifting gadgets induce goods made of plastic that end up in the water (Liu et al. 2021). The replacement of fishing equipment results in the discharge of old fishing nets into the ocean. According to statistics, at least 298 million pieces of synthetic fishing equipment are discarded annually by commerce ships and boats for fishing (Gilman et al. 2022). MP concentrations in beach soil are 539 n/kg in Switzerland and 412.50 n/kg in Shandong Province (Huang et al. 2019).

2.4 *Commercial Products*

2.4.1 Synthetic Textiles and Personal Care Products

The widespread usage of synthetic textiles and personal grooming products are major sources of soil microplastics (Galafassi et al. 2019). In artificial textiles, synthetic fibers such as ester and nylon are commonly utilized (Purwar 2016). When laundered, one synthesized fiber sweatshirt has the ability to emit 110,000 MP particles (Tian et al. 2021). In the USA, the per person consumption of MPs in these products are around 1 g each year (Galafassi et al. 2019). The imposition of prohibitions has significantly curtailed the usage of MPs in PCPP. On August 22, 2019, the European Chemicals Agency (ECHA) suggested a comprehensive limitation on the use of MPs in cosmetics, detergents, and cleaning goods in the European Union

(Guerranti et al. 2019). Following the implementation of a legislation prohibiting cosmetics that utilize plastic tiny beads, the manufacturing of MPs in EU, the amount of cosmetics has dropped from 860 tons in 2012 to 440 tons in 2020 (Liu et al. 2021).

2.4.2 Plastic Mulching

There is no apparent way to control or interrupt the massive yet invisible sources that are continuously flooding the earth and creating microplastic deposition. Mulch film, which is composed of polyvinyl chloride (PVC) and polyethylene (PE), has gained popularity in agricultural production across the world because of its many financial advantages, which include higher yield, better fruit quality, and more effective use of water (Qi et al. 2020). By 2030, the farming plastic film industry is expected to have grown from four million tons in 2016 to 5.6% annually (Huang et al. 2022). Almost 90% of the 20 million hectares of agricultural land covered with plastic mulch are in China. Additionally, the longer the ongoing mulching period, the higher the concentration of microplastics. When large pieces of plastic mulch waste are left on the surface of the field, they are exposed to UV radiation by nature, which causes decomposition and brittle (Liu et al. 2023). Plowing and other cultivation operations will also cause shearing stresses on plastic waste in fields, possibly breaking apart already delicate materials (Acharjee et al. 2023). At the same time, the plastic pieces buried in the field undergo mechanical dissolution due to factors including the chilling and melting cycle, pressure concealed in the soil or snow, and damage from organism contact. At some point, a significant quantity of MPs is created and kept in the soil, endangering both nutrition and terrestrial wildlife. However, the global use of agricultural film has increased dramatically in recent years, posing additional hidden concerns for microplastics in soil (Kim et al. 2020).

2.4.3 Microfibers from the Washing Process of the Textile Industry

Textile manufacturing and commerce are important sources of MP contamination. In the textile business, there are three basic groups of textile fiber: organic threads, artificial filaments, and rejuvenated filaments (Stanton et al. 2019). Microplastic waste from textiles has more synthetic fibers than natural fibers by more than 170%. The textile industry is anticipated to generate over forty-two million tons of artificial fibers per year, with PES accounting for around 80% (Schmutz and Som 2022). Because of the significant detrimental impact on the marine ecosystem, there is a growing awareness of synthetic textile fibers (Stanton et al. 2019). Microfibers are mostly derived from the home and commercial washing of synthetic fabrics. Microfibers originating from textile sewage were shown to be responsible for 35% of the detected MPs in coastal locations (Magni et al. 2019). Under typical residential washing conditions, washing an acrylic cloth can release approximately 700,000 tiny fibers, totaling around 6 kilograms in weight. More than 1900 fibers may be lost from a garment every wash, and all clothes discharged more than 100 fibers per liter of wastewater (Shopnil 2022).

2.5 *Littering Over 100 Threads per Liter of Sewage*

The terrestrial ecosystem contains a wide spectrum of microplastic waste due to the excessive production of plastics and haphazard or inadequate management practices (Yang et al. 2021). Similar to aquatic settings, a significant amount of plastic waste can give rise to microplastics (Napper and Thompson 2020). An estimated 6.3 billion tons of plastic waste were produced worldwide between 1950 and 2015, of which 4.97 billion tons were accumulated in landfills and natural areas (Yang et al. 2021). Unmanaged plastic waste is usually found in the ground, next to roads, or at unlawful disposal locations (Jakiel et al. 2019).

3 Distribution of Microplastic in Soil Across Globe

In the Shanghai suburbs, there are over 680,000 films of plastic gardens spread across 130 million square meters, resulting in the pervasiveness of plastic waste or its disintegration into tiny pieces on agricultural land (Liu et al. 2018). For example, 20 shallow and deep soil samples from Shanghai's crops had MP quantities of 78.0 and 62.5 particles/kg (sizes of 0.0215 mm), suggesting that raking with plastic is a major source of MP pollution (Liu et al. 2018). In agricultural regions, sewage is applied as manure that contributes to alternate major source of MP contamination in topsoil. In 2018, Ng et al. stated that the mass of MPs obtained from sludge application in Australia reached up to 104 tons per year (Xu et al. 2020). MP counts rise in 31 farming areas in Chile's central valley after sequential sludge treatments (Corradini et al. 2019). Furthermore, in three rice-fish combination sites in Shanghai, the soil had an average of 10.32 microplastic particles per kilogram, according to Li et al. (2019). Fish-eating habits, pollution from the atmosphere, and fertilizers used in agriculture were shown to be significant contributors to MP poisoning in these rice-fish co-culture systems.

Another study from Wuhan, China, reported 320–12,560 particles/kg of MPs in vegetable farmlands (Bi et al. 2023). MPs are most abundant in soils used for agriculture that has now been recorded in Yunnan tree-planted soils (Xu et al. 2020). Microplastics can produce particle toxicity in soils by causing harmful effects and oxidative stress, disrupting health balance and increasing absorption or relocation. Meanwhile, majority of plastics found in ground have a variety of substances that worsen environmental damage. The mulching films, which have been extensively contaminated in Chinese soil used for agriculture, have made extensive use of phthalic acid-based ester plasticizers (PAEs) (He et al. 2015). They are often known as "natural endocrine interfering chemicals" and have been demonstrated to negatively affect the variety of bacteria and the functioning of soil enzymes (Kumar et al. 2020). The density of PAEs in microplastics (debris crop film) and the risks of exposure across the food chain must be further examined (Qi et al. 2020). Additionally, significant bacterial population variability was seen among MPs made from different breakdown materials. In other words, MPs may establish a new biotic habitat in soils, and bacteria linked to MPs may adapt well to a surface-related existence. According to Loiseau and Sorci (2022), the presence of harmful bacteria implies that the plastic waste might be a disease vector, causing tissue damage that could encourage infection. An analysis of China's major plastic mulching film usage regions found that cotton fields in Xinjiang Province had the worst contaminating things, with a median soil vinyl film leftovers level of 259 kg/ha and a maximum soil plastic film leftovers level of 381 kg/ha in certain locations (Lian et al. 2022). Over time, a significant accumulation of microplastics is inevitable due to the large amounts of microplastic remnants in the soils of farms. This could add up to 106–109 particles/ha in the form of biosolids, assuming a median MP pollution of 104 particles/kg and a usual land treatment rate of 1–15 t/ha. This would culminate in a topsoil contaminant level of roughly 4–150 particles/kg for every use per year. However, it is not unexpected that 90% of Swiss riverbed topsoil have MPs and that European farming topsoil has contaminant levels of 670 fibers/kg (Qi et al. 2020). Additionally, compared to marine ecosystems, terrestrial environments have considerably higher concentrations of microplastics (Xu et al. 2020). Many research has been conducted to describe the microplastics' dispersion and pollution mechanisms in aquatic and coastal environments mainly in lakeside and coastal regions where human activity is considerable (Ajith et al. 2020; Veerasingam et al. 2020; Marrone et al. 2021). A figure of investigations observed that the quantity of microplastics differs at different ground lengths, ranging attaining in highly contaminated topsoil up to 7% by weight (Li et al. 2022). According to Liu et al. (2018), microplastics were found in both shallow and deep soils from a range of agricultural sites in China; yet the quantity of MPs in shallower soils was much more than that in deep soils. In addition to being a sink for MPs, soil could also serve as an incubator of them for aquatic and hydrologic life (Chia et al. 2021).

4 Transformation of Microplastic in Soil

Plastics undergo physical, chemical, and biological changes in the environment, such as deterioration and fragmentation. The early degrading processes of plastic trash are usually dominated by chemical degradation brought on by ultraviolet (UV) light, which is crucial among these three translation types (Lear et al. 2022). Phototransformation of microplastics in topsoil occurs as a result of UV irradiation at a sufficient temperature and oxygen utilization (Maddela et al. 2023). The phototransformation of MPs includes chemical processes such as chain splitting, cross-linking, the production of functional groups containing oxygen, and degradation into CO_2. These procedures may result in the production of nanoplastics ought to and even smaller microplastics (Lv et al. 2021). Additionally, microplastic disintegration is facilitated by the combined impacts of UV irradiation and mechanical friction brought on by the turbulence of soil life and agricultural operations (Sun et al. 2020). Even when MPs decompose into smaller pieces, they frequently bind with soil particles through electrostatic forces brought on by the root system liquids from roots or during digestion and waste (Tian et al. 2021). Unknown environmental issues might arise from the accumulation of plastics at the submicron level caused by the insufficient degradation of microplastics in soil. In addition to breaking down, microplastics' mobility inside the porous matrix of agricultural soils modifies their distribution. Studies have shown that microplastics with more surface functional groups and smaller sizes are more transportable in saturate porous media, such as soils (Shaniv et al. 2021). Additionally, heterogeneity, soil organic matter, surface coating, ionic strength, and cation type were all factors that affected the flexibility of MPs in soils (Boots et al. 2019; Ren et al. 2021). Despite the paucity of study on the effects of microplastics on groundwater, groundwater contamination is dangerous since it directly affects the health of people and animals (Wong et al. 2020). The growth and breakdown of plant roots may create macropores in soils, which facilitate the transport of microplastics, in addition to soil animals (Wang et al. 2022). Furthermore, plowing and harvesting are examples of agricultural practices that can cause surface and underneath the ground to invert in agricultural systems (Dang et al. 2015). Microplastics in soils are little studied in comparison to the aquatic ecology. Soils are thought to be the main source of MPs, and they can get into them by runoff from the surface, wastewater, organic waste, farming film breaking, watering, and atmospheric dissolution (Uwamungu et al. 2022). Figure 9.2 enlightens the various fates of MPs into the soil.

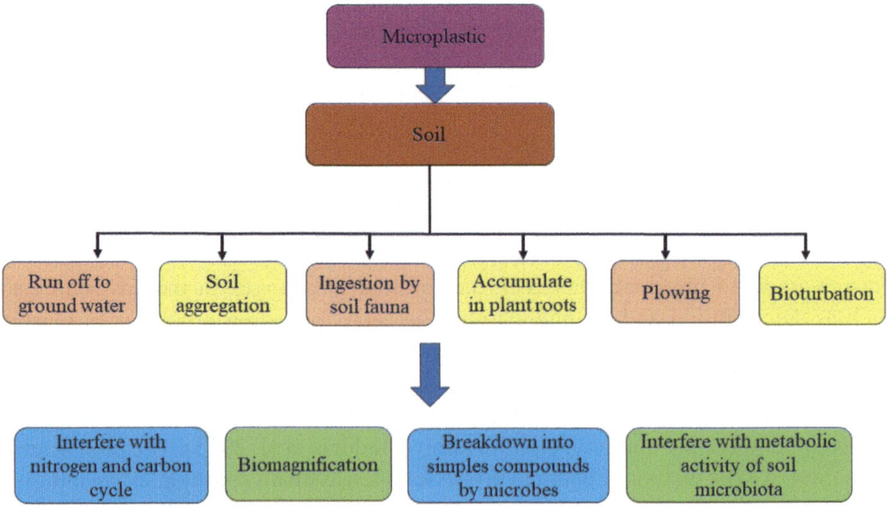

Fig. 9.2 Fate of microplastic into the soil

5 Alteration of Physicochemical Properties of Soil by Microplastics

MPs can shift the physical structures of topsoil to affect its properties due to their different qualities (Wang et al. 2022). Waste from plastics accumulates on the exterior of the soil which may have negative consequences, including alterations in nutrient and water mobility that lead to secondary salinization of the soil, declined soil pH caused by surplus nitrogen, reduced bulk soil density along with shifts to unrest, permeability and agglomeration, stimulated degradation of soil organic content, impacts on seed appearance, root length and surface, raised subsurface temperatures, and soil degradation brought on by dumps (Yin et al. 2018; Medyńska-Juraszek and Jadhav 2022; Chen et al. 2022; Riveros et al. 2022; Huang et al. 2022).

Firstly, due to their affinity for soil minerals and organic components, MPs may affect how soil aggregates; however, the exact impact will depend on the makeup, size, type, and shape of the plastic particles (Lei et al. 2018; Sajjad et al. 2022). For instance, polyethylene film reduced the proportion of big macroaggregates (>2 mm), while polyester fibers raised the proportion of water-stable aggregates (Zhang et al. 2019; Boots et al. 2019). Secondly, because plastics frequently have lower bulk densities than many minerals in the soil, MPs can promote soil aeration, which may promote root penetration. Polyethylene has been demonstrated to decrease soil pH, whereas polylactic acid may raise soil pH (Qi et al. 2020; Boots et al. 2019). Any significant changes in soil pH are anticipated to have an important consequence on the community structure and functioning of the pH-sensitive soil microbes (Sridhar et al. 2022; Peng et al. 2022).

It turned out that MPs had a bulk density that was lower than other substances existing in the soil (Chen et al. 2022). Polyester fibers (8 m) with improved capacity for holding water may be capable of maintaining soils saturated for extended tenure in regard to soil water dynamics, whereas polyester fiber lessened the development of new aggregates and lowered soil deterioration and degree of sedimentation, notably on soils that are highly erodible (Ingraffia et al. 2022). Polyethylene films (2 mm) increased the soil water loss through the enhancement of evaporation by decreasing the shrinking ratio with the increase in rate of desiccation shrinkage. The earlier process resulted from fibers' capacity to create soil clumps and entrap particles of dirt at smaller sizes (Wan et al. 2019). As in silt loam with two degree (0.1 and 0.5%, w/w), each of three diameters (10, 5 and 2 mm) of plastic residue films was subjected to natural evaporation trials, whereby loss of moisture and subsurface crack characteristics were measured. The 0.1% polymer that was less than 10 mm caused evaporation, which increased as it became bigger by creating water channels. However, because of the soil's formed waterproof barrier, the 0.5% plastic that was larger than 10 mm actually prevented water loss (Guo et al. 2022). Because of variations in the amount of compost and the ability to hold water used, MPs boosted the soil's fertility as shown by the buildup of rich molecular compounds such as humic and fulvic acids (Wang et al. 2020). Soil water changes could alleviate or aggravate drought-like situations also (Lozano et al. 2021; Lozano and Rillig 2020). This determines how pollutants migrate through fissures and into the subsurface soils, changing the soil moisture cycle in the process (Riveros et al. 2022).

Numerous researches have documented MPs' impact on topsoil nutrients (Zhang et al. 2023; Silori et al. 2023). Microplastic-carbon can be passed off as a significant anthropogenic component of high-carbon plastics including organic carbon from soil (Shen et al. 2023; Yu et al. 2023). MPs could drastically alter the carbon chain by impacting plant development and microbial actions in soil and disintegration of debris because they are carbon-based substances themselves (Kurniawan et al. 2023; Kvale 2022). Similarly, research has shown that the presence of microplastics affects enzyme activity, ultimately boosting microbial processes that promote the formation of organic phosphorus minerals and the solubilization of synthetic phosphorus, thereby increasing the availability of phosphorus in the soil (Liu et al. 2023; Yan et al. 2021). But adversely, the obtainable phosphorus reduced because of polyethylene and propylene, ascribed to the fact that maybe there was little or no mineralization (Li and Liu 2022). Similarly, nitrates and nitrites along with nitrogen were also reported to modulate with the addition of polyether sulfone, suggesting that microplastic addition might accelerate the nitrification process, which gradually reduces the ammonification (Qin et al. 2020). Microplastics have been demonstrated to have very little of an impact on key enzymes of soil which could cause the soil to degrade (Paramdeep et al. 2022). Pb-Zn-contaminated soil suffered heavy alteration sin enzymatic structures of lithosphere (Feng et al. 2022) (Fig. 9.3).

Consequently, modifications in soil structures due to shifts in soil hydrological dynamics brought on by MPs of various concentrations, kinds, size, and forms could have unpredictable effects (Shafea et al. 2023; Lehmann et al. 2020).

6 Effects on Soil Micro- and Macrobiota

Microplastics are persistent form of soil pollution that may impact development and fertility, feeding, immunity system, and enduring of the organisms present in the soil (Álvarez-Lopeztello et al. 2021; Wang et al. 2019; Guo et al. 2020). Soil and the living organisms present in soil offers a significant part in balancing nutrient cycle, providing food and maintaining biodiversity (Tikoria et al. 2022a, b). Soil micro- and macro-organisms are important components of ecosystem as they enhance soil fertility by decomposing organic matter; they also avail soil mineral nutrients for plants by mediating biogeochemical cycles for sustainable agroecosystem (Basu et al. 2021). Microfauna of soil improves soil quality by impacting microbial community or by releasing polymeric substances. Soil microorganism produce extracellular enzymes such as CAT, beta glucosidase, and phenol oxidase, and thus their enzymatic activities enhance soil quality (Tikoria et al. 2023a, b). Soil fauna such as earthworms modulate soil physical structure by remodeling soil pores (Tikoria et al. 2022c). In other words, soil organisms strongly influence the dynamic structure of

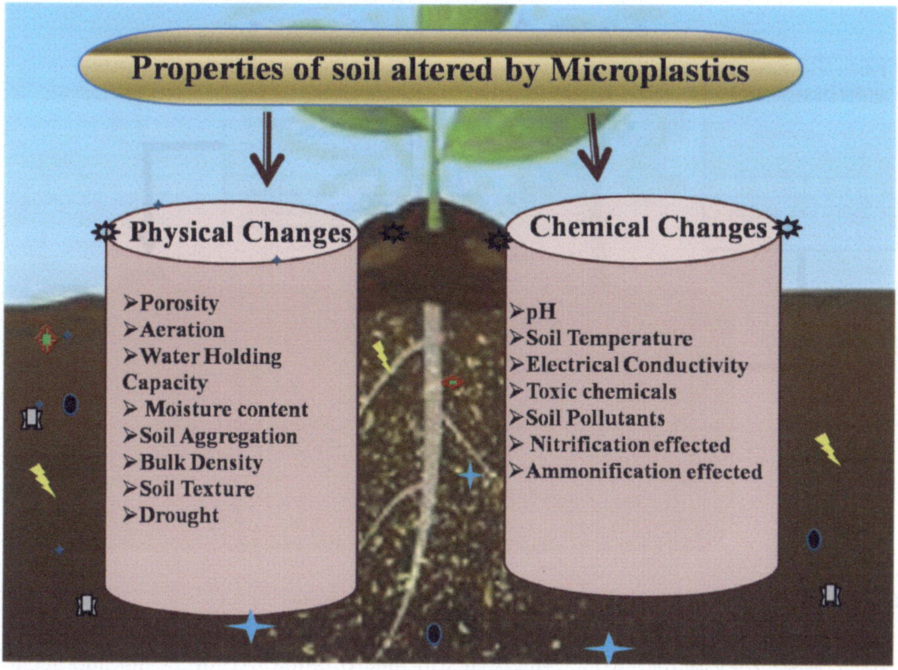

Fig. 9.3 The figure displays the key physical-chemical alterations faced by soil due to microplastics

soil. These physical properties of soil affect root growth and functions related to crop yield and survival (Yu et al. 2020a, b; Gliński and Lipiec 2018).

The microplastics are persistent in soil (Fig. 9.4). However, during degradation, microplastic undergoes various processes like thermal oxidation, UV radiation, biodegradation effects, and wind and physical abrasion. This produces variation in chemical structure of microplastic polymer (Marathe and Bank 2022, Ren and Ni 2022). Therefore, these microplastics affect soil organisms, and the extent of their effect relies on the abundance and nature of microplastic. MPs existing in soil can affect microbes that live in the soil, including fungus and bacteria by altering their microhabitat and microbial activity, spreading antibiotic genes, increasing bacterial transport, and also changing their community and function (Zhang et al. 2021; Guo et al. 2021). Plastics existing in mulch film also exert significant change in bacterial community of soil affecting various cellular process such as pathways involved in biosynthesis of secondary metabolite, membrane transport, metabolism of xenobiotics, and biodegradation process (Wu et al. 2022). With research on how soil microbial populations are affected by MPs, the observations indicated the decrease in metabolic functional capacity due to repressed catabolism in which conventional and biodegradable microplastic both equally contributed in altering the composition of soil microbial community. MPs also decrease soil microbial activities by affecting enzymes involved in soil microbial respiration, FDA, dehydrogenase activity, phosphatase activity, and P, N, or C cycles (Joos and De Tender 2022; Sun et al.

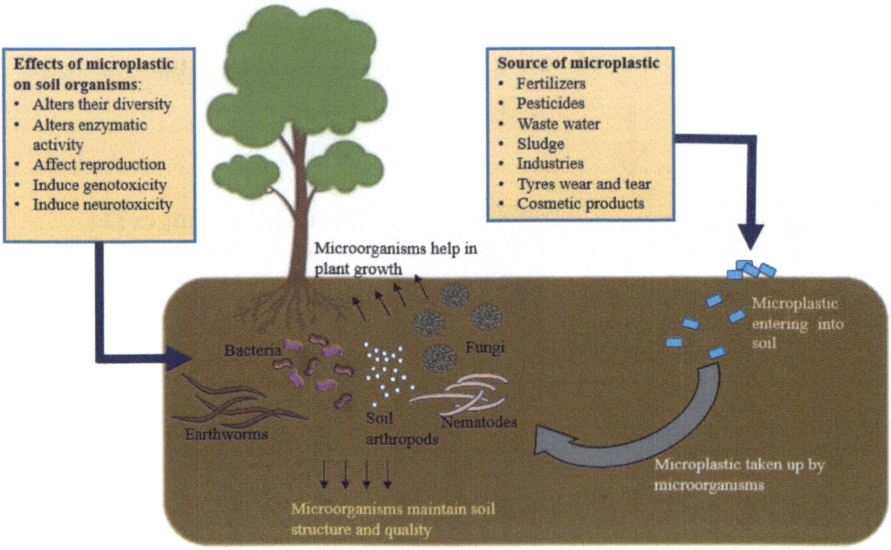

Fig. 9.4 Accumulation of microplastic in soil affect soil organisms

2022). Also, they induce toxic effects in soil macro fauna like ingesting and accumulation of microplastic, metabolic disease, genetic damage and developmental harm, pathological destruction, and harm to DNA and neurotoxicity. Also, when it combines with other materials such as metals that are heavy, antibiotics, pesticides, additives, and organic pollutants, it shows adverse effect on soil biota (Wang et al. 2022; Dhiman et al. 2021; Kaur et al. 2022). A few invertebrates in soil like oligochaeta, nematodes, isopods, and collembolan are affected by microplastics which results in their altered behavior like growth, survival rate, metabolic activity, feeding behavior, gut microbiota, and neurodegeneration. For example, *C. elegans* exposed to microplastic show poor likelihood of survival and shorter physique, downregulate gene expression, and have least life expectancy. In oligochaetes the intake of MP decreases their weight. It also affects immune system and damage tissue. In collembolan, it inhibits reproduction and growth and modifies metabolic turnover. In isopods also minor hazardous effect were observed. Microplastics also affect microarthropods by changing their litter decomposition rates and higher mass loss (Zhu et al. 2019; Barreto et al. 2020).

MPs could affect reproductive rate and cause skin lesions of earthworm also. A study shows that treatment of polystyrene on earthworm at different concentrations like 0, 0.25, 0.5, 1, and 2% has adverse effects on animal such as inhibition of expansion and a higher death rate (Cao et al. 2017). Another experiment on earthworm shows the effect of nanoplastic and microplastic fibers and shows that there can be long-term accumulation of plastic inside animals' body (Lahive et al. 2022). In another study, effect of nylon toxicity was studied on *Enchytraeus crypticus*. The effects shows that their reproduction rate was reduced (Lahive et al. 2019). In one more study, two kinds of microplastic polyethylene and biodegradable microplastic polylactic acid were tested on earthworm. The observation reveals decreased activities of POD, SOD, CAT AchE, and GST, possibly because both microplastics induced oxidative stress in earthworm (Yu et al. 2022). In a report upon earthworm with the treatment of polystyrene at 0.02% concentration shows disturbed osmoregulatory metabolism as betaine level was also decreased along with decreased level of lactate and myo-inositol, while other acid (2-hexyl-5-ethyl-furan-3-sulfonic acid) was increased at degree of respiration (Tang et al. 2023). Some more examples are given in Table 9.1.

7 Possible Remedies to Solve Microplastic Pollution

MPs that are not recyclable and tenacious will linger in the surroundings for a long time, endangering habitats over the long run (Steinmetz et al. 2016). Significant step in lowering microplastic contamination in soil is to reduce or completely do away with the consumption of plastic in the farming process. Although recyclable mulch film is crucial to farming efficiency, it is a major source of microplastics in soils used for farming. More than 30% more crops are produced as a result of these films, contributing significantly to agricultural product security in nations such as China (Huang et al. 2021). As per Zheng et al. (2022), the use of farming products film resulted in a 42.3% increase in water use efficiency. Given that water resources in many nations are declining and becoming more variable due to the effects of the changing climate, the usage of plastic mulch films is unlikely to end (Jägermeyr et al. 2016). Therefore, a practical strategy for lowering the contamination of plastic throughout the agricultural sector is the prudent use and obligatory disposal of traditional polymer mulch films. Furthermore, the introduction of poor-quality plastic film (i.e., excessive pollutant concentration) into the consumer market should be rigorously restricted by the authorities and the appropriate agencies (Ghaffour et al. 2019).

Additionally, we should encourage the long-term usage of agricultural plastic sheets. Bioplastic, a substance that may either entirely or partly destroyed by microbes, is a crucial route for the development of farming and the plastics sector. Utilizing recyclable mulching film, especially natural mulch film, is turning into a key tactic for tackling the issue of MP contamination and plastic film residual from the place of origin. Fortunately, because of the lack of resistance to mechanical stress, greater cost, and bigger manufacturing ecological impact, recyclable mulch films—especially bio-based ones—have significant restrictions for widespread application when compared to polyethylene film. Later on, it will be important to lower manufacturing costs, enhance product qualities, optimize raw resources, and manage the breakdown period of biodegradable plastic mulching film. To some extent, stronger laws and law enforcement can aid in lowering contamination from microplastics (Wong et al. 2020). As of July 2018, 127 of the 192 countries examined in the United Nations report "Legal Limits on Single-use Plastics and

Table 9.1 Effect of microplastic on the soil organisms

Soil organisms	Microplastic	Concentration	Effect	References
Soil microbial community (protozoa, bacteria, and fungi)	PS, PP, LDPE, microglass, and polyamide 12 (PA12)	1% (w/w)	Microglass had toxic effect on soil protozoa	Wiedner and Polifka (2020)
Soil microbial community	Low-density polyethylene	5% (w/w)	Altered composition of microbial community	Ya et al. (2022)
Bacterial community	PS, PLA, and PHB	0.2% and 2%	Change in diversity	Feng et al. (2022)
Soil bacterial community	Polyethylene, polyethylene terephthalate,and polyamide-6	5 g microplastic per kg of soil	Decreased bacterial network complexity was observed along with unstable connectivity, keystone species, network size, and number of modules	Shi et al. (2022)
Soil microbial community	Polyethylene, microsphere, and fibrous polypropylene	2 mg/g for PP and 20 mg/g for rest two MPs	Fibrous PP show more effect and membranous PE showed lesser effect on enzymes like dehydrogenase, alkaline phosphatase, and dehydrogenase along with altering microbial community	Yi et al. (2021)
C. Elegans	Polystyrene	55.4 cm^2/mL	Inhibited reproduction	Mueller et al. (2020)
Arbuscular mycorrhizal fungi (AMF)	PES, PLA, PET	2% (w/w)	Change in AMF structure and community was observed	Liu et al. (2023)
Soil bacterial community	Low-density polyethylene	2000 fragments per Kg soil	Decreased diversity indexes was found	Huang et al. (2019)
Eisenia fetida	Low-density polyethylene	0.1, 0.25, 1.0, 1.5 g/Kg	Surface damage of organism was observed, activity of CAT and MDA also increased, stimulation of neurotoxicity response was also observed	Chen et al. (2020)
Eisenia fetida	Polystyrenes	0–0.5% w/w	DNA damage was observed, production of juvenile was reduced	Sobhani et al. (2021)

(continued)

Table 9.1 (continued)

Soil organisms	Microplastic	Concentration	Effect	References
C. Elegans	Polystyrene microplastics	0, −100 μg/L of PS-MPs	Caused intestinal damage and oxidative stress	Yu et al. (2020a, b)
C. Elegans	Fluorescent, polystyrene spheres	1 and 5 μm MPs	At particular concentration the life span of nematodes was decreasing; defecation rhythm was also affected	Shang et al. (2020)
Achatina fulica	Polyethylene terephthalate	0.01–0.72 g kg^{-1}	Reduced food uptake, excretion, damaged intestinal villi, reduced antioxidant capacity	Song et al. (2019)
Fungal community	Polyethylene MPs	2000 pieces/kg soil	Beta diversity was affected, altered fungal community, affect biogeochemical cycles	Li et al. (2023)
Bacterial community	Polystyrene	10 mg kg^{-1}	Altered soil nitrogen availability and transformation, disturbed amino acid metabolism by bacteria	Xu et al. (2023)
Soil springtail (*Folsomia candida*)	Polyethylene MPs	0.5% to 1% MPs (w/w)	Inhibition of reproduction, changed gastrointestinal community of bacteria of organism	Ju et al. (2019)
Eisenia fetida	Polypropylene and high-density polyethylene	0.25% (w/w)	Altered activity of glutathione S-transferase CAT and SOD	Li et al. (2021)
Earthworm	PE and PP	0.02% (w/w)	Metabolic activities were altered, digestive system and immune system were damaged, change in osmoregulation process were observed	Chen et al. (2022)
Enchytraeus crypticus	Nylon and PVC	20, 50, 90, and 120 g/kg	Reduced reproduction rate	Lahive et al. (2019)
Porcellio scaber	Crumb rubber and polyester fiber	0.05%, 0.5%, and 1.5% (w/w)	Reduced AChE activity was observed, changes in immune parameter also observed	Dolar et al. (2021)
Enchytraeus crypticus	High-density polyethylene microplastic	0, 2%, 4% or 8% (w/w)	Increased oxidative stress, GST, and CAT level increased	Pflugmacher et al. (2020)

Microplastics: A Global Review of National Laws and Regulations," or almost 66% of the respondents supported some form of legislation to limit the usage of plastic bags.

Bans, tariffs, and waste management policies are among the laws and regulations enacted to improve elimination, support the reuse and recycling of materials, and advertise substitutes for plastic-based goods. Bags made of plastic, other reusable plastics, and tiny beads of plastic are the main topics of discussion. For example, 27 nations tax the production and distribution of plastic bags; meanwhile, 30 nations tax their customers for plastic bags. The use of microbeads in personal care products is limited in wealthy countries including the USA, Canada, the Netherlands, and New Zealand. It is simple and has tiny influence on individuals' life, making it a rapid and effective method of reducing the source of MPs (Zhang et al. 2019). Dispensable plastics are carefully regulated as well as application of biodegradable plastics is promoted (Wong et al. 2020). The primary source of microplastic pollution is now the extremely slow decay rate of plastic monomers, which are mostly produced from natural hydrocarbons (Shen et al. 2023). Bioplastics are polymers that are created by microorganisms and are based on organic materials like starch. It is reusable and so ecologically beneficial. Bioplastics are a sustainable alternative to plastic. The "Biotechnology Strategy Outline" and "Biomass Japan Comprehensive Strategy" were created by the Japanese government to promote the use of renewable resources, including bioplastics, stating that increasing the use of bioplastics is a serious issue.

By 2010, 20% of plastics should be derived from renewable resources, according to the "Biotechnology Strategic Outline" policy goal. Many nations forbid the direct disposal of plastic trash in landfills, which may substantially decrease the quantity of plastic waste generated. Direct garbage disposal has been forbidden in Germany from 2005. Additionally, the percentage of landfills in the USA is declining.

Additionally, nations are concentrating on employing biological and scientific approaches to eliminate MPs from sewage treatment plants (Yang et al. 2021). In order to reduce garbage made of plastic and microplastic damage, public involvement is essential.

8 Challenges and Future Perspective

In the future, many challenges have to be faced by every individual of the planet. These toxins may percolate and eventually reach water bodies/groundwater, polluting the entire ecosystem by poisoning water supplies. As mentioned earlier, novel goods may be introduced during breakdown and because of the highly reactive nature of the additives put to the things, they may have more negative impacts. Phthalates and bisphenol (commonly known as BPA) are substances that seep from plastic particles. It has been demonstrated that these substances harm the hormone systems of both vertebrates and invertebrates and are known hormone violators (Kabir et al. 2015).

The majority of research has focused on immediate effects of MPs on the soil-plant system, often spanning a few weeks or months. Based on current studies and the debate in this article, potential study priorities are indicated below: A combination of macro and micro methods is required to investigate MPs in soil. Abiotic and biotic factors must be considered when examining the connection between soil chemical reactions and MPs. The potential effects of MPs on soil metabolism of nutrients, global carbon sources, ecological equilibrium, and operation must all be examined in future studies.

Soil texture and amount of moisture are intimately related to air circulation in the soil, nutrients and water retention, and the level of farming complexity or ease. As of right now, it is known that MPs accumulate in soils of different textures, with sandy soils containing the highest concentrations of microplastics (Yu et al. 2021). Nevertheless, there is currently a dearth of study on the movement and ecological risk of MPs in soils with different textures and moisture contents.

Research on MPs in plants is now limited to quick research and a few typical species. To predict the consequences of microplastics at the individual, community, or ecosystem level, further study on their short- and long-term affects on a variety of plants is needed to reveal their mechanisms of action.

There are currently no consistent criteria for quantifying and estimating the levels of the levels of MPs in soil environments. Moreover, neither the maximum permissible values that the soil and plants can withstand nor the smallest quantities of MPs that negatively affect soil properties and development of plants have ever been determined. As a result, future studies should prioritize tracking the origins and transit pathways of microplastics.

MPs are less prevalent in deeper soil levels, yet microplastic fibers are more pliable and can penetrate deeper soil layers. It is unknown how deep MPs can penetrate the soil profile and how deep plant roots can spread in agricultural soils supporting a variety of crops. To offer particular suggestions for crops suitable for planting in soil polluted with MPs, it is necessary to look into the transportation and creation of MPs in the soil profile. In conclusion, microplastics, especially in the soil the surroundings, present a number of scientific obstacles and problems. Researchers can conduct studies to fully investigate the many risks that microplastics pose to the soil environment.

9 Conclusion

Soil systems, as major reservoirs of microplastics, absorb microplastics in a variety of methods, and actions of MPs may be altered by a variety of natural conditions. MPs may also be transported in both physical and biological processes using soil ecosystems, depending on microplastic qualities, soil environment features, and agricultural activities. MPs' effects on ecological systems are likely to continue due to continual increases in plastic manufacturing, the ongoing existence of MP fragments, and the breakdown of current plastics. As mentioned before, microplastics affect the physicochemical properties of soils. Through their interactions with biotic or abiotic techniques, MPs change the stability of soil aggregates. Additionally, by altering the primary flora on the ground or the genes and enzymes that control the C/N/P cycle, microplastics may have an impact on the nutrient cycling of the soil. The shape and size of MPs that mimic soil components have a negligible effect on circulation of water and the composition of the soil. Plant root porosity and seed chamber openings may be physically blocked by MPs. Plants are more negatively impacted by microfibers, tiny MPs, and compostable MPs overall. Further, more studies should be conducted for better understanding of microplastic effects on environment, soil-plant ecosystem, and possible methods to contain or remove such hazards from the soil.

References

Abdullahi UA, Khandaker MM, Alias N, Shaari EM, Alam MA, Badaluddin NA, Mohd KS (2021) Seaweed effects on plant growth and environmental remediation: a review. J Phytology 13:122–129

Acharjee SA, Bharali P, Gogoi B, Sorhie V, Walling B (2023) PHA-based bioplastic: a potential alternative to address microplastic pollution. Water Air Soil Pollut 234(1):21

Ajith N, Arumugam S, Parthasarathy S, Manupoori S, Janakiraman S (2020) Global distribution of microplastics and its impact on marine environment – a review. Environ Sci Pollut Res 27:25970–25986

Álvarez-Lopeztello J, Robles C, del Castillo RF (2021) Microplastic pollution in neotropical rainforest, savanna, pine plantations and pasture soils in lowland areas of Oaxaca, Mexico: preliminary results. Ecol Indic 121:107084

Barreto C, Rillig M, Lindo Z (2020) Addition of polyester in soil affects litter decomposition rates but not microarthropod communities. Soil Org 92(2):109–119

Basu S, Kumar G, Chhabra S, Prasad R (2021) Role of soil microbes in biogeochemical cycle for enhancing soil fertility. In: New and future developments in microbial biotechnology and bioengineering. Elsevier, pp 149–157

Bi D, Wang B, Li Z, Zhang Y, Ke X, Huang C, Liu W, Luo Y, Christie P, Wu L (2023) Occurrence and distribution of microplastics in coastal plain soils under three land-use types. Sci Total Environ 855:159023

Boots B, Russell CW, Green DS (2019) Effects of microplastics in soil ecosystems: above and below ground. Environ Sci Technol 53(19):11496–11506

Cao D, Wang X, Luo X, Liu G, Zheng H (2017) Effects of polystyrene microplastics on the fitness of earthworms in an agricultural soil. IOP Conf Ser Earth Environ Sci 61(1):012148

Chen Y, Liu X, Leng Y, Wang J (2020) Defense responses in earthworms (*Eisenia fetida*) exposed to low-density polyethylene microplastics in soils. Ecotoxicol Environ Saf 187:109788

Chen H, Jia Q, Sun X, Zhou X, Zhu Y, Guo Y, Ye J (2022) Quantifying microplastic stocks and flows in the urban agglomeration based on the mass balance model and source-pathway-receptor framework: revealing the role of pollution sources, weather patterns and environmental management practices. Water Res 224:119045

Chia RW, Lee JY, Kim H, Jang J (2021) Microplastic pollution in soil and groundwater: a review. Environ Chem Lett 19(6):4211–4224

Corradini F, Meza P, Eguiluz R, Casado F, Huerta-Lwanga E, Geissen V (2019) Evidence of microplastic accumulation in agricultural soils from sewage sludge disposal. Sci Total Environ 671:411–420

Dang YP, Seymour NP, Walker SR, Bell MJ, Freebairn DM (2015) Strategic tillage in no-till farming systems in Australia's northern grains-growing regions: I. Drivers and implementation. Soil Tillage Res 152:104–114
Dhiman S, Ibrahim M, Devi K, Sharma N, Kapoor N, Kaur R, Sharma N, Tikoria R, Ohri P, Mir BA, Bhardwaj R (2021) Biochar assisted remediation of toxic metals and metalloids. In: Handbook of assisted and amendment: enhanced sustainable remediation technology, pp 131–162
Dolar A, Selonen S, van Gestel CA, Perc V, Drobne D, Kokalj AJ (2021) Microplastics, chlorpyrifos and their mixtures modulate immune processes in the terrestrial crustacean *Porcellio scaber*. Sci Total Environ 772:144900
Feng X, Wang Q, Sun Y, Zhang S, Wang F (2022) Microplastics change soil properties, heavy metal availability and bacterial community in a Pb-Zn-contaminated soil. J Hazard Mater 424:127364
Galafassi S, Nizzetto L, Volta P (2019) Plastic sources: a survey across scientific and grey literature for their inventory and relative contribution to microplastics pollution in natural environments, with an emphasis on surface water. Sci Total Environ 693:133499
Ghaffour N, Soukane S, Lee JG, Kim Y, Alpatova A (2019) Membrane distillation hybrids for water production and energy efficiency enhancement: a critical review. Appl Energy 254:113698
Gilman E, Humberstone J, Wilson JR, Chassot E, Jackson A, Suuronen P (2022) Matching fishery-specific drivers of abandoned, lost and discarded fishing gear to relevant interventions. Mar Policy 141:105097
Gliński J, Lipiec J (2018) Soil physical conditions and plant roots. CRC Press/Taylor and Francis, Boca Raton
Gorman D, Gutiérrez AR, Turra A, Manzano AB, Balthazar-Silva D, Oliveira NR, Harari J (2020) Predicting the dispersal and accumulation of microplastic pellets within the estuarine and coastal waters of south-eastern Brazil using integrated rainfall data and Lagrangian particle tracking models. Front Environ Sci 8:559405
Guerranti C, Martellini T, Perra G, Scopetani C, Cincinelli A (2019) Microplastics in cosmetics: environmental issues and needs for global bans. Environ Toxicol Pharmacol 68:75–79
Guo JJ, Huang XP, Xiang L, Wang YZ, Li YW, Li H, Cai Q, Mo C, Wong MH (2020) Source, migration and toxicology of microplastics in soil. Environ Int 137:105263
Guo QQ, Xiao MR, Zhang GS (2021) The persistent impacts of polyester microfibers on soil bio-physical properties following thermal treatment. J Hazard Mater 420:126671
Guo H, Feng H, Xing X (2022) Plastic content and size together produce the dual effects on water loss and surface fissure for silt loam. Eurasian Soil Sci 55(12):1741–1748
He L, Gielen G, Bolan NS, Zhang X, Qin H, Huang H, Wang H (2015) Contamination and remediation of phthalic acid esters in agricultural soils in China: a review. Agron Sustain Dev 35:519–534
Henríquez-Hernández LA, Romero D, González-Antuña A, Gonzalez-Alzaga B, Zumbado M, Boada LD, Hernández AF, López-Flores I, Luzardo OP, Lacasaña M (2020) Biomonitoring of 45 inorganic elements measured in plasma from Spanish subjects: a cross-sectional study in Andalusian population. Sci Total Environ 706:135750
Huang Y, Zhao Y, Wang J, Zhang M, Jia W, Qin X (2019) LDPE microplastic films alter microbial community composition and enzymatic activities in soil. Environ Pollut 254:112983
Huang J, Chen H, Zheng Y, Yang Y, Zhang Y, Gao B (2021) Microplastic pollution in soils and groundwater: characteristics, analytical methods and impacts. Chem Eng J 425:131870
Huang D, Wang X, Yin L, Chen S, Tao J, Zhou W, Chen H, Zhang G, Xiao R (2022) Research progress of microplastics in soil-plant system: ecological effects and potential risks. Sci Total Environ 812:151487
Ingraffia R, Amato G, Bagarello V, Carollo FG, Giambalvo D, Iovino M, Lehmann A, Rillig MC, Frenda AS (2022) Polyester microplastic fibers affect soil physical properties and erosion as a function of soil type. Soil 8(1):421–435
Jägermeyr J, Gerten D, Schaphoff S, Heinke J, Lucht W, Rockström J (2016) Integrated crop water management might sustainably halve the global food gap. Environ Res Lett 11(2):025002

Jakiel M, Bernatek-Jakiel A, Gajda A, Filiks M, Pufelska M (2019) Spatial and temporal distribution of illegal dumping sites in the nature protected area: the Ojców National Park, Poland. J Environ Plan Manag 62(2):286–305

Joos L, De Tender C (2022) Soil under stress: the importance of soil life and how it is influenced by (micro) plastic pollution. Comput Struct Biotechnol J

Ju H, Zhu D, Qiao M (2019) Effects of polyethylene microplastics on the gut microbial community, reproduction and avoidance behaviors of the soil springtail, *Folsomia candida*. Environ Pollut 247:890–897

Jung YS, Sampath V, Prunicki M, Aguilera J, Allen H, LaBeaud D, Veidis E, Barry M, Erny B, Patel L, Akdis C (2022) Characterization and regulation of microplastic pollution for protecting planetary and human health. Environ Pollut:120442

Kabir ER, Rahman MS, Rahman I (2015) A review on endocrine disruptors and their possible impacts on human health. Environ Toxicol Pharmacol 40(1):241–258

Kaur R, Sharma N, Tikoria R, Ali M, Kour S, Kumar D, Ohri P (2022) Insights into biosynthesis and signaling of cytokinins during plant growth, development and stress tolerance. In: Auxins, cytokinins and gibberellins signaling in plants. Springer, Cham, pp 153–187

Kawecki D, Goldberg L, Nowack B (2021) Material flow analysis of plastic in organic waste in Switzerland. Soil Use Manag 37(2):277–288

Kim SW, Waldman WR, Kim TY, Rillig MC (2020) Effects of different microplastics on nematodes in the soil environment: tracking the extractable additives using an ecotoxicological approach. Environ Sci Technol 54(21):13868–13878

Kumar M, Sarma DK, Shubham S, Kumawat M, Verma V, Prakash A, Tiwari R (2020) Environmental endocrine-disrupting chemical exposure: role in non-communicable diseases. Front Public Health 8:553850

Kurniawan TA, Haider A, Ahmad HM, Mohyuddin A, Aslam HMU, Nadeem S, Javed M, Othman MHD, Goh HH, Chew KW (2023) Source, occurrence, distribution, fate and implications of microplastic pollutants in freshwater on environment: a critical review and way forward. Chemosphere:138367

Kvale K (2022) Implications of plastic pollution on global marine carbon cycling and climate. Emerging Top Life Sci 6(4):359–369

Lahive E, Walton A, Horton AA, Spurgeon DJ, Svendsen C (2019) Microplastic particles reduce reproduction in the terrestrial worm *Enchytraeus crypticus* in a soil exposure. Environ Pollut 255:113174

Lahive E, Cross R, Saarloos AI, Horton AA, Svendsen C, Hufenus R, Mitrano DM (2022) Earthworms ingest microplastic fibers and nanoplastics with effects on egestion rate and long-term retention. Sci Total Environ 807:151022

Lear G, Maday SDM, Gambarini V, Northcott G, Abbel R, Kingsbury JM, Weaver L, Wallbank JA, Pantos O (2022) Microbial abilities to degrade global environmental plastic polymer waste are overstated. Environ Res Lett 17(4):043002

Lehel J, Murphy S (2021) Microplastics in the food chain: food safety and environmental aspects. Rev Environ Contam Toxicol 259:1–49

Lehmann A, Leifheit EF, Feng L, Bergmann J, Wulf A, Rillig MC (2020) Microplastic fiber and drought effects on plants and soil are only slightly modified by arbuscular mycorrhizal fungi. Soil Ecol Lett:1–13

Lei L, Liu M, Song Y, Lu S, Hu J, Cao C, Xie B, Shi H, He D (2018) Polystyrene (nano) microplastics cause size-dependent neurotoxicity, oxidative damage and other adverse effects in Caenorhabditis elegans. Environ Sci Nano 5(8):2009–2020

Li H, Liu L (2022) Short-term effects of polyethene and polypropylene microplastics on soil phosphorus and nitrogen availability. Chemosphere 291:132984

Li Q, Wu J, Zhao X, Gu X, Ji R (2019) Separation and identification of microplastics from soil and sewage sludge. Environ Pollut 254:113076

Li B, Song W, Cheng Y, Zhang K, Tian H, Du Z, Wang J, Wang J, Zhang W, Zhu L (2021) Ecotoxicological effects of different size ranges of industrial-grade polyethylene and polypropylene microplastics on earthworms *Eisenia fetida*. Sci Total Environ 783:147007

Li S, Ding F, Flury M, Wang Z, Xu L, Li S, Jones DL, Wang J (2022) Macro-and microplastic accumulation in soil after 32 years of plastic film mulching. Environ Pollut 300:118945

Li K, Zhang M, Jia W, Xu L, Huang Y (2023) Deciphering the effects of LDPE microplastic films on diversity, composition and co-occurrence network of soil fungal community. Appl Soil Ecol 182:104716

Lian Z, Gu X, Liu L, Zhao X (2022) Diffuse phthalate acid esters losses induced from large amount of agricultural plastic film residues caused low risks for water quality in China during 1991–2017. J Hazard Mater 431:128644

Liu M, Lu S, Song Y, Lei L, Hu J, Lv W, Zhou W, Cao C, Shi H, Yang X, He D (2018) Microplastic and mesoplastic pollution in farmland soils in suburbs of Shanghai, China. Environ Pollut 242:855–862

Liu Y, Shao H, Liu J, Cao R, Shang E, Liu S, Li Y (2021) Transport and transformation of microplastics and nanoplastics in the soil environment: a critical review. Soil Use Manag 37(2):224–242

Liu X, Li Y, Yu Y, Yao H (2023) Effect of nonbiodegradable microplastics on soil respiration and enzyme activity: a meta-analysis. Appl Soil Ecol 184:104770

Loiseau C, Sorci G (2022) Can microplastics facilitate the emergence of infectious diseases? Sci Total Environ 823:153694

Long Z, Pan Z, Wang W, Ren J, Yu X, Lin L, Lin H, Chen H, Jin X (2019) Microplastic abundance, characteristics and removal in wastewater treatment plants in a coastal city of China. Water Res 155:255–265

Lozano YM, Rillig MC (2020) Effects of microplastic fibers and drought on plant communities. Environ Sci Technol 54(10):6166–6173

Lozano YM, Aguilar-Trigueros CA, Onandia G, Maaß S, Zhao T, Rillig MC (2021) Effects of microplastics and drought on soil ecosystem functions and multifunctionality. J Appl Ecol 58(5):988–996

Lv L, Yan X, Feng L, Jiang S, Lu Z, Xie H, Sun S, Chen J, Li C (2021) Challenge for the detection of microplastics in the environment. Water Environ Res 93(1):5–15

Maddela NR, Ramakrishnan B, Kadiyala T, Venkateswarlu K, Megharaj M (2023) Do microplastics and nanoplastics pose risks to biota in agricultural ecosystems? Soil Syst 7(1):19

Magni S, Binelli A, Pittura L, Avio CG, Della Torre C, Parenti CC, Gorbi S, Regoli F (2019) The fate of microplastics in an Italian wastewater treatment plant. Sci Total Environ 652:602–610

Marathe NP, Bank MS (2022) The microplastic-antibiotic resistance connection. In: Microplastic in the environment: pattern and process. Springer, Cham, pp 311–322

Marrone A, La Russa MF, Randazzo L, La Russa D, Cellini E, Pellegrino D (2021) Microplastics in the center of Mediterranean: comparison of the two Calabrian coasts and distribution from coastal areas to the open sea. Int J Environ Res Public Health 18(20):10712

Medyńska-Juraszek A, Jadhav B (2022) Influence of different microplastic forms on pH and mobility of Cu^{2+} and Pb^{2+} in soil. Molecules 27(5):1744

Mueller MT, Fueser H, Trac LN, Mayer P, Traunspurger W, Höss S (2020) Surface-related toxicity of polystyrene beads to nematodes and the role of food availability. Environ Sci Technol 54(3):1790–1798

Napper IE, Thompson RC (2020) Plastic debris in the marine environment: history and future challenges. Global Chall 4(6):1900081

Paramdeep KAUR, Singh K, Singh B (2022) Microplastics in soil: impacts and microbial diversity and degradation. Pedosphere 32(1):49–60

Peng Y, Holmstrup M, Schmidt IK, De Schrijver A, Schelfhout S, Heděnec P, Zheng H, Bachega LR, Yue K, Vesterdal L (2022) Litter quality, Mycorrhizal association and soil properties regulate effects of tree species on the soil fauna community. Geoderma 407:115570

Persson L, Carney Almroth BM, Collins CD, Cornell S, de Wit CA, Diamond ML, Fantke P, Hassellöv M, MacLeod M, Ryberg MW, Søgaard Jørgensen P (2022) Outside the safe operating space of the planetary boundary for novel entities. Environ Sci Technol 56(3):1510–1521

Petersen F, Hubbart JA (2021) The occurrence and transport of microplastics: the state of the science. Sci Total Environ 758:143936

Pflugmacher S, Huttunen JH, von Wolff MA, Penttinen OP, Kim YJ, Kim S, Mitrovic SM, Esterhuizen-Londt M (2020) *Enchytraeus crypticus* avoid soil spiked with microplastic. Toxics 8(1):10

Purwar S (2016) Application of natural dye on synthetic fabrics: a review. Int J Home Sci 2(2):283–287

Qi Y, Beriot N, Gort G, Lwanga EH, Gooren H, Yang X, Geissen V (2020) Impact of plastic mulch film debris on soil physicochemical and hydrological properties. Environ Pollut 266:115097

Qin R, Su C, Liu W, Tang L, Li X, Deng X, Wang A, Chen Z (2020) Effects of exposure to polyether sulfone microplastic on the nitrifying process and microbial community structure in aerobic granular sludge. Bioresour Technol 302:122827

Ren SY, Ni HG (2022) A method for measuring the emissions of in situ agricultural plastic film microplastics by ultraviolet and mechanical abrasion. Sci Total Environ 819:152041

Ren Z, Gui X, Xu X, Zhao L, Qiu H, Cao X (2021) Microplastics in the soil-groundwater environment: aging, migration and co-transport of contaminants–a critical review. J Hazard Mater 419:126455

Riveros G, Urrutia H, Araya J, Zagal E, Schoebitz M (2022) Microplastic pollution on the soil and its consequences on the nitrogen cycle: a review. Environ Sci Pollut Res:1–15

Sajjad M, Huang Q, Khan S, Khan MA, Yin L, Wang J, Lian F, Wang Q, Guo G (2022) Microplastics in the soil environment: a critical review. Environ Technol Innov 102408

Schmutz M, Som C (2022) Identifying the potential for circularity of industrial textile waste generated within Swiss companies. Resour Conserv Recycl 182:106132

Shafea L, Yap J, Beriot N, Felde VJ, Okoffo ED, Enyoh CE, Peth S (2023) Microplastics in agroecosystems: a review of effects on soil biota and key soil functions. J Plant Nutr Soil Sci 186(1):5–22

Shahnawaz M, Sangale MK, Ade AB, Shahnawaz M, Sangale MK, Ade AB (2019) Case studies and recent update of plastic waste degradation. In: Bioremediation technology for plastic waste. Springer, Singapore, pp 31–43

Shang X, Lu J, Feng C, Ying Y, He Y, Fang S, Lin Y, Dahlgren R, Ju J (2020) Microplastic (1 and 5 μm) exposure disturbs lifespan and intestine function in the nematode *Caenorhabditis elegans*. Sci Total Environ 705:135837

Shaniv D, Dror I, Berkowitz B (2021) Effects of particle size and surface chemistry on plastic nanoparticle transport in saturated natural porous media. Chemosphere 262:127854

Shen M, Liu S, Hu T, Zheng K, Wang Y, Long H (2023) Recent advances in the research on effects of micro/nanoplastics on carbon conversion and carbon cycle: a review. J Environ Manag 334:117529

Shi J, Sun Y, Wang X, Wang J (2022) Microplastics reduce soil microbial network complexity and ecological deterministic selection. Environ Microbiol 24(4):2157–2169

Shopnil MSR (2022) Microplastic risk characterization of plastic road. Doctoral dissertation, The University of Texas at Arlington

Silori R, Shrivastava V, Mazumder P, Mootapally C, Pandey A, Kumar M (2023) Understanding the underestimated: occurrence, distribution and interactions of microplastics in the sediment and soil of China, India and Japan. Environ Pollut 320:120978

Smith M, Love DC, Rochman CM, Neff RA (2018) Microplastics in seafood and the implications for human health. Curr Environ Health Rep 5:375–386

Sobhani Z, Panneerselvan L, Fang C, Naidu R, Megharaj M (2021) Chronic and transgenerational effects of polystyrene microplastics at environmentally relevant concentrations in earthworms (*Eisenia fetida*). Environ Toxicol Chem 40(8):2240–2246

Song Y, Cao C, Qiu R, Hu J, Liu M, Lu S, Shi H, Raley-Susman KM, He D (2019) Uptake and adverse effects of polyethylene terephthalate microplastics fibers on terrestrial snails (*Achatina fulica*) after soil exposure. Environ Pollut 250:447–455

Sridhar B, Lawrence GB, Debenport SJ, Fahey TJ, Buckley DH, Wilhelm RC, Goodale CL (2022) Watershed-scale liming reveals the short-and long-term effects of pH on the forest soil microbiome and carbon cycling. Environ Microbiol 24(12):6184–6199

Stanton T, Johnson M, Nathanail P, MacNaughtan W, Gomes RL (2019) Freshwater and airborne textile fibre populations are dominated by 'natural', not microplastic, fibres. Sci Total Environ 666:377–389

Steinmetz Z, Wollmann C, Schaefer M, Buchmann C, David J, Tröger J, Muñoz K, Frör O, Schaumann GE (2016) Plastic mulching in agriculture. Trading short-term agronomic benefits for long-term soil degradation? Sci Total Environ 550:690–705

Sun Y, Yuan J, Zhou T, Zhao Y, Yu F, Ma J (2020) Laboratory simulation of microplastics weathering and its adsorption behaviors in an aqueous environment: a systematic review. Environ Pollut 265:114864

Sun Y, Duan C, Cao N, Ding C, Huang Y, Wang J (2022) Biodegradable and conventional microplastics exhibit distinct microbiome, functionality and metabolome changes in soil. J Hazard Mater 424:127282

Tang R, Ying M, Luo Y, El-Naggar A, Palansooriya KN, Sun T, Cao Y, Diao Z, Zhang Y, Lian Y, Chen K, Yan Y, Lu X, Cai Y, Chang SX (2023) Microplastic pollution destabilized the osmoregulatory metabolism but did not affect intestinal microbial biodiversity of earthworms in soil. Environ Pollut 320:121020

Tian Y, Chen Z, Zhang J, Wang Z, Zhu Y, Wang P, Zhang T, Pu J, Sun H, Wang L (2021) An innovative evaluation method based on polymer mass detection to evaluate the contribution of microfibers from laundry process to municipal wastewater. J Hazard Mater 407:124861

Tikoria R, Kaur A, Ohri P (2022a) Potential of vermicompost extract in enhancing the biomass and bioactive components along with mitigation of Meloidogyne incognita-induced stress in tomato. Environ Sci Pollut Res 29(37):56023–56036

Tikoria R, Kaur A, Ohri P (2022b) Modulation of various phytoconstituents in tomato seedling growth and Meloidogyne incognita–induced stress alleviation by Vermicompost application. Front Environ Sci 10:891195

Tikoria R, Sharma N, Kour S, Kumar D, Ohri P (2022c) Vermicomposting: an effective alternative in integrated pest management. In: Earthworm engineering and applications. Nova Science, New York, pp 103–118

Tikoria R, Kaur A, Ohri P (2023a) Physiological, biochemical and structural changes in tomato plants by vermicompost application in different exposure periods under glass house conditions. Plant Physiol Biochem:107656

Tikoria R, Sharma N, Kour S, Kumar D, Ali M, Sharma R, Parkirti, Ohri P (2023b) Cotton oilseed cake: chemical composition and nematicidal potential. In: Ahmad F, Pandey R (eds) Oilseed cake for nematode management. CRC Press/Taylor and Francis, pp 59–69

Tsiota P, Karkanorachaki K, Syranidou E, Franchini M, Kalogerakis N (2018) Microbial degradation of HDPE secondary microplastics: preliminary results. In: Proceedings of the international conference on microplastic pollution in the Mediterranean Sea. Springer, pp 181–188

Urli S, Corte Pause F, Crociati M, Baufeld A, Monaci M, Stradaioli G (2023) Impact of microplastics and nanoplastics on livestock health: an emerging risk for reproductive efficiency. Animals 13(7):1132

Uwamungu JY, Wang Y, Shi G, Pan S, Wang Z, Wang L, Yang S (2022) Microplastic contamination in soil agro-ecosystems: a review. Environ Adv 100273

Veerasingam S, Ranjani M, Venkatachalapathy R, Bagaev A, Mukhanov V, Litvinyuk D, Verzhevskaia L, Guganathan L, Vethamony P (2020) Microplastics in different environmental compartments in India: analytical methods, distribution, associated contaminants and research needs. TrAC Trends Anal Chem 133:116071

Wan Y, Wu C, Xue Q, Hui X (2019) Effects of plastic contamination on water evaporation and desiccation cracking in soil. Sci Total Environ 654:576–582
Wang S (2023) International law-making process of combating plastic pollution: status quo, debates and prospects. Mar Policy 147:105376
Wang J, Liu X, Li Y, Powell T, Wang X, Wang G, Zhang P (2019) Microplastics as contaminants in the soil environment: a mini-review. Sci Total Environ 691:848–857
Wang T, Hu M, Song L, Yu J, Liu R, Wang S, Wang Z, Sokolova IM, Huang W, Wang Y (2020) Coastal zone use influences the spatial distribution of microplastics in Hangzhou Bay, China. Environ Pollut 266:115137
Wang F, Wang Q, Adams CA, Sun Y, Zhang S (2022) Effects of microplastics on soil properties: current knowledge and future perspectives. J Hazard Mater 424:127531
Wiedner K, Polifka S (2020) Effects of microplastic and microglass particles on soil microbial community structure in an arable soil (Chernozem). Soil 6(2):315–324
Willer H, Lernoud J, Kemper L (2018) The world of organic agriculture 2018: summary. In: The world of organic agriculture. Statistics and emerging trends 2018. Research Institute of Organic Agriculture FiBL and IFOAM-Organics International, pp 22–31
Wong JKH, Lee KK, Tang KHD, Yap PS (2020) Microplastics in the freshwater and terrestrial environments: prevalence, fates, impacts and sustainable solutions. Sci Total Environ 719:137512
Wu C, Ma Y, Wang D, Shan Y, Song X, Hu H, Ren X, Ma X, Cui J, Ma Y (2022) Integrated microbiology and metabolomics analysis reveal plastic mulch film residue affects soil microorganisms and their metabolic functions. J Hazard Mater 423:127258
Xu C, Zhang B, Gu C, Shen C, Yin S, Aamir M, Li F (2020) Are we underestimating the sources of microplastic pollution in terrestrial environment? J Hazard Mater 400:123228
Xu M, Xu Q, Wang G, Du W, Zhu J, Yin Y, Ji R, Wang X, Guo H (2023) Elevated CO_2 aggravated polystyrene microplastics effects on the rice-soil system under field conditions. Environ Pollut 316:120603
Ya H, Xing Y, Zhang T, Lv M, Jiang B (2022) LDPE microplastics affect soil microbial community and form a unique plastisphere on microplastics. Appl Soil Ecol 180:104623
Yan Y, Chen Z, Zhu F, Zhu C, Wang C, Gu C (2021) Effect of polyvinyl chloride microplastics on bacterial community and nutrient status in two agricultural soils. Bull Environ Contam Toxicol 107:602–609
Yang L, Zhang Y, Kang S, Wang Z, Wu C (2021) Microplastics in soil: a review on methods, occurrence, sources and potential risk. Sci Total Environ 780:146546
Yano KA, Geronimo FK, Reyes NJ, Kim LH (2021) Characterization and comparison of microplastic occurrence in point and non-point pollution sources. Sci Total Environ 797:148939
Yi M, Zhou S, Zhang L, Ding S (2021) The effects of three different microplastics on enzyme activities and microbial communities in soil. Water Environ Res 93(1):24–32
Yin L, Chen B, Xia B, Shi X, Qu K (2018) Polystyrene microplastics alter the behavior, energy reserve and nutritional composition of marine jacopever (Sebastes schlegelii). J Hazard Mater 360:97–105
Yu H, Fan P, Hou J, Dang Q, Cui D, Xi B, Tan W (2020a) Inhibitory effect of microplastics on soil extracellular enzymatic activities by changing soil properties and direct adsorption: an investigation at the aggregate-fraction level. Environ Pollut 267:115544
Yu Y, Chen H, Hua X, Dang Y, Han Y, Yu Z, Chen X, Ding P, Li H (2020b) Polystyrene microplastics (PS-MPs) toxicity induced oxidative stress and intestinal injury in nematode *Caenorhabditis elegans*. Sci Total Environ 726:138679
Yu L, Zhang J, Liu Y, Chen L, Tao S, Liu W (2021) Distribution characteristics of microplastics in agricultural soils from the largest vegetable production base in China. Sci Total Environ 756:143860
Yu H, Shi L, Fan P, Xi B, Tan W (2022) Effects of conventional versus biodegradable microplastic exposure on oxidative stress and gut microorganisms in earthworms: a comparison with two different soils. Chemosphere 307:135940

Yu Y, Zhang Z, Zhang Y, Jia H, Li Y, Yao H (2023) Abundances of agricultural microplastics and their contribution to the soil organic carbon pool in plastic film mulching fields of Xinjiang, China. Chemosphere 137837

Zhang GS, Zhang FX, Li XT (2019) Effects of polyester microfibers on soil physical properties: perception from a field and a pot experiment. Sci Total Environ 670:1–7

Zhang X, Li Y, Ouyang D, Lei J, Tan Q, Xie L, Li Z, Liu T, Xiao Y, Farooq TH, Wu X, Chen L, Yan W (2021) Systematical review of interactions between microplastics and microorganisms in the soil environment. J Hazard Mater 418:126288

Zhang S, Pei L, Zhao Y, Shan J, Zheng X, Xu G, Sun Y, Wang F (2023) Effects of microplastics and nitrogen deposition on soil multifunctionality, particularly C and N cycling. J Hazard Mater 131152

Zheng J, Fan J, Zhou M, Zhang F, Liao Z, Lai Z, Yan S, Guo J, Li Z, Xiang Y (2022) Ridge-furrow plastic film mulching enhances grain yield and yield stability of rainfed maize by improving resources capture and use efficiency in a semi-humid drought-prone region. Agric Water Manag 269:107654

Zhu F, Zhu C, Wang C, Gu C (2019) Occurrence and ecological impacts of microplastics in soil systems: a review. Bull Environ Contam Toxicol 102:741–749

Chapter 10
Microplastics (MPs) and Antimicrobial Resistance Gene (ARG) Dissemination: Impacts on Ecosystem Sustainability

Amritha Ajayakumar, Suma C. Surendran, H. Raja Naika, and Ashok Kumar Madikonda

Abstract Microplastics (MPs) are small plastic particles that are present in the environment and have been recognized as a significant threat to environmental sustainability. Microbial biofilms, which are complex communities of microorganisms, are known to play a crucial role in the biodegradation of microplastics. These MPs provide a surface for adhesion of antibiotic-resistant bacteria, antimicrobial-resistant genes, antibiotics, MGEs, toxins, and other chemical pollutants. This chapter gives an overview of the associations between MPs, microbial biofilms, and the spread of antibiotic resistance genes (ARGs), with a focus on their implications for environmental sustainability and human health. MPs act as microbial colonization vectors, encouraging the creation of biofilms that enable the transmission of ARGs and worsening the spread of antibiotic resistance. The implications of these interactions for environmental sustainability and human health are also discussed. The chapter also highlights concerns about the direct impact of MPs on human health. Overall, this book chapter provides a valuable contribution to the literature on microplastic pollution and MP-mediated transmission of antimicrobial resistance genes and its impacts on ecosystem sustainability.

Keywords Biofilms · Microbes · Antimicrobial substances · Sustainability · Antibiotic resistance · Gene expression

A. Ajayakumar · S. C. Surendran · A. K. Madikonda (✉)
Department of Biochemistry and Molecular Biology, School of Biological Sciences, Central University of Kerala, Kasaragod, Kerala, India

H. R. Naika
Department of Environmental Science, School of Earth Science Systems (SESS), Central University of Kerala, Kasaragod, Kerala, India

K. Khanna et al. (eds.), *Microplastics and Soil Microbiome*,
https://doi.org/10.1007/978-981-96-4978-5_10

1 Introduction

Plastics, the synthetic polymer that was referred to as the "material of 1,000 uses," have had an unstable connection with civilization since Leo Baekeland discovered them in 1907. Plastic is made of carbon, oxygen, hydrogen, chlorine, nitrogen, and silicon that are chemically linked together to form long chains of repeating units called monomers. These monomers undergo polymerization, where they bond to create large, durable, and versatile macromolecules. Polythene, consisting of long chains of ethylene monomers (C_2H_4), makes up 64% of all plastic materials (Malpass 2010). Since plastics are used in strategic industries like packaging, construction, automobile, electronics, sports, agriculture, renewable energy, and medical devices, their production has expanded to about 413.8 metric ton globally in 2023. Out of which 374.2 Mt. is fossil-based production ("Plastics – the *fast* Facts" 2024).

The Society of the Plastics Industry in 1988 developed the Resin Identification Code (RIC) system to classify plastic resins into seven groups based on their chemical composition. These include PET (polyethylene terephthalate), used for making beverage bottles and food containers; HDPE (high-density polyethylene), used in milk jugs, detergent bottles, and piping; PVC (polyvinyl chloride), seen in pipes, medical equipment, and cable insulation; LDPE (low-density polyethylene), used for making plastic bags, flexible containers, and film wraps; PP (polypropylene), commonly found in food containers, bottle caps, and straws; PS (polystyrene), used in disposable cups, food trays, and packaging foam; and other plastics, a category that includes polymers like polycarbonate (PC) and polylactic acid (PLA) that do not fit into the previous groups (Fig. 10.1). This classification system facilitates the identification and recycling of various types of plastics, contributing to sustainable

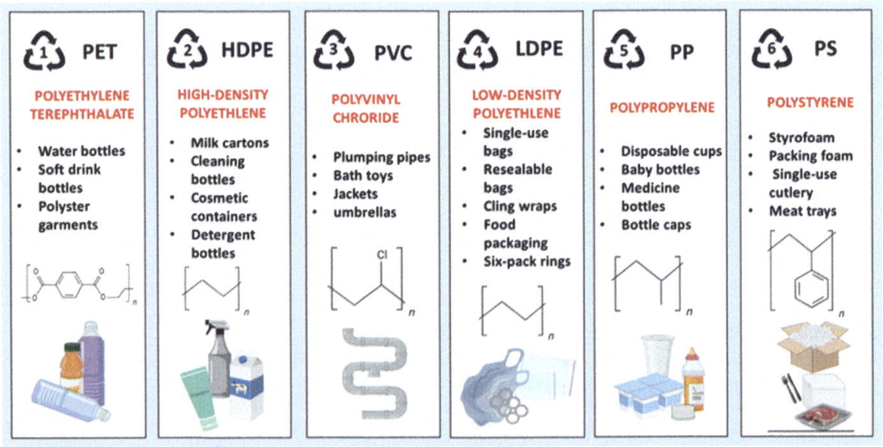

Fig. 10.1 Types of plastics and their use: polyethylene terephthalate (PET), high-density polyethylene (HDPE), polyvinyl chloride (PVC), low-density polyethylene (LDPE), polypropylene (PP), and polystyrene (PS)

waste management and material reuse (Fig. 10.1) (Serranti et al. 2011). Thompson Richard Charles described tiny plastic particles, particularly in oceans, as "microplastic" in 2004.

Microplastics (MPs) are the tiny plastic particles that are less than 5 mm in any dimension with an unknown lower limit. A lower limit for these particles is being established by the recent use of the term "nanoplastics", which generally refers to particles with dimensions of less than 100 nm (Picó and Barceló 2019). MPs are non-biodegradable particles that can persist in the environment for centuries due to their inability to decompose naturally. As a result, they accumulate in both terrestrial and aquatic ecosystems, causing long-term environmental damage. These tiny particles can contaminate soil, water, and air and cause significant hazards to plant, animal, microbial, and human health. In aquatic environments, MPs are often consumed by marine life, leading to physical harm, poisoning, and even disruption of reproductive systems. Similarly, in terrestrial ecosystems, they can affect soil quality, reduce biodiversity, and potentially enter the food chain. For humans, the risks include direct exposure through consumption of contaminated water, seafood, or other food products, which may cause numerous health problems such as inflammation, hormonal disruption, and cancer. Widespread presence of MPs and their persistence in nature have raised growing concerns about their impact on ecologies and human health, highlighting the urgent need for effective mitigation strategies (Chia et al. 2021). Huerta Lwanga et al. (2016) proved the negative impact of plastic litter on earthworms, revealing a significant reduction in their growth rate and weight. This reduction in size and development is particularly concerning, as it can severely affect the earthworms' reproductive capacity, making them less likely to reproduce and sustain their populations (Fig. 10.5). Earthworms have a critical role in soil health by providing aeration to soil and facilitating nutrient cycling, so any disruption to their growth and reproduction can have ripple effects on the entire ecosystem. The presence of plastic litter in soil not only interferes with the physical and biological processes of earthworms but also increases the potential for long-term environmental damage, as the loss of these vital organisms can degrade soil quality and disrupt agricultural productivity. Huerta Lwanga et al. (2016) study highlights the wider ecological consequences of plastic pollution, stressing the need for urgent action to address the environmental impacts of plastic waste on soil ecosystems. Recent research has shown that *Arabidopsis thaliana*, a widely used model plant in scientific studies, has the ability to accumulate both positively and negatively charged plastic particles. This phenomenon presents a serious concern for agricultural productivity, as the accumulation of these MPs in plant tissues can interfere with their growth and development. When plants absorb plastic particles, they may experience a range of negative effects, including impaired nutrient uptake, root damage, and disrupted cellular processes (Fig. 10.5). Moreover, MPs can release toxic chemicals into the plant, which may be passed along the food chain, ultimately affecting human and animal health (Fig. 10.3). The presence of these particles in the soil can also alter soil structure and fertility, leading to long-term consequences for crop yields and soil sustainability. As *Arabidopsis thaliana* serves as a model for other plants, these findings raise alarms about the broader implications of microplastic pollution on agricultural ecosystems, emphasizing the urgent need for further investigation into the environmental impacts of plastic contamination in soil and crops (Sun et al. 2020).

Studies highlight several effective methods for reducing microplastic contamination in groundwater and soil, including pyrolysis (Ni et al. 2020), the use of biodegradable plastics (Qi et al. 2020), and filtration systems integrated with chemical or biological degradation. These methods aim to mitigate the contamination caused by MPs by either removing or breaking them down in the environment. Pyrolysis, for example, involves the thermal decomposition of plastics, which can convert them into useful by-products while reducing their environmental footprint (Ni et al. 2020). The shift to biodegradable plastics is another promising solution, as these materials are designed to decompose more readily than traditional plastics, potentially preventing long-term accumulation in ecosystems (Qi et al. 2020). Plastic filtration systems, along with chemical or biodegradation techniques, can be used to capture and break down MPs in contaminated environments, preventing them from further spreading into soil and water (Chia et al. 2021). Together, these approaches represent a multi-faceted strategy to address the growing issue of plastic pollution, which, if effectively implemented, could significantly reduce the ecological as well as health risks related to microplastics. However, continued research and innovation are vital to enhance the efficiency and scalability of these methods to combat microplastic contamination on a global scale (Ni et al. 2020; Qi et al. 2020; Chia et al. 2021).

In recent years, MP pollution has become a significant concern, widely contaminating marine habitats and ecosystems (Debroy et al. 2021). This has led to an increased interaction between MPs and microorganisms, which plays vital role in the fate and effect of microplastics in the environment (Qiu et al. 2022). Microorganisms like bacteria and fungi interact with MPs primarily by utilizing the polymers or their degraded intermediates as substrates or co-substrates for growth. These microorganisms are capable of colonizing the surfaces of plastic particles, forming biofilms that can facilitate the breakdown of plastic materials (Sooriyakumar et al. 2022) (Fig. 10.2). When MPs and microbes interact, the biodegradation of plastics occurs more rapidly, as microbial enzymes help degrade the polymers, transforming them into smaller and potentially less harmful compounds (Amobonye et al. 2021). However, the biodegradation process is complex and depends on several factors like environmental conditions, nature of plastic, and the specific microbial communities involved. While this interaction can offer a potential natural solution to plastic pollution, it also raises concerns about the potential for harmful microorganisms to thrive on plastic surfaces, affecting marine life and the food chain (Debroy et al. 2021) (Fig. 10.4). The usage of plastic items promotes microbial colonization and the formation of biofilms by speeding up the release of plastic particles (Sooriyakumar et al. 2022). Additionally, these plastic fragments can serve as carriers of various pollutants and microorganisms that threaten the immune system of animals (Chen et al. 2022) (Fig. 10.2). Unusual MPs have the potential to change biogeochemical cycles in the natural environment, including the nitrogen and organic carbon cycles in the ocean (Wang et al. 2021a).

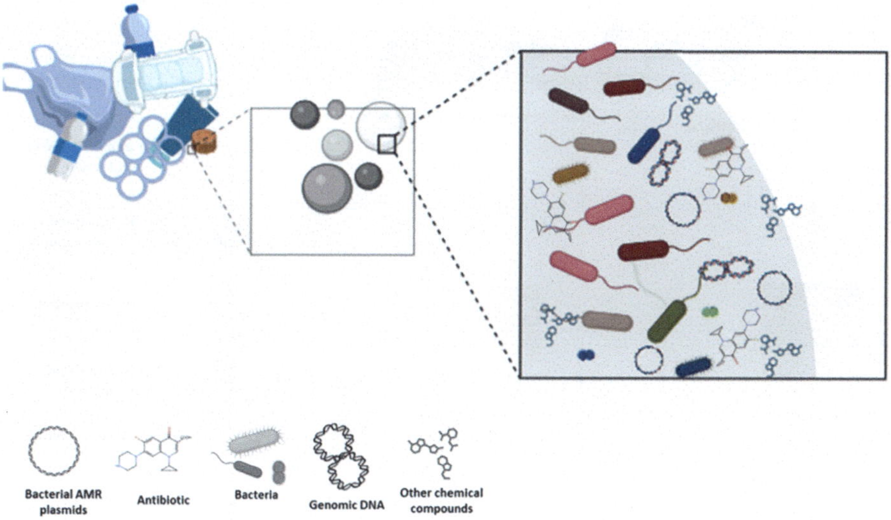

Fig. 10.2 Adsorption of ARBs, ARGs, antibiotics, MGEs, and other chemical pollutants on MP surface. The adsorption of the biological and chemical components to the MP is mediated by hydrogen bonding, electrostatic interactions, π-π interactions, hydrophobic interactions, and van der Waals forces. MPs are the carriers of adhered particles

MPs create an ideal environment for specific bacterial consortia to thrive. MP biofilms differ from those developed on natural surfaces in terms of their microbial community structure. MPs provide a new environment for biofilm development and promote the growth of pathogens (Wu et al. 2019). MPs pose a concern to ecological and human health as they are difficult to break down, move easily, and accumulate in the food chain (Rafa et al. 2024). In aquatic environments, MPs act as a "Plastisphere", a substrate for the colonization of the microbes leading to biofilm formation (Fig. 10.2). These microbes on the MP surfaces affect the degradation, sedimentation, transport, and adsorption of MPs in marine environments (Fig. 10.2) (Jia et al. 2024).

Biofilms formed on microplastics (MPs) provide a favourable environment for the colonization of pathogenic microorganisms and the conjugation of plasmids (Arias-Andres et al. 2018). These biofilms create an ideal habitat for pathogens that can harbour antibiotic resistance genes (ARGs). In some instances, ARGs are exclusively present in the pathogenic biofilm associated with MPs (Wu et al. 2019). The coexistence of ARGs and MPs in the environment has raised widespread fear due to their potential threat to human well-being. Both conventional and biodegradable MPs have been recognized as significant contributors to the increase in ARG abundance, with MPs serving as hotspots for the spread of ARGs via horizontal gene transfer (HGT) in aquatic ecosystems (Yang et al. 2019) (Fig. 10.2). Mobile genetic elements (MGEs) are central to the transmission of ARGs, and integron-integrase genes (*intI*1 and *intI*2) are commonly found in biofilms formed on plastic particles (Lu et al. 2020). This interaction between MPs and ARGs underscores the growing

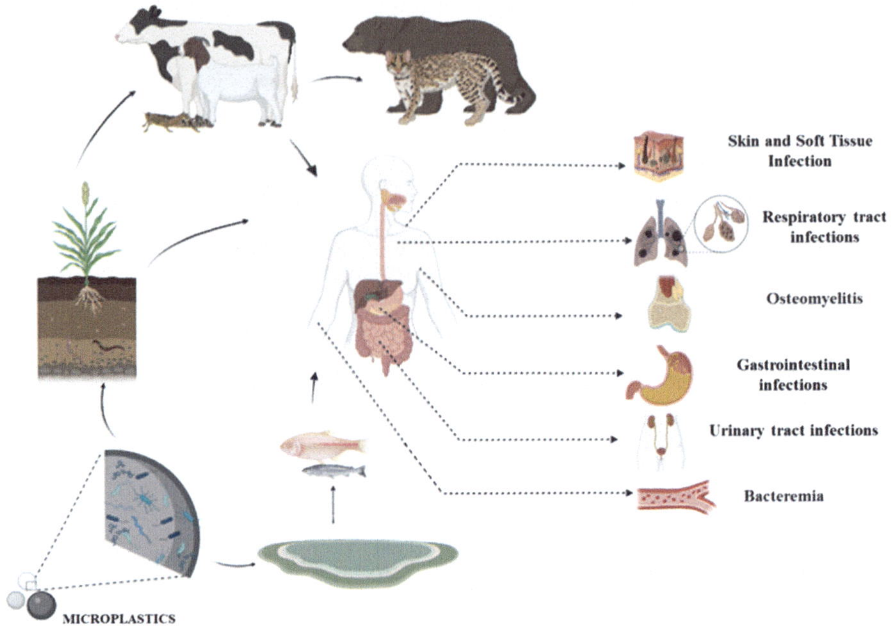

Fig. 10.3 Trophic transfer of MPs and ARGs in food chain. Aquatic organisms and earthworms are the primary consumers of the MPs. The MPs ingested by these organisms can be transferred to organisms of higher trophic level till humans. Humans consume MPs mainly through three mechanisms: ingestion, inhalation, and dermal contact. The presence of MPs can cause infections in the skin, respiratory tract, gastrointestinal tract, urinary tract, bones, and blood

risk posed by microplastic pollution as a vehicle for the dissemination of antimicrobial resistance, further complicating the challenge of managing both plastic waste and the rise of resistant pathogens.

This chapter provides an overview of the relationships between MPs, microbial biofilms, and the dissemination of antimicrobial resistance genes (ARGs), emphasizing their impact on environmental sustainability and human health. MPs serve as vectors for microbial colonization, promoting the formation of biofilms that facilitate the transmission of ARGs, exacerbating the spread of antibiotic resistance (Fig. 10.2). The chapter also explores the sorption of antibiotics onto MPs, which can persist in the environment and contribute to resistance development. MPs and ARGs are prevalent across terrestrial, aquatic, and atmospheric ecosystems. In aquatic environments, MPs act as hotspots for ARG proliferation through horizontal gene transfer (HGT), while in soil, MPs introduce resistance factors that affect soil health. Atmospheric MPs, though less studied, may spread contaminants across vast distances, further compounding the issue of antimicrobial resistance (Fig. 10.4). The chapter also highlights concerns about the direct impact of MPs on human health (Fig. 10.3). Inhalation or ingestion of MPs can lead to their accumulation in the human body, causing inflammation and facilitating the transfer of harmful

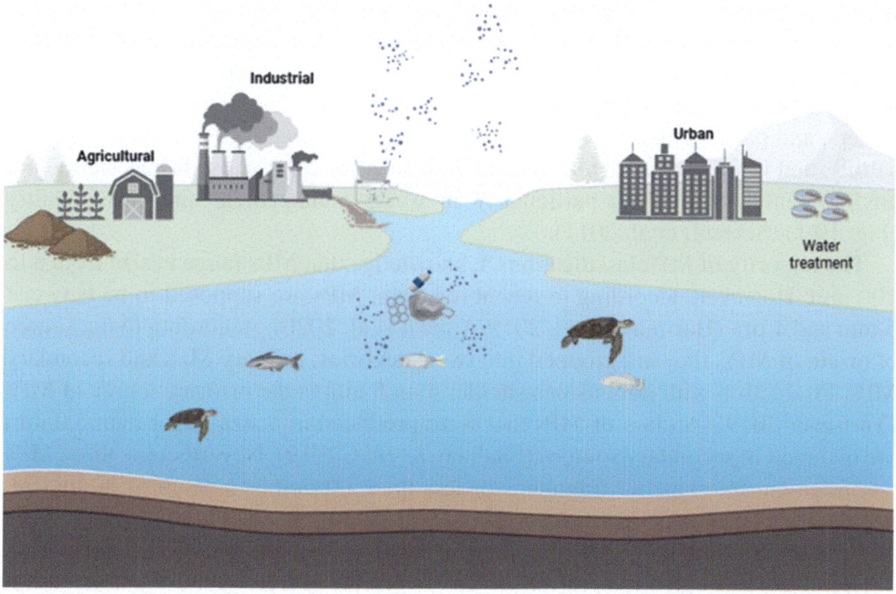

Fig. 10.4 Occurrence of MPs in terrestrial, aquatic, and atmospheric environments. The major source of MPs is agricultural, industrial, urban, and WWTPs. The MPs discharging from these sources leaches out to soil layers, underground water, as well as the aquatic environment leading to the ingestion of MPs by aquatic organisms. Also, MPs can be consumed as airborne particles

substances, including ARGs and resistant bacteria (Fig. 10.3). This poses a growing risk to human health, particularly for vulnerable populations. The chapter stresses the necessity for interdisciplinary research and solutions to address the environmental and health threats posed by MPs and antimicrobial resistance.

2 MPs in the Environment

Plastics are cost-effective, lightweight, durable, long-lasting, resistant to corrosion, and highly versatile materials. They are very useful and have strong electrical and thermal insulation properties. Currently, the most widely used synthetic plastics include polyethylene terephthalate (PET), high-density polyethylene (HDPE), polyvinyl chloride (PVC), low-density polyethylene (LDPE), polypropylene (PP), and polystyrene (PS). PET, with the chemical formula $(C_{10}H_8O_4)n$, is a durable and lightweight plastic commonly used to make food containers and beverage bottles. HDPE, represented by the formula $(C_2H_4)n$, is recognized for its strength and resistance, making it ideal for the manufacture milk jugs, detergent bottles, and pipes. PVC, with the formula $(C_2H_3C_l)n$, is a versatile material widely used in construction for pipes and fittings, as well as in medical equipment. LDPE, which shares the formula $(C_2H_4)n$ but has a more branched structure, is softer and more flexible, often found in plastic bags, film wraps, and squeeze bottles. Polypropylene (PP), with the formula $(C_3H_6)n$, is a tough, heat-resistant plastic used in automotive parts, reusable containers, and textiles. Lastly, polystyrene (PS), with the chemical formula (C_8H_8) n, is a lightweight, insulating plastic commonly used for disposable cups, food containers, and packing materials. These plastics are favoured for their durability, versatility, and resistance to chemicals, but their widespread use also brings significant environmental challenges, particularly in waste management and sustainability (Fig. 10.1) (Serranti et al. 2011).

The majority of MP classifications acknowledge that MPs range in size from 5 to 100 mm. However, according to recent research, MPs are supposed to be between 5 mm and 1 μm (Hartmann et al. 2019;Ameen et al. 2020). According to the source of origin of MPs, they are grouped into two categories: primary MPs and secondary MPs. Production with dimensions smaller than 5 mm is the primary source of MPs (Yurtsever 2019). A class of MPs that were produced at a size larger than 5.0 mm are included in secondary sources (Shahnawaz et al. 2019). Nevertheless, these MPs can fragment into smaller fragments of less than 5.0 mm as a result of biological, physical, and chemical degradation (Naik et al. 2020).

Primary microplastics (MPs) are plastic particles that are released directly to the environment through the abrasion of larger plastic objects while manufacturing, maintenance or utilization. Examples include tire erosion during driving and the shedding of synthetic fibres during washing. They can also originate from optional additives in products, such as scrubbing agents in cosmetics (Fig. 10.1). On the other hand, secondary microplastics are generated when larger plastics degrade into smaller particles due to environmental exposure over time (Kalogerakis et al. 2017). This occurs due to unintended losses, such as abandoned or lost fishing nets, which often end up in marine environments, or through the breakdown of improperly disposed waste, like plastic bags, caused by photodegradation and other weathering processes (Boucher and Friot 2017). A global report by Boucher and Friot (2017) highlights seven major sources of MPs contributing to environmental pollution. These sources include city dust, which consists of tiny plastic particles from urban areas; marine coatings used on ships, which release microplastics into the oceans, personal care products and cosmetics containing microbeads, plastic pellets used in industrial manufacture, road markings that degrade over time, synthetic textiles that shed fibres during washing, and tire wear particles generated from driving. Collectively, these sources are the cause of widespread distribution of microplastics across terrestrial and aquatic ecosystems.

Ding et al. (2020) reported that MP concentrations as high as 3410 pieces/kg could pollute agricultural soil. Plastic mulch and sewage sludge are regarded as the principal sources of MPs rain-fed agricultural land-use systems. The main source of MPs are irrigation and urea fertilizers coated with MPsin wet agricultural land-use systems (Huang et al. 2020). The possibility exists that once MPs are deposited on the soil's surface, they may be carried vertically, into soil sublayers and ultimately into deeper soil layers (Li et al. 2020), and through leaching, they may eventually enter the groundwater system. According to some research, human activities, including farming, fishing, and wastewater treatment, are connected to groundwater contamination and pollution from MPs (Chia et al. 2021).

According to recent research by Zang et al. (2020), the addition of MPs made of polyvinyl chloride, polyethylene, or polythene to soil can alter plant growth, which can then be reflected in biomass production and carbon distribution. MPs are often regarded as inert materials that are small and easily absorbed by humans or other animals and have little chemical reactivity (Sarkar et al. 2023). Oral ingestion of contaminated food or water is considered as one of the main ways of MP exposure to human. When eating marine goods, there is a significant chance of absorbing MPs. Because MPs are so tiny, they can be involuntarily consumed by lower trophic-level aquatic creatures, including plankton and filter feeders (Cole et al. 2011). MPs have been increasingly recognized for their potential harmful effects on human health. These minute plastic particles can act as vectors for toxic chemicals, disrupt cellular functions, and contribute to a range of adverse biological effects. MPs may exert toxic effects on the human body by serving as carriers for hazardous substances, including persistent organic pollutants (POPs), endocrine-disrupting chemicals, and heavy metals. These pollutants can discharge from the microplastics and enter human tissues, leading to biochemical disruptions. Moreover, MPs can generate oxidative stress, a condition due to imbalance between the production of free radicals and the body's ability to neutralize them using antioxidants (Alkadi 2020). Prolonged oxidative stress may result in cellular damage, DNA mutations, and lipid peroxidation, which are precursors to several diseases, including cardiovascular disorders, neurodegenerative disorders, and cancer (Preiser 2012). MPs are also associated with tissue damage and chronic inflammation. Studies suggest that MPs can penetrate epithelial barriers, accumulate in vital organs, and trigger immune responses. For instance, the ingestion or inhalation of microplastics has been linked to inflammatory conditions in the gastrointestinal and respiratory systems. Chronic inflammation caused by prolonged exposure to microplastics may lead to fibrosis, tissue necrosis, and an increased risk of autoimmune diseases (Fig. 10.3) (Naqash et al. 2020).

Recent research highlights that microplastics can significantly hinder plant growth through both direct and indirect mechanisms. Direct inhibitory effects include the blockage of pores in soil or plant surfaces, which disrupts water and nutrient uptake (Fig. 10.5). Microplastics can also cause root injuries and interfere with root elongation and development. Furthermore, they have been found to inhibit the expression of key plant genes associated with growth and stress response. Indirect effects arise from changes in soil properties caused by microplastics. These include altering the soil's physical structure, water retention capacity, and aeration. Microplastics can release toxic additives (e.g., plasticizers, stabilizers) into the soil, which may affect plant health. Additionally, they can disrupt the balance of soil microbial communities and fauna, including bacteria and earthworms, which are essential for nutrient cycling and soil fertility. This disruption may, in turn, impact the bioavailability of essential nutrients and contaminants, further affecting plant growth and health. Overall, the multifaceted impact of microplastics on soil ecosystems and plant health underlines the necessity for further investigation and mitigation strategies to address this emerging environmental concern (Li et al. 2022).

Fig. 10.5 Degradation of plastic used for mulching in agricultural settings. Major source of terrestrial MPs. Farming activities like ploughing, harvesting, and mulch film recycling can cause the production of MPs. It can leach to the soil layers, and organisms in the soil environment will consume these MPs

3 Microplastics and Human Health

Plastic consumption occurs among taxa within multiple trophic levels, including fishes, marine mammals, invertebrates, and fish-eating birds (Gall and Thompson 2015). MPs have emerged as significant challenges to human, animal, and environmental health. These small plastic particles can disrupt ecosystems and pose risks to various organisms. For instance, microplastics can contribute to harmful algal blooms and an increased prevalence of antibiotic-resistant bacteria. These phenomena may negatively affect other organisms and animals, disrupt ecological balance, and potentially impact entire food chains. Such disruptions can ultimately pose a risk to human health, as the effects cascade through trophic levels.

Human exposure to microplastics is through various routes, including ingestion (through MP-contaminated foods), inhalation (airborne MPs), and dermal contact (through skin exposure). This exposure elevates the concerns about potential toxic effects, bioaccumulation, and long-term health consequences, necessitating further investigation and mitigation efforts (Fig. 10.3) (Yee et al. 2021).

Studies indicated that a certain way of consumption of microplastics by humans is by ingestion (Lehner et al. 2019). The involuntary ingestion of MPs through the gastrointestinal route via food and water is also identified by the presence of diverse

MPs in human stool (Schwabl et al. 2019). Research performed in mouse models by Kim et al. (2022) identified that PS MPs exposure increased the expression level of asialoglycoprotein receptor 2 (ASGR 2) gene, thus enhancing the potential risk of gastric cancer. MPs also can reduce the lipid digestion. Lipids adsorbing onto the microplastic surface form aggregates, which in turn reduce their bioavailability for degradation. MP-adsorbed lipase reduce the lipid degradation by changing the secondary structure (Tan et al. 2020). Even the MPs ingested through the food can translocate through the circulatory system, and they can also diffuse to the tissues. The study revealed the incidence of MPs in human placental samples of patients with uneventful pregnancies (Ragusa et al. 2021).

A study by Amato-Lourenço et al. (2021) identified and characterized the existence of microplastics (MPs), specifically polypropylene (PP) and polyethylene (PE), within lung tissue. The research demonstrated that exposure to microplastics through the respiratory tract led to their accumulation in the lungs, raising concerns about potential health implications associated with inhalation of airborne microplastic particles. Inhalation is the second significant method of exposure of microplastics (Yee et al. 2021). These MPs can be retained in lungs or alveolar surfaces and can easily be translocated to different body parts as well (Stapleton 2019).

Dermal exposure is also an important route of MPs to enter the human body. The major source of MPs is cosmetic products which are used topically (Hernandez et al. 2017). Stratum corneum acts as a skin barrier which protects the skin from chemicals and microbial agents. Damage to stratum corneum facilitates the entry of MPs through the skin and which can later diffuse to the target tissues (Yee et al. 2021).

4 Microbial Biofilms and Their Associations with MPs

Microorganisms produce EPS, which includes proteins, glycoproteins, and glycolipids. This matrix surrounds the bacteria and allows to adhere to a wide range of biological and abiotic surfaces, forming biofilms (Keswani et al. 2016).

Biofilm production is an intricate, multi-step process in which bacteria transform from a free-swimming planktonic state to a surface-attached, biofilm-producing sessile lifestyle. This process involves the initial adhesion of bacteria to a surface, followed by the formation of microcolonies, synthesis of extracellular polymeric substances (EPS), and the development of mature biofilm structure. Various environmental and physiological factors influence biofilm formation (Keswani et al. 2016).

External conditions such as pH, temperature, gravitational forces, Brownian motion, hydrodynamic forces, and the properties of the surface being colonized play critical roles in determining the efficiency and structure of biofilm development. Additionally, bacterial communication through quorum sensing, which involves the production and detection of signalling molecules, regulates biofilm formation by coordinating group behaviours. Secondary messengers like cyclic-di-GMP also have vital role in this process, promoting the production of EPS and other biofilm-specific factors. These influencing factors work in conjunction to trigger a cascade of genetic and biochemical changes in bacteria, enabling them to adapt to environmental stress, enhance resistance to antimicrobials, and establish robust biofilm communities. The ability of biofilms to thrive in diverse conditions underscores their importance in both natural ecosystems and clinical settings, where they cause infections and biofouling (Rather et al. 2021).

The increased diversity of MPs enabled the production of biofilms by *Escherichia coli*, *Pseudomonas aeruginosa*, and *Vibrio cholerae*. Biofilm improves the adsorption ability of MPs to contaminants like antibiotics (Jia et al. 2024). Biofilms, surface-attached microbial colonies, can develop antibiotic resistance up to 1000 times stronger than planktonic cells (Wang et al. 2023). Biofilm generation on MPs can affect their physical and chemical properties, including surface micromorphology, roughness, charge, surface area, and density. This can have an impact on the vertical migration, weathering, and adsorption/desorption of contaminants and pathogens in maritime settings. Biofilm production also has biological consequences, such as changing microbial community structures, influencing bio-toxicity, and facilitating bio-transfer across trophic levels. Rapid colonization and aggregation of microalgae and microorganisms can alter buoyancy and sedimentation, resulting in biofilm formation. Biofilms can potentially function as a source of microplastic-degrading microorganisms (Tu et al. 2020).

5 Interaction of Antibiotics with MPs: Absorption and Adsorption

MPs can interact with antibiotics in the environment through processes of absorption and adsorption, significantly impacting their fate, transport, and bioavailability. Adsorption denotes to the surface-level attachment of antibiotic molecules superficially on MPs, influenced by factors such as the surface area, hydrophobicity, and chemical composition of MPs (e.g., polyethylene, polypropylene). On the other hand, absorption involves the dissemination of antibiotic molecules into the microplastic matrix, often dictated by the porosity and polymer characteristics of the microplastic material.

These interactions not only facilitate the accumulation and persistence of antibiotics in aquatic and terrestrial environments but also create hotspots for the proliferation of antibiotic-resistant bacteria and genes. Such interactions exacerbate ecological and public health risks, highlighting the urgent need for understanding and mitigating the combined effects of MPs and antibiotics in ecosystems. MPs have a significant adsorption impact on antibiotics and can be utilized as carriers to influence antibiotic degradation in environment. J. Wang et al. (2020a), detected that the degradation of ciprofloxacin (CIP) in the soil is inhibited by the MPs. Total nitrogen concentration in the soil also decreased when it is treated with CIP and MPs together compared to where it is only treated with CIP (Wang et al. 2020a).

Antibiotics interact differently with commercial plastics than other organic compounds due to their unique molecular structures (Yu et al. 2019; Wang et al. 2020a, b). The ability of microplastics (MPs) to absorb antibiotics is influenced by a combination of their inherent properties and the nature of their interactions with antibiotic molecules. Key mechanisms include hydrogen bonding, surface characteristics (such as roughness and functional groups), electrostatic interactions, π-π interactions, hydrophobic interactions, and van der Waals forces. These processes collectively determine the extent and efficiency of antibiotic adsorption onto MPs, as highlighted by Syranidou and Kalogerakis (2022). These interactions depend on factors like the chemical composition of the microplastics, the surrounding environmental conditions (e.g., pH, temperature, and salinity), and the structural properties of the antibiotics themselves. This complex interplay has significant implications for the environmental persistence and bioavailability of antibiotics, potentially influencing ecological balance and the spread of antibiotic resistance. Antibiotic adsorption on MPs can increase antibiotic exposure in aquatic ecosystems through long-range transport, as antibiotics and MPs share pollutant resources. MPs can transport antibiotics and ARGs in the environment, increasing their prevalence (Zhao et al. 2024).

6 MP: Antimicrobial Resistance Bacteria (ARB)/ Pathogen carrier

MPs can act as a sink for the pathogens or ARBs (Junaid et al. 2022). Microorganisms in MP biofilm differ greatly from those in surrounding water, which makes them highly distinctive (Jia et al. 2024). Wu et al. (2019) stated that MPs provide improved existence than the environment of aquatic ecosystem. *Vibrio* sp. is identified as opportunistic biofilm former in aquatic environments (Oberbeckmann et al. 2018). Pathogens such as *Klebsiella*, *Pseudomonas*, and *Salmonella* spp. are selectively abundant in biofilms with significant antibiotic resistance and possess numerous ARGs like *gyrA*, *qnrS*, and *sul1* (Silva et al. 2023;Qiu et al. 2023;Metcalf et al. 2024).

7 MP: Antimicrobial Resistance Gene (ARG) Carrier

MPs act as the vectors for the conveyance of chemical pollutants, heavy metals, ARGs, and antimicrobial-resistant bacteria (ARB) (Stapleton et al. 2023). Compared to the HGT in natural water, ARG transformation is hundred times more in MP biofilms (Wang et al. 2023). Intracellular ARGs (i-*Tet*A, i-*Tet*C, i-*Tet*O, and i-*su*l1) and extracellular ARGs (e-*Tet*A and e-*bla*TEM) are favourably adsorbed onto the surface of MPs, as demonstrated in studies by Xu et al. (2022). Intracellular ARGs can be found on chromosomes or MGEs such as ICEs, transposons, and integrons present on episomal plasmids. However, chromosomes and plasmids may transport ARGs differently (Hu et al. 2022) In 2019, Lu et al. observed 10 ARGs in the water samples and on the MPs in the recirculating aquaculture system farm, which includes *sul*1, *sul*2, *tet*G, *erm*F, and *qnrS*, out of which the sulfonamide resistance gene was identified to be the predominant one. The abundance of *intI* 1gene, which is responsible for the transport of the antimicrobial resistance cassettes, was observed more on the MP surface than in water.

Wang et al. (2021b) observed that exposure to MPs (ranging from 120 to 140 μm in size and composed of polyvinyl chloride, polyamide, polystyrene, and polyethylene) can lead to substantial rise in the abundance of both intracellular and extracellular antibiotic resistance genes (ARGs), including *tet*W, *tet*E, and *int*I1. This finding aligns with the above-mentioned studies highlighting the potential for microplastics to serve as vectors for the proliferation of ARGs in the environment, contributing to the spread of antibiotic resistance (Xu et al. 2022; Li et al. 2020). In a study, five (*tet*A, *tet*X, *sul*1, *sul*2, and *bla*TEM) ARGs were detected on MPs from the terrestrial environment soil as well (Yan et al. 2020). ARGs and MPs regularly co-existed in agricultural soils. The abundance of ARG is greater in larger MPs than in smaller MPs, and the longer the planting years, the greater the abundance and strongly aged MPs had a higher abundance (Lu et al. 2020). Studies disclosed the transfer of ARGs and MPs from soil to edible crops (Li et al. 2020).

8 Incidence of Antibiotic Resistance Genes on MPs

8.1 Terrestrial

The main sources of MPs in terrestrial environments include organic fertilizers, sewage sludge, plastic film mulching, irrigation practices, and atmospheric deposition. These sources cause the widespread dispersal of MPs in soils, potentially disturbing soil health and ecosystem functioning. Studies have shown that agricultural practices, particularly the use of plastic-based materials like mulching films and the application of sewage sludge, are significant contributors to the presence of MPs in terrestrial ecosystems (Zhang et al. 2020). Additionally, atmospheric deposition and irrigation can introduce microplastics into the soil from airborne particles and contaminated water sources (Yang et al. 2021). Soil and MPs can interact to generate unique aggregates. Soil aggregates help in the downward and horizontal transfer of contaminants such as ARGs, MPs, and heavy metals in agro-ecosystems through runoff (Wang et al. 2020b). Zhang et al. (2022b) stated that mulching farmlands have MP abundance than non-mulching farmlands, and the abundance depends on the duration of mulching (Fig. 10.5). ARGs on MPs are based on the specific soil bacteria colonizing on MP (Yang et al. 2022). Wang et al. (2023) identified that the diversity of the MPs affects the abundance of the ARGs; as the diversity of MPs increases, the richness of the ARGs in the soil environment also increases. It also stated that ARGs that encode resistance to tetracyclines, aminoglycosides, and florfenicol were significantly greater. The diversity of microplastics (MPs) has led to a significant hike of mobile genetic elements (MGEs) expressing antibiotic resistance genes (ARGs), motility genes, and virulence factor genes (VFGs) associated with human pathogenicity, posing a considerable health risk. According to Lu, Lu, and Liu (2020), the selective abundance of ARGs on microplastics is influenced by various factors, including the size of MP, weathering status of the microplastics, soil characteristics (such as years of cultivation), and specific types of ARGs. Their study found that macrolide-lincosamide-streptogramin B (MLSB) resistance genes and sulfonamide-ARGs were more likely to be enhanced on heavily weathered MPs, while aminoglycoside resistance genes and beta-lactamase resistance genes were preferentially attached to slightly weathered MPs. These findings underscore the complex interactions between microplastic characteristics and the proliferation of antibiotic resistance in the environment, raising concerns about the long-term health risks associated with these pollutants (Lu et al. 2020; Zhang et al. 2020).

8.2 Aquatic

The aquatic environment is a major reservoir for the MPs from agricultural activities, industries, and WWTPs. The MP aggregates in the aquatic environment enhance the transfer of pollutants adsorbed on the surface of MPs to the food web. Plants can adsorb MP aggregates (Yu et al. 2020), and these can be consumed by aquatic animals. Jabeen et al. (2017) identified the presence of mesoplastic and microplastics in the gut and intestine of fishes from both ocean and fresh water.

WWTPs act as one of the sinks for the ARG as well as the MPs (Cheng et al. 2021; Zhao et al. 2024). Prolonged interactions between the bacteria and antibiotics cause the generation and dissemination of ARGs. MPs can transport and exchange contaminants from various ecological compartments, even across the extended distances (Koelmans et al. 2016).

MPs have significant role in enabling the spread of antimicrobial resistance genes (ARGs) in various environments. Studies have shown that the removal of ARGs is hindered by the presence of MPs (MPs) in aerobic digestion processes. MPs not only increase the total abundance of ARGs but also alter the microbial community in sludge, potentially enhancing gene exchange between bacteria. Research on WWTP identified that, while anaerobic digestion MPs have been found to enhance the enrichment and spread of ARGs with certain types of MPs leading to an increase in ARG richness, particularly genes associated with sulfonamides, lactams, and tetracyclines (Cheng et al. 2021).

The horizontal transfer of ARGs between microorganisms is influenced by factors such as reactive oxygen species (ROS) production, permeability of cell membrane, and secretion of extracellular polymeric substances (EPS). MPs promote gene exchange and potentially increasing the frequency of antimicrobial resistance, thus exerting a selective pressure on microorganisms (Zhao et al. 2024). MPs caused the enrichment of ARGs and potential pathogenic bacteria in different sewage settings. The study found positive correlations between *tetE* and 12 bacterial genera, with *Leucobacter* carrying the targeted tet genes. The tet genes like tetW-01, tetM-01, and tetA-02 were also related to potential pathogenic bacteria. Several genera were found to be more in MP samples, potentially leading to the enrichment of antibiotic-resistant bacteria on the plastisphere. *Arcobacter*, *Hyphomicrobium*, and *Microbacterium* were identified as potential host bacteria for resistance genes. A positive network was found between MGEs, ARGs, and bacteria at the genus level. Some ARGs (*mexF*, *mexB*, *blaTEM*, and *tetW-01*) and MGEs (*intI1* and *tnpA-04*) exhibited great mobility as a result of HGT (Wang et al. 2021b).

According to Ding et al. (2020), RP4 is a conjugative antibiotic-resistant plasmid that is identified in the WWTPs and promotes the colonization of bacteria. It was observed that colonization of bacteria on the MPs is insignificant when the bacteria carrying RP4 exists alone because of the inhibition of the synthesis of conjugative pili that helps in the adhesion and colonization of the bacteria. When it co-exists with the recipients, it increases the intracellular ATP levels, thus enhancing the expression of conjugative pili synthesis and promoting the colonization of PS microplastics (Zhang et al. 2022a).

8.3 Atmospheric

MPs are occurring in remote places because of atmospheric transport. There are extensive discussions on the occurrence, effect, and transport of the MPs that have failed to observe (Evangeliou et al. 2020). β-Lactam resistance gene *bla*TEM is the abundant ARGs in the air from a global survey followed by quinolone resistance gene qepA (Li et al. 2018). According to Vianello et al. (2019), indoor atmosphere contains microplastics at concentrations ranging from 1.7 to 16.2 particles per cubic meter. Furthermore, synthetic fibres have been identified in human lung biopsies (Prata 2018). There has been a research gap on the impact of ARGs on MPs in both indoor and outdoor atmosphere (Liu et al. 2021).

9 Conclusion

This study examines the types, sources, and impact of MPs associated with ARGs on human health. MPs absorb antibiotics by physicochemical action, produce biofilm-enriched bacteria on their surfaces, enhance the abundance of ARGs, and promote their spread. MPs play a specific role in enriching ARGs and ARBs, primarily linked with antibiotics like aminoglycoside, chloramphenicol, quinolone, tetracycline, sulfonamide, macrolide, and beta-lactam. They enrich antimicrobial resistance through processes like HGT, conjugation, gene transduction, and transformation, affecting plants, fish, and humans (Junaid et al. 2022). Laboratory investigations of antibiotic adsorption and desorption in MP biofilms were only studied. So, field studies are required to identify the antibiotic adsorption and desorption in actual environments and unknown factors influencing it. The occurrence of MPs decreases the clearance of ARGs and ARBs. MPs have the capability to selectively enhance the abundance of ARGs and MGEs in the environment, as some ARGs are solely found in MP biofilms. The reasons, methods, and dynamics driving this process remain undistinguishable, but they are crucial for reducing the quantity and spread of ARGs.

Acknowledgement All the figures were created with https://biorender.com/

References

Alkadi H (2020) A review on free radicals and antioxidants. Infect Disord Drug Targets 20(1):16–26. https://doi.org/10.2174/1871526518666180628124323

Amato-Lourenço LF, Carvalho-Oliveira R, Júnior GR, dos Santos GL, Ando RA, Mauad T (2021) Presence of airborne microplastics in human lung tissue. J Hazard Mater 416:126124. https://doi.org/10.1016/j.jhazmat.2021.126124

Ameen A, Stevenson M, Blaschke AP (2020) Microplastics: are they really a threat to groundwater systems? https://doi.org/10.5194/egusphere-egu2020-9069

Amobonye A, Bhagwat P, Singh S, Pillai S (2021) Plastic biodegradation: frontline microbes and their enzymes. Sci Total Environ 759:143536. https://doi.org/10.1016/j.scitotenv.2020.143536

Arias-Andres M, Klümper U, Rojas-Jimenez K, Grossart HP (2018) Microplastic pollution increases gene exchange in aquatic ecosystems. Environ Pollut 237:253–261

Boucher J, Friot D (2017) Primary microplastics in the oceans: a global evaluation of sources. IUCN International Union for Conservation of Nature

Chen M, Yue Y, Bao X, Yu H, Tan Y, Tong B, Kumkhong S, Yu Y (2022) Microplastics as contaminants in water bodies and their threat to the aquatic animals: a mini-review. Animals 12(20):2864. https://doi.org/10.3390/ani12202864

Cheng YL, Kim J-G, Kim H-B, Choi JH, Fai Tsang Y, Baek K (2021) Occurrence and removal of microplastics in wastewater treatment plants and drinking water purification facilities: a review. Chem Eng J 410:128381. https://doi.org/10.1016/j.cej.2020.128381

Chia RW, Lee J-Y, Kim H, Jang J (2021) Microplastic pollution in soil and groundwater: a review. Environ Chem Lett 19(6):4211–4224. https://doi.org/10.1007/s10311-021-01297-6

Cole M, Lindeque P, Halsband C, Galloway TS (2011) Microplastics as contaminants in the marine environment: A review. Mar Pollut Bull 62(12):2588–2597. https://doi.org/10.1016/j.marpolbul.2011.09.025

Debroy A, George N, Mukherjee G (2021) Role of biofilms in the degradation of microplastics in aquatic environments. J Chem Technol Biotechnol 97(12):3271–3282. https://doi.org/10.1002/jctb.6978

Ding L, Zhang S, Wang X, Yang X, Zhang C, Qi Y, Guo X (2020) The occurrence and distribution characteristics of microplastics in the agricultural soils of Shaanxi Province, in north-western China. Sci Total Environ 720:137525. https://doi.org/10.1016/j.scitotenv.2020.137525

Evangeliou N, Grythe H, Klimont Z, Heyes C, Eckhardt S, Lopez-Aparicio S, Stohl A (2020) Atmospheric transport is a major pathway of microplastics to remote regions. Nat Commun 11(1):3381. https://doi.org/10.1038/s41467-020-17201-9

Gall SC, Thompson RC (2015) The impact of debris on marine life. Mar Pollut Bull 92(1–2):170–179. https://doi.org/10.1016/j.marpolbul.2014.12.041

Hartmann NB, Hüffer T, Thompson RC, Hassellöv M, Verschoor A, Daugaard AE, Rist S, Karlsson T, Brennholt N, Cole M, Herrling MP, Hess MC, Ivleva NP, Lusher AL, Wagner M (2019) Are we speaking the same language? Recommendations for a definition and categorization framework for plastic debris. Environ Sci Technol 53(3):1039–1047. https://doi.org/10.1021/acs.est.8b05297

Hernandez LM, Yousefi N, Tufenkji N (2017) Are There Nanoplastics in Your Personal Care Products? Environ Sci Technol Lett 4(7):280–285. https://doi.org/10.1021/acs.estlett.7b00187

Hu X, Waigi MG, Yang B, Gao Y (2022) Impact of plastic particles on the horizontal transfer of antibiotic resistance genes to bacterium: dependent on particle sizes and antibiotic resistance gene vector replication capacities. Environ Sci Technol 56(21):14948–14959. https://doi.org/10.1021/acs.est.2c00745

Huang Y, Liu Q, Jia W, Yan C, Wang J (2020) Agricultural plastic mulching as a source of microplastics in the terrestrial environment. Environ Pollut 260:114096. https://doi.org/10.1016/j.envpol.2020.114096

Huerta Lwanga E, Gertsen H, Gooren H, Peters P, Salánki T, van der Ploeg M, Besseling E, Koelmans AA, Geissen V (2016) Microplastics in the terrestrial ecosystem: implications for Lumbricusterrestris(Oligochaeta, Lumbricidae). Environ Sci Technol 50(5):2685–2691. https://doi.org/10.1021/acs.est.5b05478

Jabeen K, Su L, Li J, Yang D, Tong C, Mu J, Shi H (2017) Microplastics and mesoplastics in fish from coastal and fresh waters of China. Environ Pollut 221:141–149. https://doi.org/10.1016/j.envpol.2016.11.055

Jia J, Liu Q, Zhao E, Li X, Xiong X, Wu C (2024) Biofilm formation on microplastics and interactions with antibiotics, antibiotic resistance genes and pathogens in aquatic environment. Eco-Environ Health 3(4):516–528. https://doi.org/10.1016/j.eehl.2024.05.003

Junaid M, Liu X, Wu Y, Wang J (2022) Selective enrichment of antibiotic resistome and bacterial pathogens by aquatic microplastics. J Hazard Mater Adv 7:100106. https://doi.org/10.1016/j.hazadv.2022.100106

Kalogerakis N, Karkanorachaki K, Kalogerakis GC, Triantafyllidi EI, Gotsis AD, Partsinevelos P, Fava F (2017) Microplastics Generation: Onset of Fragmentation of Polyethylene Films in Marine Environment Mesocosms. Front Mar Sci 4. https://doi.org/10.3389/fmars.2017.00084

Keswani A, Oliver DM, Gutierrez T, Quilliam RS (2016) Microbial hitchhikers on marine plastic debris: Human exposure risks at bathing waters and beach environments. Mar Environ Res 118:10–19. https://doi.org/10.1016/j.marenvres.2016.04.006

Kim H, Zaheer J, Choi E-J, Kim JS (2022) Enhanced ASGR2 by microplastic exposure leads to resistance to therapy in gastric cancer. Theranostics 12(7):3217–3236. https://doi.org/10.7150/thno.73226

Koelmans AA, Bakir A, Burton GA, Janssen CR (2016) Microplastic as a vector for chemicals in the aquatic environment: critical review and model-supported reinterpretation of empirical studies. Environ Sci Technol 50(7):3315–3326. https://doi.org/10.1021/acs.est.5b06069

Lehner R, Weder C, Petri-Fink A, Rothen-Rutishauser B (2019) Emergence of nanoplastic in the environment and possible impact on human health. Environ Sci Technol 53(4):1748–1765. https://doi.org/10.1021/acs.est.8b05512

Li J, Cao J, Zhu Y, Chen Q, Shen F, Wu Y, Xu S, Fan H, Da G, Huang R, Wang J, de Jesus AL, Morawska L, Chan CK, Peccia J, Yao M (2018) Global survey of antibiotic resistance genes in air. Environ Sci Technol 52(19):10975–10984. https://doi.org/10.1021/acs.est.8b02204

Li L, Luo Y, Li R, Zhou Q, Peijnenburg WJGM, Yin N, Yang J, Tu C, Zhang Y (2020) Effective uptake of submicrometre plastics by crop plants via a crack-entry mode. Nat Sustain 3(11):929–937. https://doi.org/10.1038/s41893-020-0567-9

Li J, Yu S, Yu Y, Xu M (2022) Effects of microplastics on higher plants: a review. Bull Environ Contam Toxicol 109(2):241–265. https://doi.org/10.1007/s00128-022-03566-8

Liu Y, Liu W, Yang X, Wang J, Lin H, Yang Y (2021) Microplastics are a hotspot for antibiotic resistance genes: progress and perspective. Sci Total Environ 773:145643. https://doi.org/10.1016/j.scitotenv.2021.145643

Lu J, Zhang Y, Wu J, Luo Y (2019) Effects of microplastics on distribution of antibiotic resistance genes in recirculating aquaculture system. Ecotoxicol Environ Saf 184:109631. https://doi.org/10.1016/j.ecoenv.2019.109631

Lu X-M, Lu P-Z, Liu X-P (2020) Fate and abundance of antibiotic resistance genes on microplastics in facility vegetable soil. Sci Total Environ 709:136276. https://doi.org/10.1016/j.scitotenv.2019.136276

Malpass DB (2010) Introduction to industrial polyethylene. Wiley

Metcalf R, Messer LF, White HL, Ormsby MJ, Matallana-Surget S, Quilliam RS (2024) Evidence of interspecific plasmid uptake by pathogenic strains of Klebsiella isolated from microplastic pollution on public beaches. J Hazard Mater 461:132567. https://doi.org/10.1016/j.jhazmat.2023.132567

Naik RA, Rowles LS, Hossain AI, Yen M, Aldossary RM, Apul OG, Conkle J, Saleh NB (2020) Microplastic particle versus fiber generation during photo-transformation in simulated seawater. Sci Total Environ 736:139690. https://doi.org/10.1016/j.scitotenv.2020.139690

Naqash N, Prakash S, Kapoor D, Singh R (2020) Interaction of freshwater microplastics with biota and heavy metals: a review. Environ Chem Lett 18(6):1813–1824. https://doi.org/10.1007/s10311-020-01044-3

Ni B-J, Zhu Z-R, Li W-H, Yan X, Wei W, Xu Q, Xia Z, Dai X, Sun J (2020) Microplastics mitigation in sewage sludge through pyrolysis: the role of pyrolysis temperature. Environ Sci Technol Lett 7(12):961–967. https://doi.org/10.1021/acs.estlett.0c00740

Oberbeckmann S, Kreikemeyer B, Labrenz M (2018) Environmental factors support the formation of specific bacterial assemblages on microplastics. Front Microbiol 8:2709. https://doi.org/10.3389/fmicb.2017.02709

Picó Y, Barceló D (2019) Analysis and prevention of microplastics pollution in water: current perspectives and future directions. ACS Omega 4(4):6709–6719. https://doi.org/10.1021/acsomega.9b00222

Plastics—the fast Facts (2024). https://plasticseurope.org/knowledge-hub/plastics-the-fast-facts-2024

Prata JC (2018) Airborne microplastics: consequences to human health? Environ Pollut 234:115–126. https://doi.org/10.1016/j.envpol.2017.11.043

Preiser J (2012) Oxidative Stress. J Parenter Enter Nutr 36(2):147–154. https://doi.org/10.1177/0148607111434963

Qi R, Jones DL, Li Z, Liu Q, Yan C (2020) Behavior of microplastics and plastic film residues in the soil environment: a critical review. Sci Total Environ 703:134722. https://doi.org/10.1016/j.scitotenv.2019.134722

Qiu X, Qi Z, Ouyang Z, Liu P, Guo X (2022) Interactions between microplastics and microorganisms in the environment: modes of action and influencing factors. Gondwana Res 108:102–119. https://doi.org/10.1016/j.gr.2021.07.029

Qiu C, Zhou Y, Wang H, Chu Y, Zheng L, Chen Y, Song Y, Fang C (2023) Microplastics enrichment characteristics of antibiotic resistance genes and pathogens in landfill leachate. Chemosphere 341:140100. https://doi.org/10.1016/j.chemosphere.2023.140100

Rafa N, Ahmed B, Zohora F, Bakya J, Ahmed S, Ahmed SF, Mofijur M, Chowdhury AA, Almomani F (2024) Microplastics as carriers of toxic pollutants: source, transport, and toxicological effects. Environ Pollut 343:123190. https://doi.org/10.1016/j.envpol.2023.123190

Ragusa A, Svelato A, Santacroce C, Catalano P, Notarstefano V, Carnevali O, Papa F, Rongioletti MCA, Baiocco F, Draghi S, D'Amore E, Rinaldo D, Matta M, Giorgini E (2021) Plasticenta: first evidence of microplastics in human placenta. Environ Int 146:106274. https://doi.org/10.1016/j.envint.2020.106274

Rather MA, Gupta K, Mandal M (2021) Microbial biofilm: formation, architecture, antibiotic resistance, and control strategies. Braz J Microbiol 52(4):1701–1718. https://doi.org/10.1007/s42770-021-00624-x

Sarkar S, Diab H, Thompson J (2023) Microplastic pollution: chemical characterization and impact on wildlife. Int J Environ Res Public Health 20(3):1745. https://doi.org/10.3390/ijerph20031745

Schwabl P, Köppel S, Königshofer P, Bucsics T, Trauner M, Reiberger T, Liebmann B (2019) Detection of various microplastics in human stool. Ann Intern Med 171(7):453–457. https://doi.org/10.7326/m19-0618

Serranti S, Gargiulo A, Bonifazi G (2011) Characterization of post-consumer polyolefin wastes by hyperspectral imaging for quality control in recycling processes. Waste Manag 31(11):2217–2227. https://doi.org/10.1016/j.wasman.2011.06.007

Shahnawaz M, Sangale MK, Ade AB (2019) Bioremediation technology for plastic waste. Springer, Singapore

Silva I, Rodrigues ET, Tacão M, Henriques I (2023) Microplastics accumulate priority antibiotic-resistant pathogens: evidence from the riverine plastisphere. Environ Pollut 332:121995. https://doi.org/10.1016/j.envpol.2023.121995

Sooriyakumar P, Bolan N, Kumar M, Singh L, Yu Y, Li Y, Weralupitiya C, Vithanage M, Ramanayaka S, Sarkar B, Wang F, Gleeson DB, Zhang D, Kirkham MB, Rinklebe J, M Siddique KH (2022) Biofilm formation and its implications on the properties and fate of microplastics in aquatic environments: a review. J Hazard Mater Adv 6:100077. doi:https://doi.org/10.1016/j.hazadv.2022.100077

Stapleton PA (2019) Toxicological considerations of nano-sized plastics. AIMS Environ Sci 6(5):367–378. https://doi.org/10.3934/environsci.2019.5.367

Stapleton MJ, Ansari AJ, Hai FI (2023) Antibiotic sorption onto microplastics in water: a critical review of the factors, mechanisms and implications. Water Res 233:119790. https://doi.org/10.1016/j.watres.2023.119790

Sun X-D, Yuan X-Z, Jia Y, Feng L-J, Zhu F-P, Dong S-S, Liu J, Kong X, Tian H, Duan J-L, Ding Z, Wang S-G, Xing B (2020) Differentially charged nanoplastics demonstrate distinct accumulation in Arabidopsis thaliana. Nat Nanotechnol 15(9):755–760. https://doi.org/10.1038/s41565-020-0707-4

Syranidou E, Kalogerakis N (2022) Interactions of microplastics, antibiotics and antibiotic resistant genes within WWTPs. Sci Total Environ 804:150141. https://doi.org/10.1016/j.scitotenv.2021.150141

Tan H, Yue T, Xu Y, Zhao J, Xing B (2020) Microplastics reduce lipid digestion in simulated human gastrointestinal system. Environ Sci Technol 54(19):12285–12294. https://doi.org/10.1021/acs.est.0c02608

Tu C, Chen T, Zhou Q, Liu Y, Wei J, Waniek JJ, Luo Y (2020) Biofilm formation and its influences on the properties of microplastics as affected by exposure time and depth in the seawater. Sci Total Environ 734:139237. https://doi.org/10.1016/j.scitotenv.2020.139237

Vianello A, Jensen RL, Liu L, Vollertsen J (2019) Simulating human exposure to indoor airborne microplastics using a breathing thermal Manikin. Sci Rep 9(1):8670. https://doi.org/10.1038/s41598-019-45054-w

Wang J, Liu X, Dai Y, Ren J, Li Y, Wang X, Zhang P, Peng C (2020a) Effects of co-loading of polyethylene microplastics and ciprofloxacin on the antibiotic degradation efficiency and microbial community structure in soil. Sci Total Environ 741:140463. https://doi.org/10.1016/j.scitotenv.2020.140463

Wang W, Ge J, Yu X, Li H (2020b) Environmental fate and impacts of microplastics in soil ecosystems: progress and perspective. Sci Total Environ 708:134841. https://doi.org/10.1016/j.scitotenv.2019.134841

Wang J, Peng C, Li H, Zhang P, Liu X (2021a) The impact of microplastic-microbe interactions on animal health and biogeochemical cycles: a mini-review. Sci Total Environ 773:145697. https://doi.org/10.1016/j.scitotenv.2021.145697

Wang Z, Gao J, Zhao Y, Dai H, Jia J, Zhang D (2021b) Plastisphere enrich antibiotic resistance genes and potential pathogenic bacteria in sewage with pharmaceuticals. Sci Total Environ 768:144663. https://doi.org/10.1016/j.scitotenv.2020.144663

Wang H, Xu K, Wang J, Feng C, Chen Y, Shi J, Ding Y, Deng C, Liu X (2023) Microplastic biofilm: An important microniche that may accelerate the spread of antibiotic resistance genes via natural transformation. J Hazard Mater 459:132085. https://doi.org/10.1016/j.jhazmat.2023.132085
Wu X, Pan J, Li M, Li Y, Bartlam M, Wang Y (2019) Selective enrichment of bacterial pathogens by microplastic biofilm. Water Res 165:114979. https://doi.org/10.1016/j.watres.2019.114979
Xu C, Lu J, Shen C, Wang J, Li F (2022) Deciphering the mechanisms shaping the plastisphere antibiotic resistome on riverine microplastics. Water Res 225:119192. https://doi.org/10.1016/j.watres.2022.119192
Yan X, Yang X, Tang Z, Fu J, Chen F, Zhao Y, Ruan L, Yang Y (2020) Downward transport of naturally-aged light microplastics in natural loamy sand and the implication to the dissemination of antibiotic resistance genes. Environ Pollut 262:114270. https://doi.org/10.1016/j.envpol.2020.114270
Yang Y, Li Z, Song W, Du L, Ye C, Zhao B, Liu W, Deng D, Pan Y, Lin H, Cao X (2019) Metagenomic insights into the abundance and composition of resistance genes in aquatic environments: influence of stratification and geography. Environ Int 127:371–380. https://doi.org/10.1016/j.envint.2019.03.062
Yang L, Zhang Y, Kang S, Wang Z, Wu C (2021) Microplastics in soil: a review on methods, occurrence, sources, and potential risk. Sci Total Environ 780:146546
Yang Y, Li T, Liu P, Li H, Hu F (2022) The formation of specific bacterial communities contributes to the enrichment of antibiotic resistance genes in the soil plastisphere. J Hazard Mater 436:129247. https://doi.org/10.1016/j.jhazmat.2022.129247
Yee MS-L, Hii L-W, Looi CK, Lim W-M, Wong S-F, Kok Y-Y, Tan B-K, Wong C-Y, Leong C-O (2021) Impact of microplastics and nanoplastics on human health. Nano 11(2):496. https://doi.org/10.3390/nano11020496
Yu F, Yang C, Zhu Z, Bai X, Ma J (2019) Adsorption behavior of organic pollutants and metals on micro/nanoplastics in the aquatic environment. Sci Total Environ 694:133643. https://doi.org/10.1016/j.scitotenv.2019.133643
Yu H, Zhang X, Hu J, Peng J, Qu J (2020) Ecotoxicity of polystyrene microplastics to submerged carnivorous Utricularia vulgaris plants in freshwater ecosystems. Environ Pollut 265:114830. https://doi.org/10.1016/j.envpol.2020.114830
Yurtsever M (2019) Glitters as a source of primary microplastics: an approach to environmental responsibility and ethics. J Agric Environ Ethics 32(3):459–478. https://doi.org/10.1007/s10806-019-09785-0
Zang H, Zhou J, Marshall MR, Chadwick DR, Wen Y, Jones DL (2020) Microplastics in the agroecosystem: are they an emerging threat to the plant-soil system? Soil Biol Biochem 148:107926. https://doi.org/10.1016/j.soilbio.2020.107926
Zhang G, Chen J, Li W (2022a) Conjugative antibiotic-resistant plasmids promote bacterial colonization of microplastics in water environments. J Hazard Mater 430:128443. https://doi.org/10.1016/j.jhazmat.2022.128443
Zhang Z, Peng W, Duan C, Zhu X, Wu H, Zhang X, Fang L (2022b) Microplastics pollution from different plastic mulching years accentuate soil microbial nutrient limitations. Gondwana Res 108:91–101. https://doi.org/10.1016/j.gr.2021.07.028
Zhao K, Li C, Li F (2024) Research progress on the origin, fate, impacts and harm of microplastics and antibiotic resistance genes in wastewater treatment plants. Sci Rep 14(1):9719. https://doi.org/10.1038/s41598-024-60458-z

Chapter 11
Role of Microorganisms in Biodegradation of Microplastics

Madhu Chandel, Sonia Sharma, Harpreet Kaur, and Madhu Raina

Abstract The usage of plastics is increasing as a result of its uses in the economy, human health, and aesthetics. Polyethylene, polypropylene, polyvinylchloride, and polyethylene terephthalate make up the majority of microplastics (MPs). Microplastics are pieces of any kind of plastic that are fewer than 5 mm in length. They infiltrate natural environments through a number of sources, such as cosmetics, clothing, food packaging, and industrial processes. Microplastics have a high chance of being ingested, incorporated into, and accumulating in the bodies and tissues of numerous creatures due to the slow rate of plastics' degradation. It is known that primary and secondary microplastics remain in the environment at high concentrations, especially in aquatic and marine habitats, where they contribute to water pollution. Clothing and textiles account for 35% of all ocean microplastics, mostly as a result of the deterioration of polyester-, acrylic-, or nylon-based clothing, which frequently occurs after washing. Microplastics, however, also build up in the atmosphere and terrestrial ecosystems. Microplastics have been shown to lower the weight of earthworms and the viability of soil ecosystems in terrestrial ecosystems. MPs are thought to be hazardous to both humans and wildlife because of their bioaccumulation and persistence characteristics. The current chapter highlights the potential of several microorganisms in the biodegradation of microplastics. The two possible approaches to reduce the "plastic waste" are (1) to develop biodegradable commodity plastics from fossil fuel and/or renewable resources building blocks (hydrobiodegradable) or reengineering of full carbon backbone commodity polymers (oxo-biodegradable) and (2) to identify potential micro-algae and its toxins to develop protocol to effectively biodegrade the polymeric materials.

Keywords Microplastic · Biodegradation · Microorganisms

M. Chandel (✉) · S. Sharma · H. Kaur
P.G. Department of Botany, Khalsa College Amritsar, Amritsar, Punjab, India

M. Raina
Department of Botany, Sri Venkateswara College, University of Delhi,
New Delhi, Delhi, India

K. Khanna et al. (eds.), *Microplastics and Soil Microbiome*,
https://doi.org/10.1007/978-981-96-4978-5_11

1 Introduction

Microplastics can come in many different forms, such as bigger plastic pieces that have broken up, resin pellets used in the production of plastic, or microbeads, which are tiny, produced plastic beads used in cosmetic and health products. Microplastics (MPs) are possible contaminants of growing concern due to their widespread occurrence in all environmental compartments, including the atmospheric, terrestrial, and aquatic ones (Klein et al. 2018; Blair et al. 2019). They have been described as plastic particles that come in a variety of primary and secondary types, sizes, shapes, and polymer compositions, and range in size from 1 to 5 mm. Due to their bioaccumulation and persistence traits, MPs are believed to be dangerous to both humans and wildlife. MPs include extra dangerous components that might be added during production (such as colours, plasticizers, and flame retardants). Moreover, once being released into the environment, MPs may take longer to degrade (Wagner and Lambert 2018). Microplastics can affect an organism's stomach and digestive system over time and slow down its pace of feeding (Gregory 2009). Also, the prolonged presence of microplastics in the environment stunts the growth of some plants, changes the function of some microbial enzymes, and raises biological mortality rates. (Shen et al. 2021). It is impossible to disregard the issues caused by their extensive use notwithstanding their benefits. In order to biodegrade microplastics, microorganisms including bacteria, fungi, and algae come into play. It is possible to carry out this type of biodegradation both aerobically and anaerobically without the use of heat energy.

Microorganisms may degrade a variety of chemicals, including MPs, and they have an inherent aptitude to adapt in any environment (Brooks et al. 2023). For bioremediation, they use contaminants and dangerous materials as food sources. Bioremediation has many advantages over traditional cleanup methods. Because bioremediation only relies on the typical metabolic processes of living systems, the environmental damage is kept to a minimum. Also, because living things are like little factories with all the components needed to degrade harmful compounds, bioremediation does not require massive equipment that uses fossil fuels and produces pollutants. Hence, bioremediation is more efficient and cost-effective. According to Kumar Sen and Raut and others (Kumar et al. 2018; Qi et al. 2020; Restrepo-Florez et al. 2014) microbes to degrade MPs will accelerate biodegradation without endangering the environment, making it a promising and risk-free method to promote natural bioremediation and influence the cleaning of natural ecosystems (Shah et al. 2024). Only 21% of plastic has been recycled or burned, while the remaining 79% has weathered into "microplastics," (Crew et al. 2020). Microplastics (MPs), which make up 92.4% of plastic waste, are mostly made of polystyrene (PS), polyethylene (PE), and polypropylene (PP), according to statistical study (Carr et al. 2016; Santana et al. 2016). Now, MPs are distributed around the globe and have contaminated lakes, rivers, and other farmed places (Duis and Coors 2016; Eerkes-Medrano et al. 2015). MP pollution has effectively become a global issue due to the prevalence of MP pollution detections, and MPs' ultimate fate is a source of significant concern (Bakir et al. 2014).

2 Microorganisms in Biodegradation of Microplastics

The availability of the substrate, the surface properties, the shape, and the molecular weight of the polymers are all important elements that affect the highly complex biodegradation process of polymers. Despite these variations, the overall method for plastic biodegradation starts with the production of particular plastic-degrading enzymes that differ depending on the microbe and the polymer material being degraded. The hydroperoxidation/hydrolysis of the bonds is then carried out by these enzymes after their participation in adsorption on the plastic surface. Short-lived degradation intermediates are then released as a result, and eventually, the tricyclic acid (TCA) cycle produces the final byproducts of the biodegradation.

3 Microplastic Degradation with Bacterial Species

Cacciari and group (1993) carried out the initial study on microplastics biodegrading bacteria by combining *Pseudomonas chlororaphis*, *Pseudomonas stutzeri*, and *Vibrio* sp. to digest polypropylene. The inclusion of starch was also said to improve biodegradation ability in the same study. Strains *Rhodococcus rhodochrous* ATCC 29672 and *Nocardia asteroides* GK were observed for their potential to degrade polyethylene. The degradation of oxidation products in the polymer film's surface was measured by FTIR measurements and was found to be associated with the formation of protein and polysaccharides, which can be attributed to the growth of the microorganisms (Bonhomme 2003). Sivan and group (2011) isolated a biofilm-producing strain (C208) of *Rhodococcus ruber* that degraded polyethylene at a rate of 0.86% per week. According to Arkatkar et al. and Fontanella et al., a combination of *Bacillus subtilis*, *B. flexus*, *Pseudomonas stutzeri*, and *Rhodococcus rhodochrous* was used to biodegrade polypropylene (Arkatkar et al. 2009; Foekema et al. 2013). Low-density polyethylene was effectively biodegraded over the course of 2 months by strains such *Microbacterium paraoxydans* and *Pseudomonas aeruginosa*, which demonstrated nearly 61.0% and 50.5% breakdown, respectively (Rajandas et al. 2012). Jeon and Kim reported microplastic degradation by *Chelatococcus sp* E1 isolated from compost with 44.5% efficiency after 80 days of incubation (Jeon et al. 2021). There have also been reports of microorganisms with microplastics biodegradation properties being present in the gut microflora of numerous arthropods. Yang et al. (2014) studied the microplastic degradation by two bacterial species viz. *Enterobacter asburiae* YT1 and *Bacillus* sp. YP1 isolated from gut of plastic eating waxworms which showed polypropylene degradation efficiency of 10.7 and 6.1% after 28 days of incubation, respectively (Yang et al. 2014). In an another study isolated a bacterial strain *Exiguobacterium* sp. from the guts of mealworms which was able to degrade polystyrene after 60 days of incubation (Yang et al. 2018). Jeon and Kim studied the efficiency *of Stenotrophomonas panacihumi* PA3-2 isolated from municipal solid waste to degrade polypropylene at the rate of 20.3% after incubation of 90 days (Jeon and Kim 2013). Aravinthan et al. (2016) studied the ability of *Bacillus flexus* and *pseudomonas azotoformans* to degrade UV treated polymers with the efficiency of 22.7% after 12 months. Muenmee et al. (Muenmee et al. 2016) isolated *Nitrosomonas* sp., *nitrobacter* sp., *Burkholderia* sp. and *Pseudomonas* sp. from municipal solid waste and after incubation of 90 days observed the potential of these microorganisms to degrade high density polyethylene (15–20%) and polypropylene (5%–9%). Kowalczyk isolated *Achromobacter xylosoxidans*, which was able to effect on the structure of high-density polyethylene (Kowalczyk et al. 2017). Auta et al. (2017) reported microplastic degradation ability of *Bacillus cereus* and *Sporosarcina globispora* isolated from mangrove sediments in Peninsular Malaysia with 12% and 11% degradation percentage respectively. According to Auta et al., *B. gottheilii* caused microplastics to lose weight to varying degrees in polyethylene, polyethylene terephthalate, polypropylene, and polystyrene, with percentages of 6.2%, 3.0%, 3.6%, and 5.8%, respectively (Auta et al. 2017). In a study by Auta *Bacillus* and *Rhodococcus* bacterial strains isolated from mangrove sediments had polypropylene breakdown efficiencies of 4.0 and 6.4%, respectively, after 40 days of incubation (Auta 2018). *Bacillus thuringiensis* and *Bacillus licheniformis* isolated from compost were found associated with polypropylene and poly-L-lactide degradation with percentages of 12% and 10% respectively after 15 days of incubation (Jain et al. 2023). Huerta Lwanga et al. also looked into the low-density polyethylene degradation caused by the earthworm (*Lumbricus terrestris*) (Lwanga et al. 2018). Actinobacteria and Firmicutes were among the genera of gut isolates that were investigated independently and found to be capable of breaking down low-density polyethylene microplastics and releasing volatile substances like eicosane, docosane, and tricosane. *Bacillus* sp. and *Paenibacillus* sp. along with other bacteria were able to cut the dry weight of microplastics by 14.7% in just 60 days (Park and Kim 2019). Alkyl cycloalkanes, isoprenoid hydrocarbons, alkanes, and branched aliphatic compounds were all capable of being degraded by a number of marine hydrocarbonoclastic bacteria, including *Alcanivorax borkumensis* (Davoudi 2023). Within 120 days, a combination of *Enterobacter* and *Pseudomonas* from cow dung increased weight reduction by up to 15% (Skariyachan et al. 2021). According to Matjai et al., 21% of the various bacterial genera connected to microplastic breakdown belonged to Pseudomonas, 15% to Bacillus, and 17% were a combination of these two genera (Mataji et al. 2020). The biodegradation of LDPE (low-density polyethylene) microplastics was studied using *B. subtilis* V8, *P. aminophilus* B1 4-, *P. putida* C 25, *P. aeruginosa* V1, and *A. calcoaceticus* V4. With increased electrical conductivity, *P. aeruginosa* V1 demonstrated the most significant degradation when compared to the other tested cultures (CO_2 evolution of 8.86 g.l-l and percentage weight loss of 18.21%). *B. subtilis* V8, *A. calcoaceticus* V4, and *P. putida* C 2–5 and *P. aminophilus* B1 also demonstrated significant degradation (CO_2 evolution and percentage weight loss of 8.10 g.l^{-1}, 16.12%; 7.21 g. l^{-1},15.44%; 5.76 g.l^{-}, 13.30%; 5.62 g.l^{-1}, 11.72%) (Pathak 2023). Both pure cultures and microbial consortiums have been used to evaluate bacteria for their ability to break down microplastics. In particular, bacterial consortium exhibit higher effectiveness and community stability. The consortium's bacterial isolates work together in a symbiotic, synergistic, and mutualistic way that increases their resistance to the treatment procedure and their activity (Zhu et al. 2022; Salinas et al. 2024; Torena et al. 2021; Singh 2024).

4 Microplastic Degradation with Fungal Species

Fungus is mostly saprotrophs, opportunistic parasites, or obligate parasites. They are extremely adaptable and may flourish in a variety of aquatic and terrestrial ecosystems and in a wide range of climatic conditions. They produce a wide variety of extracellular enzymes and natural biosurfactants like hydrophobins that can break down complex polymers into simple monomers, making them a source of electrons and carbons for microorganisms, facilitating the degradation and mineralization of complex pollutants (Ortúzar et al. 2022). The various genera of fungus are reported to be associated with the degradation of different types of polymers such as polyethylene, polypropylene, polyethylene terephthalate and polyester polyurethane which use microplastics as sole carbon source after degradation by extracellular enzymes (Ibrahim et al. 2021; Novak et al. 2024; Devi and Kumar 2024; Sheikh et al. 2025; Paço et al. 2017; Oliveira et al. 2020). They promote the formation of different types of chemical bonds (having carbonyl, carboxyl, and ester functional groups) and decrease their hydrophobicity. Chandra and Rustgi (1997) reported fungal mediated degradation of low density polyethylene after 28 days of incubation period (Chandra and Rustgi 1997). Volke-Sepulveda et al. studied *Penicillium pinophilum* ATCC 11, 797 strain for its ability to degrade low density polyethylene powder after 31 months of incubation of the fungus (Volke-Sepúlveda et al. 2002). Ibrahim et al. reported 100% degradation of polyester polyurethane by fungal species viz. *Fusarium solani, Spicaria spp, Alternaria solani* and *Aspergillus flavus* isolated from soil and wall paint coated with polyurethane (Ibrahim et al. 2021). Nowak et al. reported degradation of low density polyethylene by different fungal species viz. *Aspergillus niger, Aspergillus terreus, Aureobasidium pullulans, Paecilomyces varioti, Penicillium funiculosum, Penicillium ochrochloron, Scopulariopsis brevicaulis, Trichoderma viride* isolated from waste dump (Novak et al. 2024). Devi et al. studied the potential of *Aspergillus flavus* VRKPT2 isolated from marine coastal area to degrade high density polyethylene after 30 days of incubation (Devi and Kumar 2024). *Zalerion maritimum* isolated from marine sediments found to be degrading polyethylene pellets when incubated for 28 days. Two potential fungal strains, namely, *Penicillium oxalicum* NS4 (KU559906) and *Penicillium chrysogenum* NS10 (KU559907) had been isolated from soil sample collected from a plastic dumping ground near IISER Kolkata, Mohanpur campus and identified to have plastic degrading abilities. The fungal isolate NS10 degrade HDPE and LDPE sheets and the weight loss was 17.06%, 48.00%, 58.598%, and 19.32%, 33.33%, 34.35% after 30, 60, and 90 days of incubation respectively. After 30, 60, and 90 days of incubation, the fungus isolate NS4 showed the overall percentage of weight loss in HDPE and LDPE sheets which was 24.18, 43.73, and 55.34% respectively. The findings show that both isolates can degrade HDPE rather effectively, which is more difficult to break down than LDPE, proving that both fungal isolates have the capacity to break down plastic (Ojha et al. 2017). The marine fungus *Zalerion maritimum* was also able to reduce the weight of PE microplastic by 43% within 14 days (Paço et al. 2017). Using both morphological criteria and molecular methods, the two *Aspergillus* isolates that degraded polythene the most effectively were identified as *Aspergillus terreus* strain MANGF1/WL and *Aspergillus sydowii* strain PNPF15/TS and among the two strain PNPF15/TS was found most effective with a weight reduction of 94% within 60 days (Sangale et al. 2019). Ekanayaka et al. provide a review regarding various taxa of fungi belonging to different classes of fungi isolated from varying habitats with the potential to degrade plastics (Ekanayaka et al. 2022). There were 109 different fungal isolates found in all from 12 different eco-geographical areas along India's west coast.

5 Microplastic Degradation with Algae

Comparing to the bacterial systems which may be considered as a biological pollutant due to endotoxins and requirement of a rich carbon source for growth, microalgae is a potential candidate since they do not contain endotoxins and organic carbon sources are not required under photoautotrophic conditions (Moog et al. 2019; Chia et al. 2020; Manzi et al. 2022). Moreover *Ideonella sakaiensis* and other microbes used for PETase generation do not adapt well to marine habitats where the accumulation of most plastic waste occurs (Yan et al. 2021). The endocrine disruptor bisphenol A (BPA), which is discharged into the environment, was examined on the green alga *Chlorella fusca*. Under constant lighting of 18 W/m2, *Chlorella fusca* was able to eliminate practically all BPA in the concentration range of 10–80 mM for 168 hours (Hirooka et al. 2005). Algal biofilms were discovered by Suseela and Toppo on the surface of underwater PE plastic waste in the Uttar Pradesh region (Suseela and Toppo 2007). A total of 15 algal biofilm consisting of *Chaetophora*, *Cooleochaete scutata*, *Coleochaete soluta*, *Aphanochaete*, *Gloeotaenium*, *Oedogonium*, *Oocystis*, *Oscillatoria*, *Phormidium*, *Chroococcus*, *Aphanoteche*, *Fragillaria*, *Cocconis*, *Navicula* and *Cymbella* formed on the surface of polythene-type plastic submerged in water damaged the plastic polymer surfaces.

The exopolysaccharides (EPS) that cyanobacteria and green algae produce around their cells or filaments make it easier for them to attach to a substrate, create the perfect environment for microalgae, and promote the breakdown of numerous polymers (Kumar et al. 2021; Gu et al. 2024; Vinagre et al. 2020). Research on BPA was done by Gulnaz and Dincer (2009). The outcomes of the biodegradation experiment demonstrated that *Chlorella vulgaris* was capable of quickly degrading the bisphenol A (BPA) compound. 10 algae species were found by Sharma et al. to possibly decompose polythene polymers (Sharma et al. 2014). They were referred to as *Amphora ovalis*, *Chlorella vulgaris*, *Closterium costatum*, *Microcystis aeruginosa*,

Monaraphidium contortum, *Navicula cuspidata*, *Oscillatoria tenuis*, *Phormidium tenue*, *Scendesmus acuminatus*, and *Selanestrum minutum*. A green microalga (*Scenedesmus dimorphus*), blue green algae (*Anabaena spiroides*), and diatom (*Navicula pupula*) were studied for their ability to deteriorate polyethylene. The highest percentage of degradation was obtained from *Anabaena spiroides* treatment (8.18%) (Kumar et al. 2022). Sarmah and Rout conducted scanning electron microscopy and optical microscopy study on cyanobacterial species viz. *Phormidium lucidum* and *Oscillatoria subbrevis* isolated from submerged polyethylene carry bags in domestic sewage water (Sarmah and Rout 2018). The results showed significant disruption of polyethylene structure. In a study Aldaby and Mawad found potential of *Oscillatoria* sp. and *Chlorella sp* to degrade pyrene in the concentrations of 10 and 30 mg/L (Aldaby et al. 2023). Percent inhibition of 95% and 78.7% was reported for *Oscillatoria* sp. and *Chlorella sp* respectively. Plastic degradating ability of *Spirulina* sp. incubated for 112 days in liquid media containing microsized PP and PET separately, was studied by Khoironi et al. (2019). Carbon Reduction of 48.61% and 36.7% was reported for PET and PP respectively, which was based on the results of Energy Dispersive X-ray spectroscopy (EDX) analysis. Whereas on the basis of tensile strength analysis, PET plastic tensile value decreased about 0.9939 MPa/day and PP tensile strength was decreased about 0.1977 MPa/day. Sanniyasi et al. isolated *Uronema africanum* Borge (microalga), from a waste plastic bag collected from a domestic waste dumping site in a freshwater lake (Sanniyasi et al. 2021). This microalga was further treated with the LDPE sheet in BBM culture medium. The results based on light microscopy, dark field microscopy, GC–MS, FT-IR, SEM and AFM showed that the microalga had initiated degradation of LDPE sheet within 30 days of incubation. Govindan et al., reported the potential of consortium of *Chlorella* sp. and *Cyanobacteria* sp. to deteriorate low-density polyethene (LDPE sample) which was further confirmed by UV-Spec, FESEM, EDX, CHNO, FTIR and DSC analysis (Govindan et al. 2024). Hadiyanto and co-workers conducted a study on Microalgae and microplastic interactions in two 1 L glass bioreactors containing *D. salina* with oxium microplastics and oxidized HDPE at various concentrations (100 mg/500 mL, 200 mg/500 mL, and 300 mg/500 mL) for 15 d (Hadiyanto et al. 2022). Researchers concluded that oxium plastics have a faster biodegradation ability owing to the addition of additives to the plastic. However, oxidation pre-treatment with H_2O_2 on HDPE plastic can also accelerate the plastic degradation process.

6 Conclusion

Biodegradation of plastic waste can be carried out by microorganisms including bacteria, fungi, and algae. These microorganisms can reduce the weight of plastic samples tested under certain conditions. The biodegradation rate of PE bags carried out by fungi was also higher than bacteria. The fungus *Aspergillus glaucus* was able to reduce the weight of the plastic sample by 28%. Meanwhile, the highest reduction rate of plastic by bacteria was *Pseudomonas* sp. with a plastic weight reduction of 20.54%. However, the reduction rate of plastic PE cups was higher in the bacteria *Moraxella* sp. compared to fungi. In plastic biodegradation, bacteria convert plastic polymers into dimers and oligomers which will be used by them as energy and carbon sources for their cell growth. Meanwhile, the optimum biodegradation of LDPE was carried out by fungi and algae. Also, the biodegradation of PY can be carried out effectively by algae. Algae will colonize the plastic surface and produce enzymes that will weaken the bonds in plastic polymers. Algae will use plastic polymers as their carbon sources and then further convert them into the algal cell biomass, CO_2 and H_2O. The algal biomass can be used in biorefinery for various goals such as biofuel production.

Plastic waste accumulation is one of the serious environmental issues in the world. Plastic waste pollution in the aquatic environment threatens many marine organisms as well as causes bioaccumulation. This problem should be resolved immediately. Some conventional methods in overcoming this problem have been carried out, such as landfills and incineration. However, this method still releases side effects for the environment. Biodegradation is considered an appropriate method of handling plastic waste. Biodegradation is carried out using plastic degrading microbes such as bacteria, fungi, and algae. They can produce various enzymes catalyzing the breakdown of plastic polymers into simpler fragments. The fragments can be assimilated as their carbon and energy sources as well as decreasing the amount of plastic waste polluting the environment. Still, there are great opportunities for research on the optimization of various kinds of plastic degrading microorganisms and their applications on a large scale.

References

Aldaby ES, Mahmoud AH, El-Bery HM, Ali MM, Shoreit AA, Mawad AM (2023) Microalgal upgrading of the fermentative biohydrogen produced from *Bacillus coagulans* via non-pretreated plant biomass. Microb Cell Factories 22:190

Aravinthan A, Arkatkar A, Juwarkar AA, Doble M (2016) Synergistic growth of *bacillus* and *pseudomonas* and its degradation potential on pretreated polypropylene. Prep Biochem Biotechnol 46:109–115

Arkatkar A, Arutchelvi J, Bhaduri S, Uppara PV, Doble M (2009) Degradation of unpretreated and thermally pretreated polypropylene by soil consortia. Inter Biodeter Biodegr 63:106–111

Auta HS (2018) Bioremediation of microplastic using microbes isolated from mangrove sediments. Doctoral dissertation, University of Malaya, Malaysia

Auta HS, Emenike CU, Fauziah SH (2017) Screening of *Bacillus* strains isolated from mangrove ecosystems in peninsular Malaysia for microplastic degradation. Environ Pollut 231:1552–1559

Auta SH, Emenike CU, Fauziah SH (2017) Screening for polypropylene degradation potential of bacteria isolated from mangrove ecosystems in Peninsular Malaysia. Int J Biosci Biochem Bioinforma 7:245–251

Bakir A, Rowland SJ, Thompson RC (2014) Transport of persistent organic pollutants by microplastics in estuarine conditions. Estuar Coast Shelf Sci 140:14–21

Blair RM, Waldron S, Gauchotte-Lindsay C (2019) Average daily flow of microplastics through a tertiary wastewater treatment plant over a ten-month period. Water Res 163:114909

Bonhomme S (2003) Environmental biodegradation of polyethylene. Polym Degrad Stab 81:441–452

Brooks JM, Stewart CJ, Haberstroh CJ, Arias ME (2023) Characteristics and fate of plastic pollution in urban stormwater ponds. Environmen Pollut 320:121052

Cacciari I, Quatrini P, Zirletta G, Mincione E, Vinciguerra V, Lupattelli P, Giovannozzi Sermanni G (1993) Isotactic polypropylene biodegradation by a microbial community: physicochemical characterization of metabolites produced. Appl Environmen Microbiol 59:3695–3700

Carr SA, Liu J, Tesoro AG (2016) Transport and fate of microplastic particles in wastewater treatment plants. Water Res 91:174–182

Chandra R, Rustgi R (1997) Biodegradation of maleated linear low-density polyethylene and starch blends. Polym Degrad Stab 56:185–202

Chia WY, Tang DYY, Khoo KS, Lup ANK, Chew KW (2020) Nature's fight against plastic pollution: algae for plastic biodegradation and bioplastics production. Environ Sci Ecotechnology 4:100065

Crew A, Gregory-Eaves I, Ricciardi A (2020) Distribution, abundance, and diversity of microplastics in the upper St. Lawrence River. Environ Pollut 260:113994

Davoudi H (2023) Toxicity of microplastics, graphene quantum dots and molybdenum disulfide to the freshwater organism Daphnia magna. Doctoral dissertation, University of Birmingham

Devi SS, Kumar AB (2024) Impacts of mesoplastics and microplastics on marine biodiversity. In: Meso-and microplastic risk assessment in marine environments. Elsevier, pp 309–340

Duis K, Coors A (2016) Microplastics in the aquatic and terrestrial environment: sources (with a specific focus on personal care products), fate and effects. Environ Sci Eur 28:2

Eerkes-Medrano D, Thompson RC, Aldridge DC (2015) Microplastics in freshwater systems: a review of the emerging threats, identification of knowledge gaps and prioritisation of research needs. Water Res 75:63–82

Ekanayaka AH, Tibpromma S, Dai D, Xu R, Suwannarach N, Stephenson SL et al (2022) A review of the fungi that degrade plastic. J Fungi 8:772

Foekema EM, De Gruijter C, Mergia MT, Van Franeker JA, Murk AJ, Koelmans AA (2013) Plastic in North Sea fish. Environ Sci Technol 47:8818–8824

Govindan K, Fard FSN, Asgari F, Sorooshian S, Mina H (2024) A bi-objective location-routing model for the healthcare waste management in the era of logistics 4.0 under uncertainty. Int J Prod Econ 276:109342

Gregory MR (2009) Environmental implications of plastic debris in marine settings – entanglement, ingestion, smothering, hangers-on, hitch-hiking and alien invasions. Philos Trans R Soc B: Biol Sci 364:2013–2025

Gu JD, Ford T, Mitton B, Mitchell R (2024) Research on biodeteriotration of plastics. Int Biodeterior Biodegradation 186:105699

Gulnaz O, Dıncer S (2009) Biodegradation of bisphenol A by *Chlorella vulgaris* and *Aeromonas hydrophilia*. J Appl Bio Sci 3:79–84

Hadiyanto H, Khoironi A, Dianratri I, Huda K, Suherman S, Muhammad F (2022) Biodegradation of oxidized high-density polyethylene and oxo-degradable plastic using microalgae *Dunaliella salina*. Environ Pollut Bioavailab 34:469–481

Hirooka T, Nagase H, Uchida K, Hiroshige Y, Ehara Y, Nishikawa JI et al (2005) Biodegradation of bisphenol A and disappearance of its estrogenic activity by the green alga *Chlorella fusca* var. vacuolata. Environ Toxicol Chem 24:1896–1901

Ibrahim YS, Hamzah SR, Khalik WMAWM, Yusof KMKK, Anuar ST (2021) Spatiotemporal microplastic occurrence study of setiu wetland, South China Sea. Sci Total Environ 788:147809

Jain R, Gaur A, Suravajhala R, Chauhan U, Pant M, Tripathi V, Pant G (2023) Microplastic pollution: understanding microbial degradation and strategies for pollutant reduction. Sci Total Environ 905:167098

Jeon J, Kim J (2013) Cold stress signaling networks in Arabidopsis. J Plant Biol 56:69–76

Jeon S, Lee DK, Jeong J, Yang SI, Kim JS, Kim J, Cho WS (2021) The reactive oxygen species as pathogenic factors of fragmented microplastics to macrophages. Environ Pollut 281:117006

Khoironi A et al (2019) Evaluation of the interaction among microalgae *spirulina* sp, plastics polyethylene terephthalate and polypropylene in freshwater environment. J Ecol Eng 20:161–173
Klein S, Dimzon IK, Eubeler J, Knepper TP (2018) Analysis, occurrence, and degradation of microplastics in the aqueous environment. In: Wagner MSL (ed) Freshwater microplastics. The handbook of environmental chemistry, vol 58. Springer
Kowalczyk N, Blake N, Charko F, Quek Y (2017) Microplastics in the maribyrnong and yarra rivers, Melbourne, Australia. Port Phillip EcoCentre, Clean Bay Blueprint, pp 1–36
Kumar VE, Ravikumar G, Jeyasanta KI (2018) Occurrence of microplastics in fishes from two landing sites in Tuticorin, south east coast of India. Mar Pollut Bull 135:889–894
Kumar M, Chen H, Sarsaiya S, Qin S, Liu H, Awasthi MK et al (2021) Current research trends on micro-and nano-plastics as an emerging threat to global environment: a review. J Hazard Mater 409:124967
Kumar R, Manna C, Padha S, Verma A, Sharma P, Dhar A et al (2022) Micro (nano) plastics pollution and human health: how plastics can induce carcinogenesis to humans? Chemosphere 298:134267
Lwanga EH, Thapa B, Yang X, Gertsen H, Salánki T, Geissen V, Garbeva P (2018) Decay of low-density polyethylene by bacteria extracted from earthworm's guts: a potential for soil restoration. Sci Total Environ 624:753–757
Manzi HP, Abou-Shanab RA, Jeon BH, Wang J, Salama ES (2022) Algae: a frontline photosynthetic organism in the microplastic catastrophe. Trends Plant Sci 27:1159–1172
Mataji A, Taleshi MS, Balimoghaddas E (2020) Distribution and characterization of microplastics in surface waters and the southern Caspian Sea coasts sediments. Arch Environ Contam Toxicol 78:86–93
Moog D, Schmitt J, Senger J, Zarzycki J, Rexer KH, Linne U et al (2019) Using a marine microalga as a chassis for polyethylene terephthalate (PET) degradation. Microb Cell Factories 18:1–15
Muenmee S, Chiemchaisri W, Chiemchaisri C (2016) Enhancement of biodegradation of plastic wastes via methane oxidation in semi-aerobic landfill. Int Biodeterior Biodegradation 113:244–255
Novak AB, Gerstenbacher CM, Lord KS, Rotjan RD (2024) Microplastic abundance and accumulation patterns in eelgrass (*Zostera marina* L.) meadows throughout coastal Massachusetts, USA. Environ Monit Assess 196:775
Ojha N, Pradhan N, Singh S, Barla A, Shrivastava A, Khatua P et al (2017) Evaluation of HDPE and LDPE degradation by fungus, implemented by statistical optimization. Sci Rep 7:39515
Oliveira AR, Sardinha-Silva A, Andrews PL, Green D, Cooke GM, Hall S et al (2020) Microplastics presence in cultured and wild-caught cuttlefish, Sepia officinalis. Mar Pollut Bull 160:111553
Ortúzar M, Esterhuizen M, Olicón-Hernández DR, González-López J, Aranda E (2022) Pharmaceutical pollution in aquatic environments: a concise review of environmental impacts and bioremediation systems. Front Microbiol 13:869332
Paço A, Duarte K, da Costa JP, Santos PS, Pereira R, Pereira ME et al (2017) Biodegradation of polyethylene microplastics by the marine fungus *Zalerion maritimum*. Sci Total Environ 586:10–15
Park SY, Kim CG (2019) Biodegradation of micro-polyethylene particles by bacterial colonization of a mixed microbial consortium isolated from a landfill site. Chemosphere 222:527–533
Pathak VM (2023) Exploitation of bacterial strains for microplastics (LDPE) biodegradation. Chemosphere 316:137845
Qi R, Jones DL, Li Z, Liu Q, Yan C (2020) Behavior of microplastics and plastic film residues in the soil environment: a critical review. Sci Total Environ 703:134722
Rajandas H, Parimannan S, Sathasivam K, Ravichandran M, Yin LS (2012) A novel FTIR-ATR spectroscopy based technique for the estimation of low-density polyethylene biodegradation. Polym Test 31:1094–1099
Restrepo-Florez JM, Bassi A, Thompson MR (2014) Microbial degradation and deterioration of polyethylene- a review. Int Biodeterior Biodegradation 88:83–90
Salinas CX, Palacios E, Pozo K, Torres M, Rebolledo L, Gómez V et al (2024) Marine litter pollution in a subantarctic beach of the Strait of Magellan, Punta Arenas. Chile Mar Pollut Bull 202:116313

Sangale MK, Shahnawaz M, Ade AB (2019) Potential of fungi isolated from the dumping sites mangrove rhizosphere soil to degrade polythene. Sci Rep 9:5390
Sanniyasi E, Gopal RK, Gunasekar DK, Raj PP (2021) Biodegradation of low-density polyethylene (LDPE) sheet by microalga, *Uronema africanum* Borge. Scient Rep 11:17233
Santana MFM, Ascer LG, Custódio MR, Moreira FT, Turra A (2016) Microplastic contamination in natural mussel beds from a Brazilian urbanized coastal region: rapid evaluation through bioassessment. Mar Pollut Bull 106:183–189
Sarmah P, Rout J (2018) Efficient biodegradation of low-density polyethylene by cyanobacteria isolated from submerged polyethylene surface in domestic sewage water. Environ Sci Pollut Res 25:33508–33520
Shah T, Khan Z, Asad M, Imran A, Niazi MBK, Dewil R et al (2024) Straw incorporation into microplastic-contaminated soil can reduce greenhouse gas emissions by enhancing soil enzyme activities and microbial community structure. J Environ Manag 351:119616
Sharma BK, Moser BR, Vermillion KE, Doll KM, Rajagopalan N (2014) Production, characterization and fuel properties of alternative diesel fuel from pyrolysis of waste plastic grocery bags. Fuel Process Technol 122:79–90
Sheikh HI, Sai P, John A (2025) Identifying microplastic contamination in marine biota. In: Sustainable management of urban plastic waste through circular economic approaches. CRC Press, pp 12–26
Shen M, Zeng Z, Wen X, Ren X, Zeng G, Zhang Y, Xiao R (2021) Presence of microplastics in drinking water from freshwater sources: the investigation in Changsha, China. Environ Sci Pollut Res 28:42313–42324
Singh S (2024) Recent advances in treatment of microplastics in wastewater. In: Detection and treatment of emerging contaminants in wastewater, pp 55–72
Sivan A (2011) New perspectives in plastic biodegradation. Curr Opin Biotechnol 22:422–426
Skariyachan S, Taskeen N, Kishore AP, Krishna BV, Naidu G (2021) Novel consortia of *Enterobacter* and *Pseudomonas* formulated from cow dung exhibited enhanced biodegradation of polyethylene and polypropylene. J Environ Manag 284:112030
Suseela MR, Toppo K (2007) Algal biofilms on polythene and its possible degradation. Curr Sci 92:113891
Torena P, Alvarez-Cuenca M, Reza M (2021) Biodegradation of polyethylene terephthalate microplastics by bacterial communities from activated sludge. Can J Chem Eng 99:69–82
Vinagre PA, Simas T, Cruz E, Pinori E, Svenson J (2020) Marine biofouling: a European database for the marine renewable energy sector. J Mar Sci Eng 8:495
Volke-Sepúlveda T, Saucedo-Castañeda G, Gutiérrez-Rojas M, Manzur A, Favela-Torres E (2002) Thermally treated low density polyethylene biodegradation by *Penicillium pinophilum* and *Aspergillus niger*. J Appl Polym Sci 83:305–314
Wagner M, Lambert S (2018) Freshwater microplastics. Emerg Env Chem 58. https://doi.org/10.1007/978-3-319-61615-5_1
Yan Y, Zhu F, Zhu C, Chen Z, Liu S, Wang C, Gu C (2021) Dibutyl phthalate release from polyvinyl chloride microplastics: influence of plastic properties and environmental factors. Water Res 204:117597
Yang J, Yang Y, Wu WM, Zhao J, Jiang L (2014) Evidence of polyethylene biodegradation by bacterial strains from the guts of plastic-eating waxworms. Environ Sci Technol 48(23):13776–13784
Yang X, Bento CP, Chen H, Zhang H, Xue S, Lwanga EH et al (2018) Influence of microplastic addition on glyphosate decay and soil microbial activities in Chinese loess soil. Environ Pollut 242:338–347
Zhu J, Liu S, Wang H, Wang D, Zhu Y, Wang J et al (2022) Microplastic particles alter wheat rhizosphere soil microbial community composition and function. J Hazard Mater 436:129176

The manufacturer's authorised representative in the EU is Springer Nature Customer Service Centre GmbH, Europaplatz 3, 69115 Heidelberg, Germany. If you have any concerns regarding our products, please contact ProductSafety@springernature.com

Printed and bound by CPI Group (UK) Ltd, Croydon, CR0 4YY
01/10/2026
02225930-0003